THIRD EDITION

INVESTIGATING PHYSICAL SYSTEMS

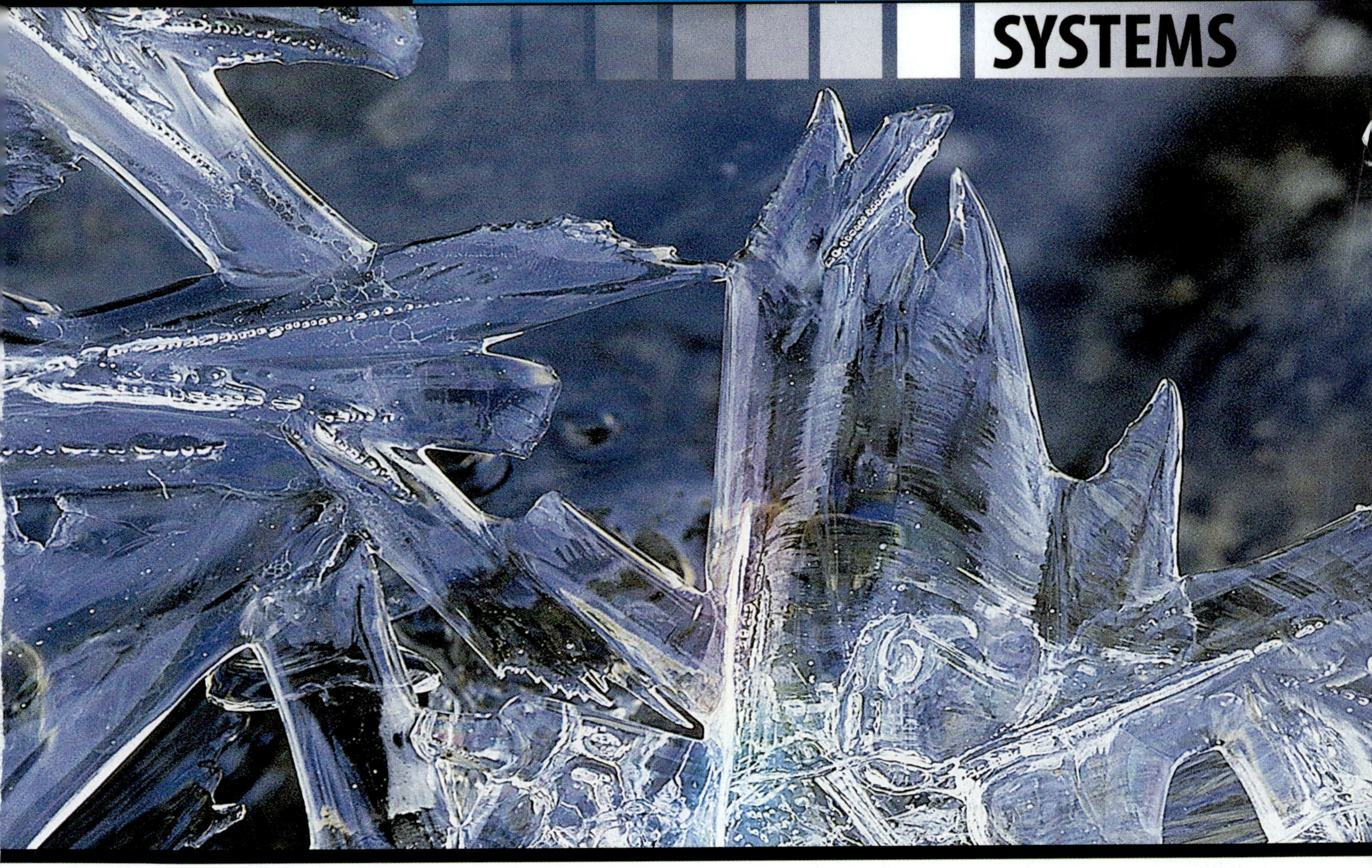

BSCS SCIENCE & TECHNOLOGY

KENDALL/HUNT PUBLISHING COMPANY
4050 Westmark Drive   Dubuque, Iowa 52004

BSCS

NATIONAL SCIENCE FOUNDATION

BSCS Development Team, First Edition

These titles and dates indicate the primary area of responsibility for each person and the years he or she worked on the project. Everyone on the project team contributed in numerous ways to create this curriculum.

Rodger W. Bybee, *Principal Investigator (1988–92)*
Janet Carlson Powell, *Project Director (1988–92)*
Kathrine A. Backe, *Staff Associate, Implementation (1991–92)*
Wilbur C. Bergquist, *Staff Associate, Evaluation (1991-92)*
Deirdre Binkley-Jones, *Project Secretary (1992)*
Jan Chatlain Girard, *Art Coordinator (1989–92)*
Sariya Jarasviroj, *Production Assistant (1992)*
Terri Johnston, *Project Secretary (1991–92)*
Donald E. Maxwell, *Staff Associate, Staff Development (1990–92)*
Mary E. McMillan, *Staff Associate, Curriculum Development (1991–92)*
Josina Romero-O'Connell, *Staff Associate, Curriculum Development (1991–92)*
Teresa Powell, *Project Secretary (1989–92)*
Judith Martin Rhode, *Research Assistant (1989–91), Staff Associate, Curriculum Development (1992)*
Joe Ramsey, *Production Assistant (1992)*
William C. Robertson, *Staff Associate, Curriculum Development (1989–91)*
Nancy Smalls, *Project Secretary, Graphics (1990–92)*
Jenny Stricker, *Staff Associate, Curriculum Development (1992)*
Pamela Van Scotter, *Staff Associate, Editing (1990–92)*
Lee B. Welsh, *Production Coordinator (1989–92)*
Yvonne Wise, *Project Secretary (1989–92)*

BSCS Revision Team, Third Edition

Janet Carlson Powell, *Project Director*
Pamela Van Scotter, *Project Director*
Melissa Richie, *Curriculum Developer*
Steve Getty, *Curriculum Developer*
Barbara Perrin, *Production Manager*
Richard Bascobert, *Editor*
Stacey Luce, *Manuscript Specialist*
Lisa Rasmussen, *Graphic Designer*
Diane Gionfriddo, *Photo Researcher*
Raphaela Connor, *Administrative Assistant*
Nicole Knapp, *Researcher*

BSCS Administrative Staff

Carlo Parravano, *Chair, Board of Directors*
Rodger W. Bybee, *Executive Director*
Janet Carlson Powell, *Associate Director and Chief Science Education Officer*
Larry Satkowiak, *Associate Director and Chief Operating Officer*
Pamela Van Scotter, *Director, The BSCS Center for Curriculum Development*
Nancy Landes, *Director, The BSCS Center for Professional Development*
Ted Lamb, *Director, The BSCS Center for Research and Evaluation*
Marcia Mitchell, *Director of Finance*

Reviewers, Third Edition

Clyde R. Burnett, *Fritz Peak Observatory, Rollinsville, Colorado (Science Content)*
Deb Hannigan, *BSCS, Colorado Springs*
Joe Taylor, *BSCS, Colorado Springs*

continued on page 589.

Cover image © 2004 Eyewire
SciLinks® is owned and provided by the National Science Teachers Association. All rights reserved.
0-7575-0105-2

Copyright © 2005 by BSCS.

Copyright © 1994, 1999 by BSCS as *BSCS Middle School Science & Technology: Diversity and Limits*.
All rights reserved.

No part of this work may be reproduced or transmitted in any form or by any means, electronic or mechanical, including photocopying and recording, or by any information storage or retrieval system, without permission in writing. For permissions and other rights under this copyright, contact BSCS, 5415 Mark Dabling Blvd., Colorado Springs, CO 80918-3842.

This material is based on work originally supported by the National Science Foundation under Grant No. MDR 8855657. Any opinions, findings, conclusions, or recommendations expressed in this publication are those of the authors and do not necessarily reflect the views of the granting agency.

Printed in the United States of America.

2 3 4 5 6 7 8 9 10   08 07 06

Contents

	Preface	viii
	Program Overview	ix
	Scope and Sequence	xii

Introduction	**WHAT WILL THIS PROGRAM BE LIKE?**	1
Chapter 1	**A Learning Journey**	3
ENGAGE	Investigation: A Big Ball of String	4
EXPLORE	Investigation: Learning about One Another through Data Collection	5
EXPLORE	Investigation: Rainbow Colors	8
EXPLORE	Investigation: Meet Al, Marie, Isaac, and Rosalind	11
EXPLAIN	Reading: Learning and Working Cooperatively	18
ELABORATE/EVALUATE	Connections: What's in Here?	25
Chapter 2	**Science Safety**	29
ENGAGE	Investigation: Science Is . . . Technology Is . . .	30
EXPLORE	Connections: Thinking about Safety	32
EXPLAIN	Connections: Science Safety Contracts	33
ELABORATE	Investigation: Cooperating for Safety's Sake	34
EVALUATE	Investigation: Science Safety	36

UNIT 1

	EXPLORING RANGES OF LIMITS AND DIVERSITY	41
	Cooperative Learning Overview	42
Chapter 3	**Identifying Limits and Diversity**	45
ENGAGE/EXPLORE	Investigation: Star Tracers	46
EXPLORE	Investigation: Threading the Needle	50
EXPLORE	Connections: How Do You Spell Success?	53
EXPLAIN	Investigation: If at First You Don't Succeed	54
EXPLAIN	Reading: Doing It All the Same	56
ELABORATE	Investigation: Seeing the World around You	60
EVALUATE	Investigation: Light, Lenses, and the Eye	63

Chapter 4	**Ranges of Limits and Diversity**		75
ENGAGE/EXPLORE	Investigation: An Invasion of the A-maize-ing Popcorans		76
EXPLAIN	Reading: Diversity Is Part of Our Natural World		79
EXPLAIN	Investigation: A Diversity of Popcorn		82
EXPLAIN	Connections: More on the Meaning of the Bell-Shaped Curve		84
ELABORATE	Reading: The Value of the Bell-Shaped Curve		85
ELABORATE	Investigation: The Bell-Shaped Curve and You		88
EVALUATE	Investigation: A Flag of a Different Color		90
Chapter 5	**Using Limits to Set Standards**		99
ENGAGE	Investigation: What Do You Really Know about TV?		100
EXPLORE	Investigation: Taking a Closer Look at TV Pictures		101
EXPLORE	Connections: The Ultimate TV		103
EXPLORE	Reading: TV Pictures and Color TV		103
EXPLORE/EXPLAIN	Investigation: A Learning Adventure		107
EXPLORE/EXPLAIN	Connections: Your Experiences at the Stations		123
EXPLORE/EXPLAIN	Investigation: The Optimal TV Viewing Distance		123
EXPLAIN	Reading: Setting Standards and Human Factors		125
ELABORATE/EVALUATE	Connections: TV for a New Millennium		127
Chapter 6	**Using Diversity to Set Standards**		133
ENGAGE/EXPLORE	Investigation: Watching *The Final Factor*		134
EXPLORE	Reading: The Three Phases of Stopping		135
EXPLORE	Investigation: Your Personal Reaction Time		139
EXPLORE	Investigation: Determining Reaction Distances and Perception Distances		144
EXPLAIN	Connections: Total Stopping Distances		149
EXPLAIN	Investigation: Life in the Fast Lane		153
ELABORATE	Investigation: Setting Speed Limits		162
ELABORATE	Reading: The Bell-Shaped Curve and Setting Standards		165
ELABORATE/EVALUATE	Investigation: Don't Drink and Drive		168
Chapter 7	**Evaluating Your Understanding of Limits and Diversity**		175
ENGAGE	Connections: Shhh . . . Hush . . . QUIET!		176
EXPLORE/EXPLAIN	Reading: What Have You Learned in Unit 1?		177
ELABORATE/EVALUATE	Investigation: How Much Noise Is Too Much Noise?		179

UNIT 2

WHY ARE THINGS DIFFERENT? — 183
Cooperative Learning Overview — 184

Chapter 8 — Properties of the Material World — 187
- **ENGAGE/EXPLORE** — Investigation: Bounce That Ball — 188
- **EXPLAIN** — Reading: How Things Are Different — 192
- **ELABORATE** — Investigation: How Can I Learn More about That Property? — 198
- **EVALUATE** — Connections: Is the Most Always the Best? — 204

Chapter 9 — Scientific Explanations Are Ancient History — 207
- **ENGAGE** — Connections: What Happens to Light inside a Prism? — 208
- **EXPLORE** — Investigation: Strange Phenomena — 209
- **EXPLAIN** — Reading: Elements of Explanations — 214
- **ELABORATE/EVALUATE** — Connections: Thinking like the Ancients — 217

Chapter 10 — Using Scientific Models to Answer Questions — 219
- **ENGAGE\EXPLORE** — Investigation: Mystery Box — 220
- **EXPLAIN** — Reading: Another Explanation — 224
- **ELABORATE** — Investigation: How Well Does the Particle Model Work? — 232
- **ELABORATE** — Connections: Particle Movement—Improving the Model — 236
- **EVALUATE** — Investigation: Applying the New Model — 239

Chapter 11 — Using Models to Test and Predict — 247
- **ENGAGE** — Connections: Shaping Models — 248
- **EXPLORE** — Investigation: Gloop — 251
- **EXPLAIN** — Reading: More on Models — 258
- **ELABORATE** — Investigation: Leak-Free Models — 265
- **ELABORATE\EVALUATE** — Investigation: A Penny's Worth of Water — 272
- **EVALUATE** — Connections: Properties and Models in Review — 280

UNIT 3

HOW DOES TECHNOLOGY ADDRESS DIVERSITY AND INFLUENCE LIMITS? — 283
Cooperative Learning Overview — 284

Chapter 12 — Consumer Concerns — 287
- **ENGAGE/EXPLORE** — Investigation: Putting Paper Towels to the Test — 288
- **EXPLORE** — Reading: Paper Towel Consumers — 291
- **EXPLORE** — Connections: Comparing Ratings — 301

EXPLAIN	**Reading:** Why Products Fit		302
ELABORATE	**Connections:** Do You Understand Criteria and Constraints?		309
ELABORATE	**Investigation:** Part of Your Complete Breakfast		310
EVALUATE	**Connections:** Evaluating Your Understanding of Criteria and Constraints		312

Chapter 13 — Your Designing Ways — 317

ENGAGE	**Investigation:** Small-Scale Boats	318
EXPLORE	**Reading:** Is a Boat a Boat?	324
EXPLORE	**Investigation:** Sails, Propellers, and Gas	327
EXPLORE	**Investigation:** Anchors Away!	338
EXPLAIN	**Connections:** Technological Problem Solving	340
ELABORATE	**Investigation:** Toys for Tots	347
EVALUATE	**Connections:** Human Factors as a Design Constraint	354

Chapter 14 — Why Are There So Many Products That Do the Same Thing? — 359

ENGAGE	**Investigation:** One Problem, Different Decisions	360
EXPLORE	**Investigation:** Designing with Shapes	361
EXPLAIN	**Reading:** Similarity and Diversity in Designs	368
ELABORATE	**Investigation:** Ideas That Fly	373
EVALUATE	**Connections:** Explaining Design Diversity	378

Chapter 15 — Masters of Design — 381

EXPLORE/EXPLAIN	**Reading:** Let's Talk Technology—Again	382
ELABORATE	**Connections:** Evaluating Your Environment	383
ELABORATE	**Investigation:** Enabling the Physically Challenged	385
EVALUATE	**Connections:** A Science and Technology Gazette	388

UNIT 4 — LIMITS OF ENERGY IN SYSTEMS — 395
Cooperative Learning Overview — 396

Chapter 16 — Exploring Energy in Systems — 399

ENGAGE\EXPLORE	**Investigation:** What Do You Already Know about Energy?	400
EXPLAIN	**Reading:** Thinking More about Energy	404
ELABORATE	**Investigation:** Heat In, Heat Out	411
ELABORATE	**Connections:** What If Energy from the Sun Were Blocked?	414
EVALUATE	**Connections:** Tracing the Flow of Energy	416

Chapter 17 — Solar Sources: Energy from the Sun — 419

ENGAGE	**Investigation:** Sun Seekers	420

EXPLORE	**Connections:** What Is Skin Cancer?	426
EXPLORE	**Investigation:** Collecting the Sun	433
EXPLAIN	**Reading:** Systems for Solving Problems	437
ELABORATE	**Investigation:** Hotter and Hotter	439
EVALUATE	**Connections:** Water-Heating Systems	443

Chapter 18	**Energy Benefits and Costs**	447
ENGAGE	**Investigation:** Generating Electricity	448
EXPLORE	**Connections:** Doing Things in Reverse	454
EXPLORE	**Investigation:** Paperclip Lifters	456
EXPLAIN	**Reading:** Benefits and Costs	461
ELABORATE	**Connections:** Wind Farms	466
ELABORATE	**Investigation:** A Variety of Systems	468
EVALUATE	**Connections:** Which System Do We Use?	477

Chapter 19	**The Power to Choose**	511
ENGAGE	**Connections:** Toast	512
EXPLORE	**Investigation:** Cans of Energy	512
EXPLAIN	**Reading:** Complex Technological Systems	515
ELABORATE	**Connections:** It's Not a Simple Matter	524
ELABORATE	**Investigation:** Ways to Improve Our Use of Energy	527
ELABORATE	**Connections:** What Is My Choice?	530
EVALUATE	**Connections:** The Global Picture	535

HOW TO:

1	Make a T-Chart	540
2	Construct a Data Table	544
3	Construct a Bar Graph	553
4	Have a Brainstorming Session	559
5	Determine Averages	561
6	Conduct a Research Project	564
7	Read a Thermometer	572
8	Conduct a Fair Test	574
9	Use a Bunsen Burner	578
	Glossary	581
	Acknowledgments	589
	Index	593

Preface

Welcome to *BSCS Science & Technology*, third edition! We hope you enjoy your learning journey. We developed this science program specifically for students like you. In the process, we considered students, their teachers, and schools. We communicated with more than 20,000 students across the United States and in South America.

Our program has been a great success with students, teachers, and parents. Students are enjoying science as a way of learning about their world. Teachers, although working harder in this program than in other science programs, know it is worth it when they see their students truly thinking and understanding. Parents are pleased because their children are enjoying science and are successful at it.

There are, however, always ways to make things better, and during the past five years, there also have been advances in science and technology that we want to share with you. In this third edition, there are some new investigations, readings, connections, and sidelights.

These activities provide you with opportunities to explore in more depth some of the science content. These changes also reflect the most current understanding of various concepts in science. We also added more artwork to better explain some ideas that need more than words alone. The new artwork also makes the book more enjoyable to look at. LaurelTech, a design firm in New Hampshire, has created an exciting and fun design, and Kendall/Hunt, a publishing company in Iowa, has worked hard to bring you the highest quality book.

If you have comments or questions about the program, please write to us at the following address. We would enjoy hearing from you.

BSCS
Attention: MSP
5415 Mark Dabling Blvd.
Colorado Springs, CO 80918

Program Overview

BSCS Science & Technology is probably unlike other science programs you have used in school. This overview briefly describes some of the key features of the program. You will learn more about these special features of the program when you complete Chapters 1 and 2 of the Introduction.

Themes

The organization of information in this program might be different from other science programs that you have used. That is because we have organized each level, or year, around a unifying theme. You can think of a unifying theme as a common idea that keeps coming up to tie other ideas together. Writers use a unifying theme, called a story line, when they write novels. Often music has a unifying theme that you hear over and over in the piece of music; Beethoven's Fifth Symphony is a famous example.

We use a different unifying theme at each level, but each theme has the same purpose: to provide a thread that ties together many scientific and technological ideas. In *Investigating Earth Systems*, the unifying theme is patterns of change; in *Investigating Physical Systems*, the unifying theme is diversity and limits; and in *Investigating Life Systems*, the theme is systems and change. We use a different curriculum emphasis and focus question for each unit. The scope and sequence chart shows how we organized all of these parts to form a complete program that extends over three years. You can use the framework to orient yourself and see where you have been and where you are going.

The Five "Es"

As you begin using your textbook, you will notice that each investigation, connections activity, and reading has a word that begins with an E to introduce it: *Engage, Explore, Explain, Elaborate, Evaluate*. These words make up five phases that you will use as you learn an idea. These phases describe what you are doing as a learner and what your teacher is doing as a teacher. Generally, you will go through each phase approximately once per chapter. For example, when you are doing an **engage** activity, you will be thinking about a new idea. Then you will **explore** that idea in one or more activities. Next, you will begin to **explain** an idea by constructing an explanation for the idea. Then you will **elaborate** your understanding of the idea, usually by doing another activity. Finally you and your teacher will **evaluate** your understanding of the idea. If your evaluation shows that you are successful in understanding the idea, it is time to be engaged in a new idea and to go through the phases for that idea. If your evaluation shows that you do not entirely understand the idea, you and your teacher will know to work through some more activities. You will learn more about the five Es in Chapter 1 A Learning Journey.

Cooperative Learning

We have incorporated cooperative learning strategies into about two-thirds of the program for a variety of reasons. One reason is that cooperative learning gives you a chance to learn and practice how to work successfully with others. The skills that you gain will be important to you in many different aspects of school and life. In addition, by using cooperative learning strategies, you will be working in a way that is similar to the way professional scientists and engineers work.

The Characters

We use four characters in this book. Al, Marie, Isaac, and Rosalind ("Ros"). These characters

1. provide a concrete method for demonstrating the value of different learning styles and ways of being smart,
2. teach some of the history of science, and
3. provide a type of positive role model for cooperative learning and doing science.

Al, Marie, Isaac, and Ros are introduced and described thoroughly in Chapter 1 of the Introduction. You learn about the strengths of each character's learning style, the historical reference for the character's name, and a bit about the character's personality. In the text, the characters provide examples of why science is something everyone can do, because they are a diverse group of learners and each contributes something positive to the group.

Questions

There are two primary places you will find questions in this book: in readings as Stop and Think questions and at the end of investigations as Wrap Ups. At first, some of the questions may seem hard to answer because you have to think carefully about your answers—you cannot just copy the answer out of the book. One of our goals for this program is to increase your ability to think critically. You will notice that many of the questions in the book have more than one answer that can be considered correct. That is because we tried to write questions that you can answer in a variety of ways as long as you provide support for your answer. If this is a new way of answering questions for you, you might feel frustrated at first, but eventually you might find that you enjoy learning this way. Using this questioning method places you more in charge of your own learning.

In addition to the questions that we ask you, there are many opportunities for you to ask questions. Science often begins with a question, and as you have more practice thinking like a scientist, you will have more questions of your own. Keep track of these questions and share them with your teacher and classmates. Although there are some questions in science that we cannot answer (at least not yet), your teacher will help you discover ways of finding answers to some of your questions.

Assessment

Because the topics, themes, and questions in this book are different from most other science programs, it makes sense to include different ways of assessing your success and progress. In many programs, you are assessed only by your performance on quizzes and tests at the end of a chapter or unit. In this program, we recommend that teachers use a variety of assessment strategies that include daily notebooks, checklists, performance tests, short-answer tests, and portfolios to measure how much you have learned and to identify areas that you might want to focus on in the future. So don't be surprised if you have "tests" that don't remind you of the tests you are used to taking.

Safety

As in any science program, safety is a concern for everyone who uses *BSCS Science & Technology*. Chapter 2 Science Safety will introduce you to the important safety rules that you will need to follow while in the science classroom. We have made every effort to alert you, the learner, to potentially dangerous situations or materials. We have marked those places with the following symbol:

In addition, your teacher should tell you about the safe behaviors that you should use in your science classroom. It is your responsibility to follow all safety warnings, rules, and procedures to avoid possible injury to yourself and others.

SCOPE AND SEQUENCE

UNIT	1	2	3	4
Investigating Earth Systems				
Unifying Theme	Patterns of Change			
Curriculum Emphasis	Personal dimensions of science and technology	The nature of scientific explanations	Technological problem solving	Science and technology in society

UNIT	1	2	3	4
Investigating Physical Systems				
Unifying Theme	Diversity and Limits			
Curriculum Emphasis	Personal dimensions of science and technology	The nature of scientific explanations	Technological problem solving	Technological problem solving and society

UNIT	1	2	3	4
Investigating Life Systems				
Unifying Theme	Systems and Change			
Curriculum Emphasis	Personal dimensions of science and technology	The nature of scientific explanations	Scientific explanations and social issues	Science and technology in society

INVESTIGATING PHYSICAL SYSTEMS

BSCS SCIENCE & TECHNOLOGY

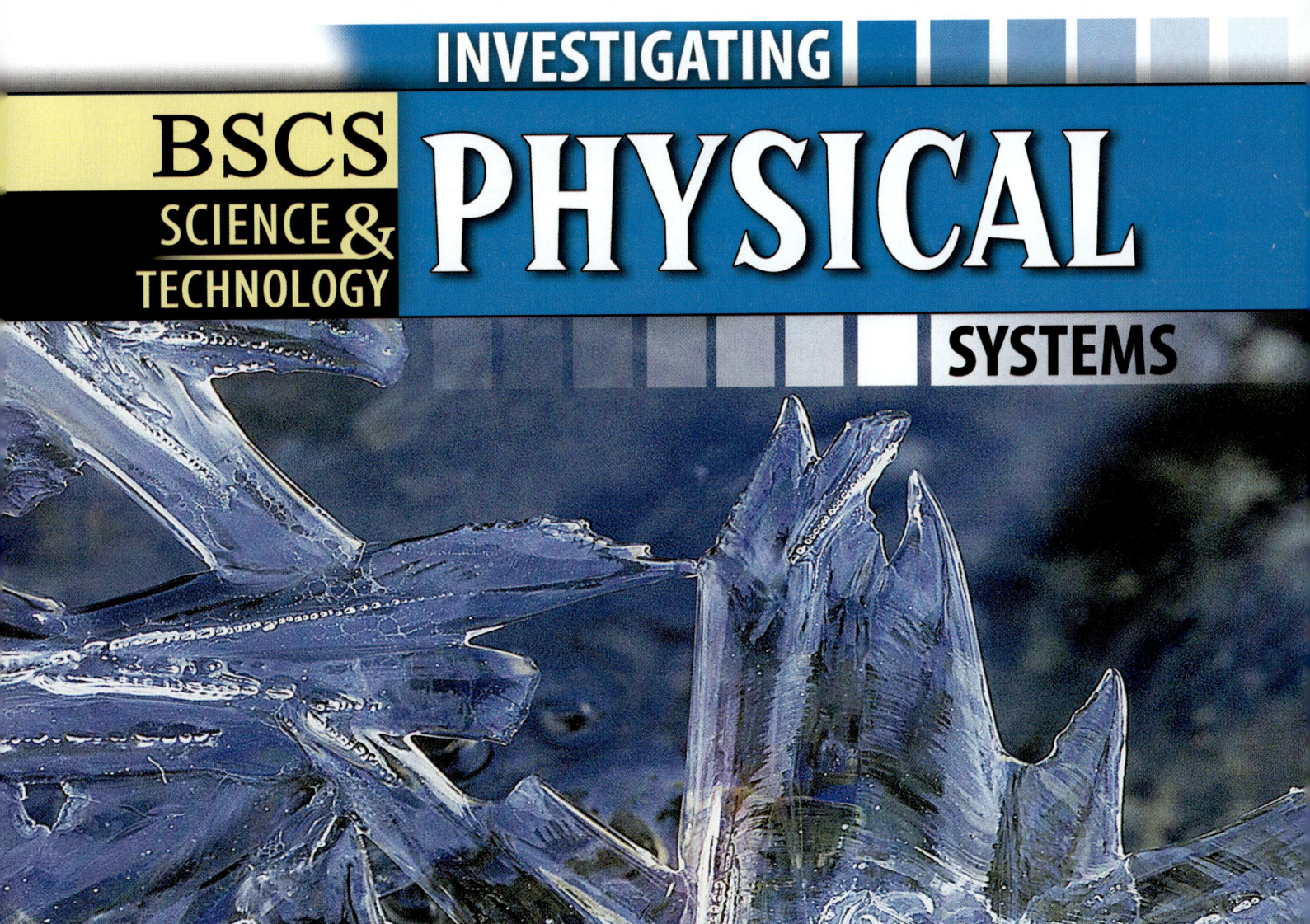

Introduction

WHAT WILL THIS PROGRAM BE LIKE?

Chapter

1. A Learning Journey
2. Science Safety

Your learning journey in science and technology continues. This year will be filled with questions, exploration, and discoveries as you investigate the physical systems in the world around you. If you used *BSCS Science & Technology* last year, you are familiar with some aspects of the road ahead. If you did not, this unit will introduce the program so that you can make the most of this year. If you keep an open, inquiring mind and ask the best of yourself and your classmates, you will experience a learning journey that can continue for a lifetime.

CHAPTER 1

A Learning Journey

By this time in your school experience, you probably have worked both in small cooperative groups, in large classroom groups, and individually. Which way do you best like to work? Why? In this chapter, you will explore different aspects of cooperative learning and individual work. During the year, you will have the opportunity to work in a variety of settings with different people, just as these people at NASA do everyday. You also will have the opportunity to work independently.

ENGAGE
- A Big Ball of String

EXPLORE
- Learning about One Another through Data Collection
- Rainbow Colors
- Meet Al, Marie, Isaac, and Rosalind

EXPLAIN
- Learning and Working Cooperatively

ELABORATE
EVALUATE
- What's in Here?

ENGAGE investigation

A Big Ball of String

What can you do with a big ball of string? In this activity, you will explore one particular task that you and your classmates can accomplish using a big ball of string.

Process and Procedure

1. Listen as your teacher describes a task that the principal would like your science class to accomplish.
2. Complete the task in the best way you know how.

Wrap Up

Participate in a class discussion of your experiences as you attempted to complete this task.

Learning about One Another through Data Collection

Any time a group of curious people gets together, the people want to know more about each other. Sometimes this is not an easy task. In this investigation, you will use the methods of science to learn more about the people in your classroom. You will work together often this year, and you may find this information useful. As you work together, think about what you learned about your classmates, what your classmates have in common, and also what opportunities there are to learn new things from your classmates.

Work individually first and then in teams of two. You will need to sit together at a table or push your desks together to form a table. As you work together, think about ways to share ideas. Be sure to *use your teammate's name.*

FIGURE 1.1 How do you decide what types of food, colors, and music you like?

CHAPTER 1 A Learning Journey

Materials for Each Team of Two:
- assorted art supplies

Process and Procedure

1. Move into your groups.
2. Answer the following questions in your notebook:
   → Do this step by yourself.
   a. On what day of the week were you born?
   b. What are your two favorite types of food?
   c. What is your favorite musical group?
   d. Which after-school activities do you enjoy?
   e. What good books have you read recently?
   f. If you could have a part-time job, what would you like to do?
   g. What would you like to learn more about in science class?
   h. If you could travel anywhere in the world, where would you go?
   i. If you could study with a scientist, with whom would you like to study?
   j. What is the best movie you have ever seen?
   k. If you could speak another language, what language would you choose?
   l. If you were elected president, what is the first thing you would do?
   m. Where would you like to live 10 years from now?
   n. If you were able to choose your own first name, what name would you select?
3. Compare your answers with your partner's.
4. Look at the letter your teacher assigned to you.
   → This letter represents the question you will be collecting, analyzing, and presenting data about.
5. Participate in a class discussion about how to systematically collect the data you need.

FIGURE 1.2 Have you ever wanted to learn another language? How might that be useful?

6. Collect the data you need according to the method your class has decided on.

7. Create a chart in which to compile the responses that you have to your question. Then write 3 sentences about your data.
    → Notebook entry: Each of you should record the chart and the sentences.

FIGURE 1.3 Do you enjoy trying different types of food?

CHAPTER 1 A Learning Journey   7

8. With your partner, decide on a way to present the information you collected to the rest of the class.
   → Your presentation must include something to look at. You may use any of the materials available to create your visual display.
9. Work on your presentation with your partner.
10. As a team, present your visual display to the class and listen to other teams' presentations.

Wrap Up

Discuss these questions with your partner. Then record your own responses to these questions in your notebook. Be prepared to participate in a class discussion.

1. What did you do to learn your teammate's name?
2. What do the data that your team collected represent?
3. What did you learn about your classmates that surprised you?
4. Based on the data presented, what are some things that your classmates have in common?
5. What other questions do you have about your classmates?

Rainbow Colors

The chemicals that you will use in this investigation can make purple cabbage water change color. You will try to produce as many different colors as possible and then determine the patterns associated with the appearance of each color. As you look for patterns, also watch for strategies that make it easier to accomplish this task.

Materials for Each Student:
- 1 pair of safety goggles
- 1 dropper of purple cabbage water
- 3 droppers of different solutions
- 1 reaction tray

Process and Procedure
1. Put on your goggles.
2. Mix 3 drops of the purple cabbage water with 3 drops of 1 of the solutions.
    → Notebook entry: Record today's date and the title of the activity. Then record the solution's letter and the color you obtain.
3. Repeat the procedure with each of the other solutions.
    → Be careful not to mix solutions.
4. Identify how many different colors you can obtain with the purple cabbage water and the 3 solutions with which you began, using only one solution at a time.

You will begin this investigation by yourself and then work in teams of four. When working with others in your team, try to *use their names* as you *listen and share your ideas* with each other.

FIGURE 1.4 Have you ever played on a field hockey team? What strategies do teams use to make teamwork easier?

5. Form a group with 3 other people as your teacher assigns.
   ➡ Notebook entry: Record the names of the people in your group.
6. Share your information with each other so that you know what each person has discovered already.
   ➡ Notebook entry: Record the information in the same way for every person you talk to.
7. Develop a plan to find out how many colors it is possible to create by mixing cabbage water with each of the solutions.
   ➡ You can consult other groups, but you cannot use any more droppers with cabbage water or solutions.
8. Use your plan to collect as much information as you can about the variety of colors that are possible when combining cabbage water with the solutions.
   ➡ Notebook entry: Keep a record of the information you collect. Look for patterns in your information. Be prepared to contribute to a class discussion.

Wrap Up

Answer these questions in your notebook and then be ready to share your answers with the class.

1. How many different colors did you make by yourself?
2. How many different colors did your group make?
3. How many different colors did your class make altogether?
4. Which solutions gave similar colors? Which ones resulted in very different colors?
5. Describe the relationship between the solutions and the pattern of colors you observed.

6. List three reasons why working in a team is an advantage.
7. List three reasons why working in a team is a disadvantage.

Meet Al, Marie, Isaac and Rosalind

Take a moment to look in your book and see whether you can find all four characters who will be joining you in *BSCS Science & Technology: Investigating Physical Science*. In this investigation, you will learn about the scientists for whom these characters are named and why we have included them in the program.

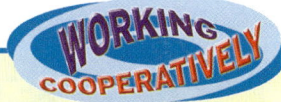

In this activity, you will work in teams of four. As you work, practice letting other people know your thoughts and ideas and continue to use your teammates' names.

CHAPTER 1 A Learning Journey

Materials for the Entire Class:
- poster board
- magazines
- glue
- scissors
- assorted art supplies

Process and Procedure

1. Make an entry in your notebook indicating today's date and the title and purpose of the investigation.
2. Participate in the count off.
   → Notebook entry: Write down whether your number is one, two, three, or four.
3. Locate the reading Getting to Know You that follows the Wrap Up for this activity. Each person on the team should read about one character. Divide the reading as follows:

   1s—read about Al

   2s—read about Marie

   3s—read about Isaac

   4s—read about Rosalind
4. When you have finished the reading, create teams that include 1 person who has been assigned each number. Copy the chart in Figure 1.5 into your notebook.

FIGURE 1.5 Copy this chart into your notebook and fill it in as your team discusses each character.

Character	Named After	Born	Died	Area of Science
Al				
Marie				
Isaac				
Rosalind				

➡ You should be in a team that has a person whose is number one, a person whose is number two, and so on.

5. Take turns telling the rest of your teammates about the scientist and character you just read about. Fill in the chart as you learn about each scientist and character.
6. Read the sets of words in the chart in Figure 1.6 and think about how they apply to you.
7. Think about which set of words best describes you.
    ➡ You may not identify with each entry in a column, but think about which column best fits you.

FIGURE 1.6 Think about each set of words. Which set best describes you?

Chart of Learning Styles

Type W	Type X	Type Y	Type Z
appreciates facts more than ideas	likes to learn by reading	likes to think about many things at once	likes to learn by experiment
likes to have exact directions	likes to think about what makes an idea important	likes to work without thinking about time limits	likes to discover new information
uses numbers effectively	knows self well	is perceptive about others' intentions and feelings	enjoys musical expression
likes to do one thing at a time	uses language effectively	appreciates ideas more than facts	likes to be independent and to compete
likes to find out how things are useful	likes to know what is happening next	enjoys working in a group	enjoys using tools or gadgets

→ Notebook entry: Write down which set of words you chose—W, X, Y, or Z—and why.

8. Find 3 classmates who chose the same set of words that you did. Take time to learn everyone's name and find out what other things you have in common.

9. In this new team of 4, use the materials that your teacher provides to create a collage to represent the traits and ideas that describe your group.
   → In addition to images, you may want to use words or phrases.

10. Display your collage alongside other collages that depict the same set of words.

Getting to Know You

We have developed four characters to serve 2 primary purposes in the curriculum. The characters show the different ways people learn, the various talents that people have, and some important events in the history of science. The characters do not represent actual animals or people, although they are named for real scientists.

Meet Al

Al is named after Alfred Wegener, a German geophysicist and meteorologist who was born on 1 November 1880. Dr. Wegener made many expeditions to Greenland to test his ideas about meteorology and geophysics. On his last expedition, Dr. Wegener and an Eskimo companion named Rasmus Villumien brought supplies to a science station in the center of Greenland. Without these supplies, the people at the station would not have been able to survive the winter. A few days later, on his fiftieth birthday, Dr. Wegener and Villumien left to return to the coast by crossing 250 miles of Greenland with two sleds and 17 dogs. As the two men left the science station, the other scientists noted that the dogs

FIGURE 1.7 Al would like you to know about Alfred Wegener, a German geophysicist and meteorologist.

were worn out and the men were racing with death. Wegener and his colleague were never seen alive again. Six months later, Wegener's body was discovered carefully buried beside the trail halfway between the science station and the west coast station, but Villumien was never found. The search team assumed that he lost his way and fell into a crevasse.

Dr. Wegener is most noted for his work on the origins of the continents and oceans; he developed the theory of continental drift. Wegener believed that all of the continents originally had been one landmass. He also claimed that the movement of the continents caused great changes in climate. When Wegener proposed his ideas in 1912, other scientists scoffed at him. In the 1930s and 1940s, professors used Wegener's ideas as an example of scientific blundering. It was not until the 1950s that Wegener's ideas were reexamined and taken seriously. In the 1960s, a number of ecological discoveries completely supported his theory.

In *BSCS Science & Technology*, Al is someone who does science for the fun of it. He would rather learn by doing than by reading about things. He likes knowing how science and technology connect to his life. He also likes using his imagination and being part of a group.

Meet Marie

Marie is named for Marie Curie. Dr. Curie was born in Poland in 1867. She moved from Poland to France in 1891 to study science at the Sorbonne because she could not find satisfying work in her homeland. At this university, she met the man who became her husband, Pierre Curie. He was already a well-known scientist. They worked together on many projects until Pierre was killed in an accident in 1906. After her husband's death, Dr. Curie continued the research on her own.

FIGURE 1.8 Marie would like you to know about Marie Curie, a Polish physicist and chemist.

Marie Curie was quite a trendsetter for her time. She is most famous for her research on radioactivity. She won two Nobel Prizes, one for physics and one for chemistry. Her first Nobel Prize, in 1903, was credited to Marie, Pierre, and their colleague Antoine Becquerel. The second prize, which Dr. Curie won in 1911, was for her discovery of the new elements radium and polonium (named after her homeland of Poland), and for her study of the chemical properties of radium. She is the only person ever awarded Nobel Prizes in both physics and chemistry. Although Dr. Curie made many scientific discoveries, earned two Nobel Prizes, and taught at a prestigious university, she was never elected to the Academy of Sciences because, at that time, the academy did not elect women. Marie lived until 1934.

In this curriculum, Marie is the one who wants to understand. It is important for her to know how everything connects. She enjoys reading and writing. She always asks *why* and gets frustrated when her *why* questions cannot be answered. She understands herself well.

Topic: Scientists' Biographies
Go to www.scilinks.org
Code: physical16

Meet Isaac

Isaac is named after Sir Isaac Newton, an English physicist, astronomer, and mathematician who was born on 25 December 1642. He lived until he was 75 years old. During his long life, he caused some trouble in the scientific community because he questioned some ideas that had been around for many years. For instance, his theory of gravity is well accepted today, but it was not 300 years ago. Newton developed the notion of gravity while living in isolation in the country. He moved there to avoid the outbreak of bubonic plague in Cambridge, England.

He also invented the branch of mathematics known as calculus. And he developed an explanation for the nature of light and color. Other scientists did not accept Newton's

FIGURE 1.9 Isaac would like you to know about Isaac Newton, an English physicist, astronomer, and mathematician.

A Big Ball of String, Rainbow Colors, and Learning about Each Other through Data Collection are three very different activities that you have just completed. One purpose of this unit is to introduce you to some key ideas that you will need throughout this year, and we designed these activities to introduce these ideas to you.

One aspect of this science program that you probably are familiar with is **cooperative learning**. In many activities, you will work cooperatively with other students to complete a task. This means that you will be doing more than just working in a group. You will be sharing ideas and incorporating the ideas of others as you work. Cooperative learning is not group work; it is teamwork. If you have worked cooperatively in the past, this unit will be a review of what cooperative learning is and how it works. If you haven't worked cooperatively before, this unit will help you become familiar enough with cooperative learning to begin using it in Unit 1.

Not all of the activities in this science book will be for cooperative teams. The ones that are cooperative are activities that require a lot of materials, or that require sharing ideas to reach a goal.

4. In which of the activities that you just completed did you work cooperatively?
5. Why would you say that the activities you listed in Question 4 were cooperative?

Social Skills

In some of these investigations, you came across a section called Working Cooperatively. In all the cooperative activities in this book, the Working Cooperatively section will tell you to try to practice a specific action, such as learning names. These actions are called **social skills**.

Social skills are always a part of cooperative learning. In this program, we use two types of social skills—activity skills and unit skills.

The activity skills are specific to each investigation. These skills usually have to do with how the cooperative team functions together. The activity skills that you will see in this book are as follows:

- Use your teammate's name.
- Move into your teams quickly and quietly.
- Listen and share your ideas.
- Stay with your team.
- Show caring and respect for others and their ideas.
- Share your thoughts and ideas.
- Let others finish without interrupting them.
- Look at the person speaking to you.
- Choose an explanation that includes the ideas of all teammates.
- Respect the working environment of others.

6. How would you describe the purpose of these activity skills to a friend in a different class?
7. What skills did the investigation Rainbow Colors ask you to practice?

In addition to the activity skills, you will have one unit skill to practice throughout a particular unit. The unit skills are similar to the skill of sharing ideas in Rainbow Colors. Unit skills concentrate more on communicating, trusting, and solving or avoiding conflict in your team. These skills tend to be harder to practice because they don't always feel natural at first. By the end of this year, however, if you have used the unit skills, they probably will feel natural in many team situations. The unit skills that you will practice in this book are as follows:

ideas at the time. Newton was a sensitive person, and the criticism hurt him greatly. In fact he did not always want to publish his ideas, and his friends had to plead with him to do so. In 1705, Queen Anne knighted him. Newton enjoyed exploring ideas in many areas, not just science and mathematics.

Isaac in *BSCS Science & Technology* likes to know facts, lots of them. But he does not particularly care whether he understands how the facts fit together. Isaac understands science books very well and has a good memory for factual information. He is always on time and helps others stay on schedule.

Meet Rosalind

The character Rosalind is named after Rosalind Franklin, a British chemist and molecular biologist who lived from 1920 until 1958, when she died of cancer. Rosalind was born in London and attended St. Paul's Girls' School where she received an excellent background in physics and chemistry. She entered Cambridge University in 1938. Between 1947 and 1950, she worked in a lab in Paris. In this lab, she worked with Jacques Mering and learned X-ray diffraction techniques.

In 1951, she left France to begin a three-year research fellowship at King's College. Even though she had poor equipment to work with, Dr. Franklin was able to develop a system for taking high-resolution photographs of single fibers of DNA (deoxyribonucleic acid). This technique helped reveal the shape of DNA. Her work contributed greatly to the work of James Watson and Francis Crick who developed the idea that DNA is shaped like two spirals (a double helix). Franklin also studied the structure of coal, the tobacco mosaic virus, and the polio virus, which helped people better understand the properties of each.

FIGURE 1.10 Ros would like you to know about Rosalind Franklin, a British chemist and molecular biologist.

In this program, Rosalind loves computers and other technological gadgets as well as music and musical instruments. She enjoys doing experiments and trying to figure out how to do the most precise experiment possible. She would rather be *doing* science than talking about ideas.

Wrap Up

1. What are some similarities among the scientists listed in your chart?
2. What are some differences among the scientists listed in your chart?
3. Each character likes to learn in a unique way. Write down the name of each character and two characteristics of that character's style of learning.
4. Which character are you most similar to in terms of how you like to learn? Why?
5. When working in a group, why is it beneficial to have people with different talents?

Learning and Working Cooperatively

Think about the differences in the characters that you just met. Do you think these characters learn in the same way or learn in different ways? Understanding *how* you learn can make learning more fun. It also can help you be more successful. Scientists who study how people learn are called cognitive psychologists. When we wrote this program, we used ideas from these scientists because we

FIGURE 1.11 Look at this photograph and the rest of the photographs in this reading. How are the people working cooperatively?

wanted to develop a program that helped everyone learn better. As you continue the program, you will notice that each investigation, reading, or connection activity has a word that begins with an E associated with it. There are five E words altogether: Engage, Explore, Explain, Elaborate, and Evaluate. This series of words helps describe an effective way for students to learn.

For each big idea that we present, you will experience that idea in a series of ways. First, the idea must **engage** you. To do this, we might ask you some interesting question to find out what you already know about an idea. Or we might provide you with a simple and fun activity to generate your interest. After you are interested in the idea, then you need an opportunity to **explore** it in more detail. In the explore activities, you will have an opportunity to investigate the idea more closely and

CHAPTER 1 A Learning Journey

thoroughly and to ask more questions. After you have explored the idea and have had more experience with it, then you begin to develop an **explanation** for what you have been exploring. Sometimes explain activities are readings and sometimes they are investigations. After you have an explanation, then you need to **elaborate** on that understanding. In the elaborate activities, you apply your understanding to new settings or learn more about some specific aspect of the idea. Finally, you are able to **evaluate** what you have learned.

When we present another big idea, we begin the cycle again. You will notice that in each chapter, we go through this cycle one time. Sometimes there are several explore, explain, or elaborate activities, for example, and sometimes we combine an engage and an explore activity or an elaborate and evaluate activity. During the year as you learn, you will notice that rather than explaining things all the time, your teacher is there to help *you* do the learning and the explaining. Cognitive scientists know that this is an effective way for students to really understand science. This way of learning also accommodates the different strengths and styles of each student.

STOP & THINK

1. What do you think is the difference between an engage activity and an explain activity?
2. What do think is the difference between an explore and an evaluate activity?
3. Even though the entire cycle of activities is important for learning, which E activity do you think Rosalind would like best? What about Isaac? What about you?

Unit 1—Show caring and respect for others and their ideas.
Unit 2—Be open to others' ideas.
Unit 3—Disagree with the idea, not the person.
Unit 4—Choose an explanation that includes the ideas of all Team Members.

These skills involve a lot more than using names or moving into your groups. Even if an activity does not specifically require that you use the unit skill, you still should try to practice it.

Now think again about the activities you just completed. Answer these questions about these activities.

8. When you worked in groups, did you work as a team? That is, was everyone participating?
9. Make a list of jobs that you saw people doing as they worked on the first three activities.

Roles

In cooperative learning, one key idea is for everyone to be involved. One way to avoid problems about who will do what is to have assigned jobs. In this program, every Working Cooperatively section for every cooperative investigation will tell you which of four jobs are necessary—Manager, Communicator, Tracker, or Team Member. Each member of a team will have a chance to do all four of these jobs several times during the year.

10. Study the job descriptions in Figure 1.12. Describe in one phrase or sentence what the major responsibilities of each job are.

STOP & THINK

11. Describe how using any or all of these jobs might have helped you avoid confusion as you worked on the first three activities.

As you probably guessed, the Working Cooperatively section of each investigation will provide you with important information about cooperative learning.

Processing

At the end of some of the cooperative learning investigations, you will have the opportunity as a team to discuss how well you functioned as a team or to discuss how well you practiced the activity and unit skills. Your

FIGURE 1.12 As you work in cooperative teams in science class this year, you will use these roles.

ROLES AND DUTIES

Manager:
- Pick up, distribute, and return materials.
- Report shortages and damage of materials to the teacher.
- May leave the team to collect materials.

Communicator:
- Maintain communications with other teams.
- Seek help from other teams or the teacher.
- May leave the team to talk with other teams or the teacher.

Tracker:
- Help the team keep track of procedural steps, deadlines, and times.
- Make sure that each member is aware of the directions for an activity.

Team Member:
- Help in team cleanup.
- Summarize the team's thoughts or activities.
- Act as spokesperson for a team.
- Help others remember their job descriptions.

chance to do this usually comes in the Wrap Up section. It is important for you to realize that you are working cooperatively to benefit your learning and not just to make the work easier. In the end, you alone are responsible for understanding whatever you accomplished as a team. If each team member does not understand what happened, then the team's work is not finished.

12. List three things that team members could do to ensure that everyone understands what happened during an investigation.

connections

ELABORATE
EVALUATE

What's in Here?

To help you learn the features of this book before you begin to use it, you will complete a scavenger hunt. You will locate and answer questions about certain features of *BSCS Science & Technology: Investigating Physical Systems.* Work on both parts according to your teacher's directions.

Part A—Scavenger Hunt

Answer the following questions in your notebook as your teacher directs:

1. What is the chapter opening art for Chapter 17?
2. What is the social skill listed in Working Cooperatively for the investigation Small-Scale Boats?
3. What is one of the elaborate activities in Chapter 15?
4. On what page does the reading Paper Towel Consumers begin?
5. What is the title of one of the Sidelights in Chapter 5?
6. What does the Caution in the investigation Light, Lenses, and the Eye warn you about?
7. In the investigation Taking a Closer Look at TV Pictures, what roles are listed?
8. How many Stop and Think sections are in the explain activity in Chapter 11?
9. On what page is the connections activity Human Factors as a Design Constraint?
10. How does your glossary define the term "cones"?
11. List all of the materials for the investigation A Penny's Worth of Water.
12. In which phase of the instructional cycle (engage, explore, explain, elaborate, or evaluate) is the investigation A Penny's Worth of Water?
13. What is the title of the engage activity for Chapter 9?
14. What roles are listed in the Working Cooperatively section for the investigation Toys for Tots?
15. What system is being explained in Complex Technological Systems?

Part B—Discussion Questions

Work on these questions according to your teacher's instructions.

1. In general, who received more points for this activity, the individuals or the teams?
2. Explain your answer for Question 1.
3. Which group of students learned more during this activity?
4. Explain your answer to Question 3.
5. Do you think it would be fair for your teacher to grade everyone on this investigation?
6. What are the advantages of working as a team?
7. What might be some disadvantages of working as a team?

Chapter 2

Science Safety

Safety is an important concern in our lives, and it will be especially important in your science classrooms in the years ahead. This chapter will help you understand the basics of science safety and will provide a foundation on which you can build in the future.

ENGAGE — Science Is . . . Technology Is . . .

EXPLORE — Thinking about Safety

EXPLAIN — Science Safety Contracts

ELABORATE — Cooperating for Safety's Sake

EVALUATE — Science Safety

ENGAGE investigation

Science Is . . . Technology Is . . .

The title of this program uses the words science and technology. What do these words mean to you? What do these words mean to your teammates? In this investigation, you will begin to develop answers to those two questions.

Process and Procedure

1. Make an entry in your notebook that indicates the date, your name, the purpose of the investigation, who is on your team, and what each person's job is for the day.

Work in combined teams of two to create teams of four. Your teacher will assign each of you a role so that you have a Tracker, a Manager, a Communicator, and a Team Member. Work on the social skill *Move into your teams quickly and quietly*. Form a circle or arrange your desks in a circle.

2. The Tracker should read the directions to the team. The Team Member should say the sentence "Science is _____." filling in the blank.

3. Then the Communicator should repeat the sentence said by the Team Member and add something to the sentence.

4. Continue in this way until each person has added two things to the sentence.
    → The Tracker and Manager should go next. You may want to take notes in your notebook to help you remember the sentence.

5. After everyone has added two things to the sentence, each person should write a summary of what the group said about science.
    → Notebook entry: Read each other's summaries and develop a summary sentence.

6. After you agree on a summary, the Communicator should visit two other groups, compare summaries, and report back to the team.

7. Repeat Steps 2–6 using the sentence "Technology is _____."

Wrap Up

Answer these questions. Be sure everyone is prepared to contribute to a class discussion.

1. How was your team's definition of science similar to other teams' definitions? How was it different?

2. How was your team's definition of technology similar to other teams' definitions? How was it different?

3. How easy was it to move into your groups quickly and quietly?

4. Did the teams around you move into their groups quickly and quietly?

EXPLORE connections

Thinking about Safety

In the scene below, Al, Marie, Ros, and Isaac are carrying out an investigation. We can see that this investigation involves some traditional lab equipment.

You might have had experience working in the laboratory. But whether you have or not, you probably have a good idea about what would be safe or unsafe in a lab. Study the scene in Figure 2.1. Then, in your notebook, record every unsafe action that you see. Write beside each action why you think it is unsafe. When you are finished, think of several rules that would help the characters avoid their unsafe behaviors.

Topic: Science Safety
Go to www.scilinks.org
Code: physical32

FIGURE 2.1 What unsafe actions do you see in this drawing?

connections EXPLAIN

Science Safety Contracts

Think about the school building that you are in right now. What are some general and some specific safety concerns that you might have about the building and about the school activities that take place in the building? What are some concerns that might come up about safety, specifically in the science classroom? What are some things that you can do to reduce the possibility of accidents and unsafe conditions? What are some rules that all students in the science classroom might follow to ensure everyone's safety?

After you have discussed these questions with your classmates and have generated a list of classroom safety rules, work in teams to create a poster that depicts one of these rules. Your teacher will help you display your posters around the classroom as reminders.

Carefully read over the safety contracts and letters that your teacher distributes to you. Take them home and discuss them with your parent or guardian. Be sure to have him or her sign the appropriate forms before you return them to your teacher. You have to turn this in before the teacher can allow you to participate in any science labs.

CHAPTER 2 Science Safety

ELABORATE  investigation

Cooperating for Safety's Sake

Conducting science safely and cooperatively is a rewarding experience. In this activity, you will work with four chemicals that react when you combine them. Your task during the investigation is to determine what effect each chemical has when combined with the other chemicals.

Materials for Each Team of Two:
- 2 pairs of safety goggles
- 1 medicine cup of chemical A
- 1 medicine cup of chemical B
- 1 medicine cup of chemical C
- 1 dropper bottle of chemical D
- 1 teaspoon
- 6 empty resealable plastic bags
- 1 resealable plastic bag with a mixture of chemicals A, B, and C

You will work in your team of two. The roles you need to use are a Manager and a Communicator/Tracker. You should try to use your teammate's name as you share your ideas. Be sure to work on the social skill *Listen and share your ideas* during this investigation.

Process and Procedure

1. Review the procedure.
   → The Tracker should lead this review.
   → Notebook entry: Record the date, title, and purpose of this investigation, and the name of your teammate. Also create a list or chart to record your initial observations.

2. When you think that you understand the activity, obtain the materials.
   → This is the Manager's job.

3. Leave the bag closed and record in your notebook as many observations as you can about each of the chemicals.
   → Share your observations with your teammate.

4. Add 20 drops of chemical D to the bag of chemicals.
   → After you have added chemical D, press the air out of the bag and seal it.

5. Pass the bag back and forth to your partner and describe all of your observations when the bag comes to you.
   → Notebook entry: Record your own and your teammate's observations.

6. Think about these 2 questions and discuss them with your partner.
   - Could you tell which chemical caused which result?
   - How might you determine which chemical caused which result?
   → Notebook entry: Record your responses to the questions. Make sure you listen and share ideas.

7. Develop a plan to determine which chemical caused which result.

8. Follow your plan.
   → Use safe procedures at all times. Be sure to wear your safety goggles.
9. Dispose of the contents of your plastic bags down the drain and rinse out each bag.

Wrap Up

Discuss these questions as a team. Then record your own answers in your notebook.

1. What did chemical A do?
2. What did chemical B do?
3. What did chemical C do?
4. What did chemical D do?
5. Explain how you decided on your answers to the first four questions.
6. What safety precautions did you take and why were they important?
7. Rate how effective you were in listening and sharing your ideas. Use a scale from 1 (*not at all*) to 10 (*very often*).

Science Safety

Cooperative Learning! Safety! Are you ready for Unit 1? In this investigation, you will play a game that gives you the chance to use cooperative learning as a way to review your understanding of safety. As you play the game, you will have the opportunity to look ahead and see what is coming in science.

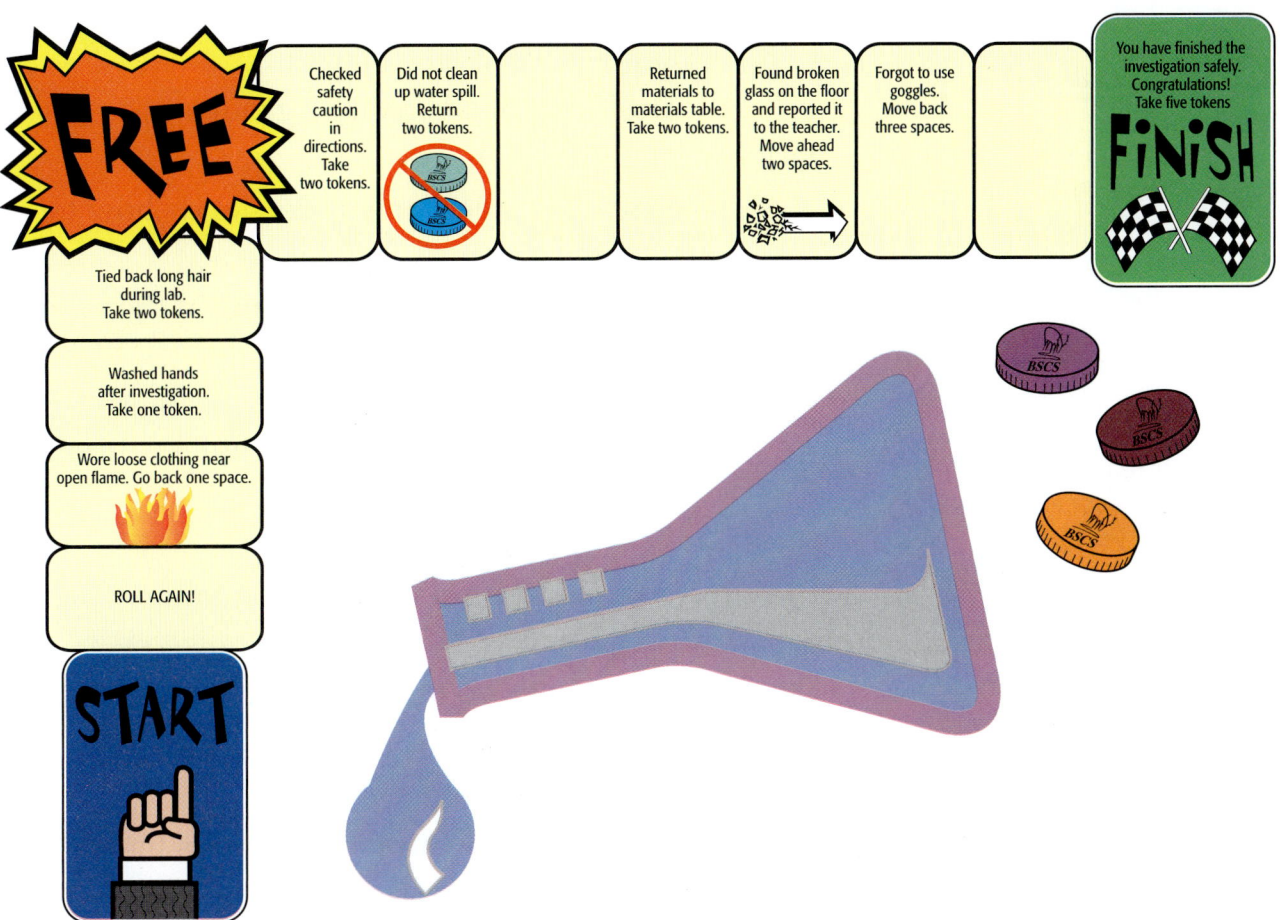

Materials for Each Team of Four:
- 1 die or spinner
- 1 game board
- 1 set of student game rules
- 2 cups for holding tokens
- 100 tokens
- 4 player pieces

Chapter 2 Science Safety

You will work cooperatively in teams of four. Push your desks together to form a table, or work at a large table or on the floor. All Team Members will need equal access to a game board. Use the roles of Manager, Communicator, Tracker, and Team Member. As you play, practice the social skill *Use your teammates' names*.

Process and Procedure

1. Get in your team of 4.
   → This team of 4 is really 2 groups of 2.
2. Read all the procedures.
   → Make sure everyone understands how to play the game.
3. Place token cups on the game board in the marked locations.
4. Select a playing piece and place it on "Start."
   → Each Team Member should have a different playing piece.
5. Take 1 token.
   → Do this before beginning the game.
6. Begin to play.
   → Play should rotate clockwise with players taking turns.
7. Roll the die or spin the spinner.
8. Move your piece the number of spaces indicated.
9. Follow the directions that are on the space on which you land.
   → If you land on a space that sends you to "time out," you lose a turn. When it is again your turn, return to the space that sent you to "time out" and roll the die or spin the spinner. You will collect and lose tokens as you move around the game board.
10. When you reach the last square "Completed Science Investigation Safely," take 5 tokens.
    → Do this only if you are the first person to finish. The game is then over.
11. Count the tokens you collected.
    → The winner is the individual who has the most tokens. This is not necessarily the first player to complete the game.

Wrap Up

Discuss the following questions as a team. Record your responses in your science notebook. Be prepared to describe your team's process if the teacher calls on you.

1. On a scale of 1–10, rate your team for how well the Team Members understand science safety.
2. Which of the activities that you previewed look the most exciting to you?
3. Describe your team's success in using each other's names.

UNIT 1
EXPLORING RANGES OF LIMITS AND DIVERSITY

Chapter

3 Identifying Limits and Diversity

4 Ranges of Limits and Diversity

5 Using Limits to Set Standards

6 Using Diversity to Set Standards

7 Evaluating Your Understanding of Limits and Diversity

As individuals, we experience limits and diversity in the things we do every day. The human form, amazing as it is, has limits of its own. As humans, we are diverse in the way we work within our limits and sometimes surpass a previous limit. What do you think are some of your limits?

In this unit, you and your classmates will explore your abilities to do various things. In the process, you will be exploring a range of limits and diversity. You will use your results to set standards.

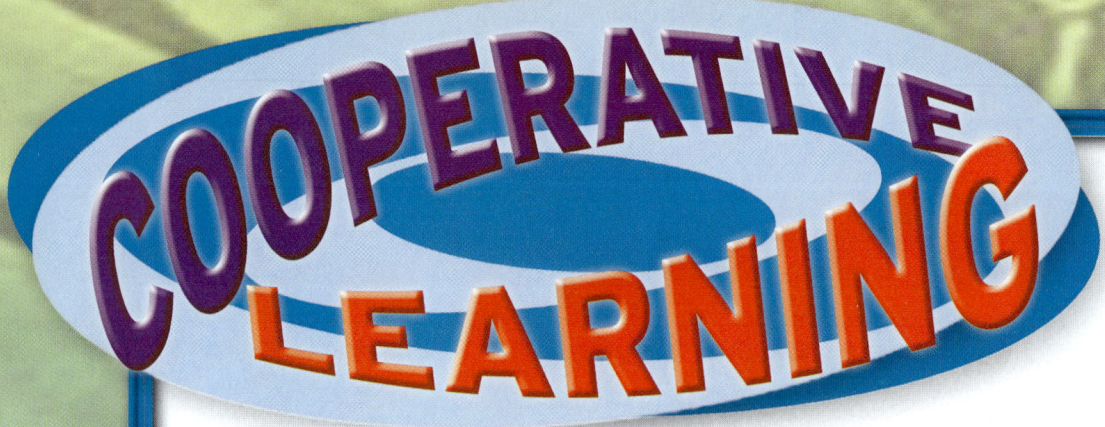

In this character scene, the characters are assembling themselves into a cooperative team. It appears that they know something about cooperative learning. Perhaps this is because they worked cooperatively the previous year or perhaps because they have just finished working on the Introductory Unit. In either case, what they are doing merits some study.

First, the members of the team have made name tags and role tags for themselves. Your teacher might ask you to do the same. Name tags help your teammates and other members of the class learn your name. Role tags help identify what your responsibilities are.

The roles you will use in this book are Communicator, Manager, Tracker, and Team Member. Second, as Ros points out, you are supposed to assume the role of a Team Member in all cooperative activities. You also might have to assume another role during certain investigations. Refer frequently to the list of role descriptions to make sure you understand the duties of the roles you assume.

Finally, Al mentions that it is time to make a T-chart. During this unit, you will work on one skill during each cooperative activity. The unit skill, as Al mentioned, is *Show caring and respect for others and their ideas*. Before you begin the first chapter, you need to discuss this skill and create a T-chart according to your teacher's instructions. If you have never made a T-chart, read How to Make a T-chart (How To #1) at the end of the book. You will need to refer to this T-chart to help you practice the skill. Even if the activity does not specifically mention this skill, you should try to practice it. Most cooperative activities will mention another skill that you should practice in addition to the unit skill.

CHAPTER 3

Identifying Limits and Diversity

Do you play a musical instrument? Think about the range of instruments in an orchestra. The instruments create sound in a variety of ways and each instrument also has its own set of limits. What is the highest note a musician can create on an oboe?

In this book, we will use the term **limit** to mean boundary, or something that you cannot go beyond. For example, you are familiar with city limits, the points at which a city begins and ends. You have probably noticed that you and your classmates have different abilities to do certain things. We will use the term **diversity** to describe this variation in individual abilities. You and your classmates have a diversity of limits to what you can do.

In this chapter, you will explore your limits and the diversity among your classmates by performing three different tasks.

ENGAGE / EXPLORE
- Star Tracers

EXPLORE
- Threading the Needle
- How Do You Spell Success?

EXPLAIN
- If at First You Don't Succeed
- Doing It All the Same

ELABORATE
- Seeing the World around You

EVALUATE
- Light, Lenses, and the Eye

45

ENGAGE EXPLORE

investigation

Star Tracers

How well do your brain and hands communicate with each other when your eyes are not getting the messages they usually get? How quickly can you adjust to such a situation? In this investigation, you will attempt to complete a challenging task. You and your teammates will measure how well you each accomplish this task and then share your results with the class.

Materials for Each Team of Two:
- one 5 × 7-in. mirror
- 1 cardboard box to make a visual shield
- 1 pair of scissors
- 1 stopwatch or clock with a second hand
- 10 copies of the Star Pattern Test Sheet

Process and Procedure

Part A—The Social Skill

1. Prepare your notebook for this investigation.
   → Notebook entry: Record the title and the purpose of the investigation, and the date.
2. Write your teammate's first and last name in your notebook.
3. Discuss why it is important to use your teammate's name.
4. Record two of the best reasons you discussed.

Part B—The Activity

1. Copy the column headings from Figure 3.1 into your notebook. Copy them across the top of the page so that you can record information from this activity.

Work cooperatively with a partner. You will need the roles of Communicator and Manager. Each of you will be a Team Member. Check the description of roles in Chapter 1 on page ** to be certain of your duties. Move your desks together so you are facing each other or sit at a table across from each other. Work on the unit skill *Show caring and respect for others and their ideas* and the activity skill *Use your teammate's name.*

46 CHAPTER 3 Identifying Limits and Diversity

Hubble snapshot captures life cycle of stars

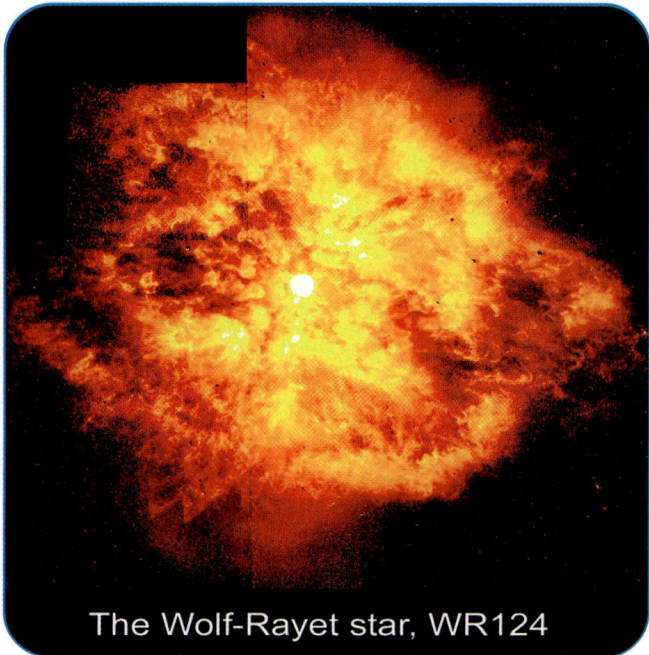
The Wolf-Rayet star, WR124

→ This is called a data table. When scientists conduct investigations, they record their information in this organized way. You will construct many data tables in the future, so this is an important skill to learn. If you want to learn more about constructing data tables, refer to How to Construct a Data Table, How To #2.

2. Obtain all the materials you will need for this investigation.

    → This is the Manager's role.

3. Cut out two opposite sides of your cardboard box and place the box on the table or desk in front of you like a tunnel.

Person attempting the task	Trial #	Time	Describe difficulty if any	Success? YES or NO

FIGURE 3.1 Make a data table just like this one on a page in your notebook. You will use the table to record your team's data from Star Tracers.

CHAPTER 3 Identifying Limits and Diversity 47

→ This will serve as a visual shield to keep you from seeing directly what your hands are doing.

4. Sit behind the shield and write your name on a copy of the Star Pattern Test Sheet, and then slide it into the tunnel.
   → The Manager will do this first.

5. Place the mirror in front of the tunnel and position it so the person doing the tracing can see his or her own hands and the star in the mirror.
   → The Communicator should position the mirror for the Manager, who will be the first tracer. You will take turns doing this activity. The person doing the tracing should not be able to observe his or her hands directly, only a reflection of them (see Figure 3.2).

6. When your teammate is ready to trace, say "Begin," and start timing.
   → To begin timing, the Communicator either notes the starting time on a clock or starts the stopwatch.

FIGURE 3.2 This is the correct way to use your Star Tracer's setup.

7. When you hear the word "begin," use a pen or pencil to begin tracing the star pattern. Try to stay between the two borders of the star.
   → Watch your hands in the mirror.

8. When you complete your first attempt at tracing the star, say "stop."
   → The Communicator should note the stopping time on the clock or stopwatch.

9. Calculate the total time it took your teammate to trace the star.
   → The Communicator will do this.

STOP: Are you using each other's names?

10. Fill in the columns of your data table with the appropriate information from this tracing.
    → You and your teammate will need to decide what you will count as an error and how you will determine whether each tracing was a success.

11. Attempt to trace the star 4 more times, using a new test sheet each time.
    → The Manager traces the pattern 4 more times. After each attempt, the Communicator records the necessary information under the appropriate headings in the data table.

12. After 5 tries, trade places with your teammate.
    → The Communicator now tries to trace the star 5 times, while the Manager records the appropriate information in the data table.

13. Analyze your data.
    → This task should be simpler because you have organized your data into a data table. To analyze your data, do the following:

    a. Count the number of times each person was successful in tracing the star.

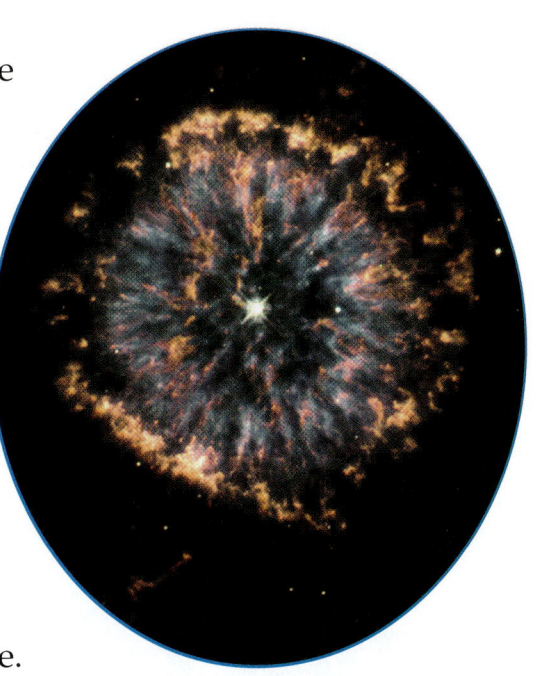

Chapter 3 Identifying Limits and Diversity   49

b. Count the number of times each person was unsuccessful in tracing the star.

c. Calculate the successful attempts and the unsuccessful attempts for your team.

14. Record each of these totals.
    → Notebook entry: Record this in your notebook following your data table.

Wrap Up

Write the words "Wrap Up" in your notebook. Discuss these questions with your teammate, and then write your own answers in your notebook. Be certain you can explain your answers during a class discussion.

1. Did anyone on your team trace the star successfully the first time he or she tried the task?

2. Describe your experience of trying to trace the star by looking in a mirror.

3. What limits did you experience as you tried to trace the star?

4. As a team, rate how successful you were at using each other's names. Pick one of these words to describe your success: excellent, good, fair, poor.

EXPLORE  investigation

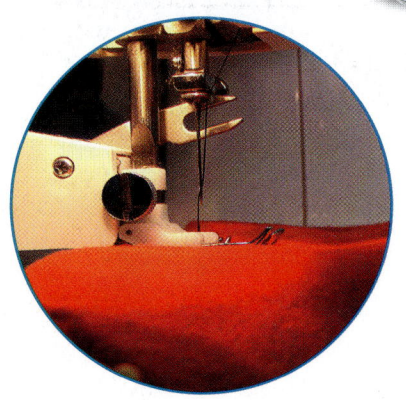

Threading the Needle

In Star Tracers, your ability to trace the star was limited because you couldn't see your hands. Some of your classmates may have been quite successful, while others were probably less successful. In other words, there was a diversity of limits within your class. Will the same people who were successful in Star Tracers be successful in another investigation that places a different limit on their vision? This investigation gives you a chance to find out.

Materials for Each Team of Two:
- 1 eyebolt
- 1 bolt
- 2 rulers
- 2 sheets of graph paper

Process and Procedure

1. Construct a data table to use with this activity.
   → Follow the steps in How To #2, How to Construct a Data Table. Each person will need a data table in her or his notebook.

2. Collect the materials you will need for this investigation.
   → This is the Manager's role.

3. Stand up. Hold the eyebolt in one hand at arm's length in front of you with your elbow slightly bent. Hold the bolt in your other hand with your arm down at your side.
   → The Communicator will do the task first, while the Manager records the results of each attempt in your data table.

4. Close your right eye. Try to get the bolt through the hole of the eyebolt in one motion.
   → You have 5 tries.

5. Open your right eye and close your left eye. Try to get the bolt through the hole of the eyebolt in one motion.
   → You have 5 tries.

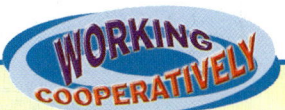

Work cooperatively in your teams of two. Use the roles of Communicator and Manager as well as Team Member. Continue to practice the unit skill and the activity skill *Use your teammate's name*. Push your desks together, side by side, or sit beside each other at a table. Clear enough space by your desks or table so you can stretch out your arms.

FIGURE 3.3 This is one way to thread the eyebolt.

6. Keep both eyes open. Try to get the bolt through the hole of the eyebolt in one motion.
   → You have 5 tries.
7. Record which attempts were successful and which were not successful for each set of attempts described in Steps 4 through 6.
   → Notebook entry: Record this in your data table. Help your partner by recording data for him or her.
8. Switch places and repeat Steps 3 through 7.
   → Don't forget to record the data.
9. Exchange and record the information in your data tables.
   → Each person should have all your team's attempts recorded in his or her own data table.
10. Return your materials to the appropriate location.
11. Analyze your data.
    → Calculate these totals for both people:
       a. The number of successes with the right eye closed.
       b. The number of successes with the left eye closed.
       c. The number of successes with both eyes open.
       d. The total number of successes.
    → Notebook entry: Record this information in your notebook following your data tables.
12. Enter each Team Member's individual success on the class data table.
    → Your teacher will provide this information on the board or on an overhead transparency.
13. After discussing your data, make a bar graph of your results.
    → See How To #3, How to Construct a Bar Graph, in the back of this book.

Wrap Up

Write the words "Wrap Up" in your notebook. Discuss these questions with your team and record your answers. Be sure to write each answer in complete sentences and be able to explain your answers.

1. Did you need to change your data table during this activity?
2. Did the data table make it easy or difficult to analyze the data once you had completed the procedure?
3. Compare how many times you were successful with either eye closed and with both eyes open.
4. Describe the limits that influenced your ability to perform this task successfully.
5. How much diversity was there in your team's results?
6. Describe how closely your team's results match the results of other teams in your class.

How Do You Spell Success?

Discuss your answers to these questions with the rest of the class.

1. Describe how your team determined which Team Member had the most success during the Star Tracers investigation.
2. How many different definitions of success were there in Star Tracers?
3. How did each team define success in Threading the Needle?

4. Explain whether you think it is fair to compare all the teams and say that the team with the most successful attempts in Star Tracers or Threading the Needle is the best team in the class.

5. Develop a class definition of a successful attempt at threading the bolt through the eye and record that definition in your notebook. Remember that you need to account for both technique and your definition of s successful attempt.

EXPLAIN  investigation

If at First You Don't Succeed

In this investigation, you will use your class's definition of success to repeat the Threading the Needle investigation. You will use the same materials and basic procedure, except you must adhere to class decisions. The object is to do everything the same way as everyone else.

Materials for Each Team of Two:
- 1 eyebolt
- 1 bolt
- 2 rulers
- 2 pieces of graph paper

Process and Procedure

Part A—The Social Skill

1. Construct a T-chart with two columns.
    → Notebook entry: Record the date and the title of this activity, as well as the purpose of the activity before constructing the T-chart.

Work cooperatively in your teams of two. In addition to Team Members, you will need a Communicator and a Manager. Set up your working environment as you did in Threading the Needle, but this time practice the social skill *Move into your teams quickly and quietly.*

➡ If necessary, review How To #1, How to Construct a T-Chart.

2. Label one column "Sounds Like" and the other column "Looks Like."
3. Discuss the social skill for this activity and then, as a team, fill in the columns based on your discussion.

Part B—The Activity

1. Create a data table.
   ➡ You may use the format for Threading the Needle or try a new data table format if your last one did not work well for you.
2. Refer back to the procedure for Threading the Needle and repeat Steps 2 through 11.
   ➡ This time, use the class definition of success from the previous connections activity to complete the task and to measure success in your team. You do not need to do the Wrap Up for Threading the Needle again.
3. Enter your new data on the class data table.
4. Construct a graph of the new class data.
   ➡ Refer to How To #3, How to Construct a Bar Graph, if necessary.
5. Compare this class graph with the class graph your teacher constructed for Threading the Needle.
6. Help your teacher construct a graph of the class data for this investigation.
   ➡ Use your graph and offer suggestions. Correct your graph if necessary.

Wrap Up

Discuss these questions as a team. Record your own answers in your notebook. Be ready to discuss the questions with the rest of the class.

1. Did you change the format of your data table? If you did, describe what changes you made and why.
2. This time, did your number of successes in threading the eyebolt differ from your partner's? If so, how do you explain the difference?
3. List any new limits you think were due to how the class defined success.
4. Describe the differences between the first class graph and the second.
   - Have the heights of the bars and the differences between the curve over the bars changed?
   - Which graph shows the most diversity among your classmates?
5. Decide which graph gives you the most reliable information about class success in threading a bolt through an eyebolt and explain why you think the graph you selected is more reliable.
6. List specific strategies that your team could use to improve moving into your group more quickly and quietly.

Doing It All the Same

When you first measured success at different tasks, chances are you agreed with Al's approach. Now that you have compared the results of people who do things differently, however, you might agree more with Isaac.

The different class graphs show that when you agree on the same way of doing things, it makes a difference in your

results. It also shows a difference in the pattern of the graphs. In fact, the graph that provided the most reliable information was the one you constructed after you agreed on a class definition of a successful attempt. That was because you eliminated the differences in the techniques that you and your classmates used to thread the eyebolt, and you had agreed on what you could consider a success. By eliminating the different techniques, you tested each student's ability to accomplish the task within given limits. The first time you completed Threading the Needle, the members of team A might have held the eyebolt at their side, while the members of team B might have held the eyebolt in front of their bodies. The difference in the success rate between team A and team B might result from their different abilities, but it also may be due to the different way they held the eyebolt. Without keeping the position of the bolt consistent, you have no way of knowing for sure.

To be certain that your results reflect only the differences you wish to test, you must consider all parts of your experiment that can change or "vary." This includes the position of the bolt and the distance of the bolt from your body. Anything that can affect the results of your experiment is called a **variable**.

You can control variables in an experiment by keeping all of the variables constant except for the one that you want to test. When you do this, you can compare your results with others who do the same experiment. To do this, you must decide how to keep all variables the same for everyone. For example, in Star Tracers, the class might decide that each team should remain silent during an attempt and everyone should arrange their desks so the amount of light is constant from team to team. Then the only variable left would be each student's ability to trace the star. That is exactly the variable you wanted to test.

1. What variables affected your measurement of class success the first time you did the eyebolt activity?
2. Which variables did your class decide were important to keep constant?
3. How did you control variables the second time you did the eyebolt activity?

One way to control variables is to agree on one class technique. But we still have more work to do. Think again of the star tracing experiment. One team might have allowed its members to cross the border of the pattern in a successful attempt. Another team might have required its members to complete the tracing without crossing the border at all. The second team would have fewer successes. Without a standard definition of how to measure success, you might never make a fair comparison between teams.

With a standard definition, each team will measure success in exactly the same way.

A standard definition of how you measure something is called an **operational definition**. You can develop operational definitions of techniques for most scientific investigations. Once you do, someone else can repeat the experiment and make fair comparisons between your work and theirs.

4. Read the following operational definitions and decide which one best defines a successful attempt in Star Tracers.
   a. You are successful in tracing the star pattern when you draw the star pattern well.
   b. You are successful in tracing the star pattern when you draw the star pattern without crossing either border.
   c. You are successful in tracing the star pattern when you draw the star pattern in one minute without touching either border.
5. Explain why you chose the operational definition you did.
6. How would you explain to a friend what an operational definition is?

Identifying and controlling variables and deciding on operational definitions are important skills in conducting investigations in science. With time and practice, your ability to use these skills increases.

ELABORATE investigation

Seeing the World around You

During the previous investigations, you explored diversity among students in your class as they accomplished small tasks. You discovered that people have different limits in accomplishing certain tasks. You tried your hand at recording and analyzing data in data tables. You made a graph of your data that showed your results. You learned about controlling variables and using operational definitions.

This investigation will bring this knowledge and these skills together. You will construct data tables, graph data, and explore the limits of peripheral vision. **Peripheral vision** is defined as the ability to see at the edge of your field of vision without moving your head or your eyes. Because all of you will test your peripheral vision and compare your results, be certain to make the investigation fair. That means use a common operational definition and control variables.

Materials for Each Team of Two:
- any materials you need from those your teacher provides
- graph paper

Process and Procedure

1. Construct a data table for this investigation.
   → Remember, this means reading through the entire procedure first.
   → Notebook entry: Record the date, the name, and the purpose of this investigation.

2. Determine what procedure you will use for measuring the edge of your field of vision without moving your head or your eyes.
   → You will compare the results of all the teams in your class. Read the Background Information following this procedure to learn more about what you will measure. As you decide on your procedure, consider the following:
   - If you plan to compare results, you will have to make the test fair. What can you do to ensure a fair test?
   - If you are going to compare results, you will need to think of a way to compare the same type of data. How will you make sure you are comparing similar results?

   STOP: Are you showing caring and respect for the ideas of your Team Members?

3. Visit the materials station, carefully look at what is available, and make a list of the items.
   → This is the Manager's role.

4. Decide what materials you will need.
   → Both Team Members should review the list and decide together.

5. Experiment with the materials to figure out a way to measure peripheral vision and to see what affects peripheral vision.

WORKING COOPERATIVELY

Work cooperatively in your teams of two. One person will be the Manager. The other will be the Communicator. Push your desks together. Try to be the team that moves most quickly and quietly.

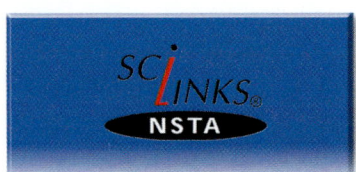

Topic: The Eye
Go to www.scilinks.org
Code: physical62

6. Conduct your investigation.
   → Notebook entry: Record your data in your data table.
7. Record your data on the class data table.
8. Construct a class graph to show the number of people who have each specific peripheral vision measurement.
   → Use your class data table to construct the class graph.

Background Information

If you look straight ahead and see something moving to one side of your head, you are seeing that object with your peripheral vision. When you see just a shadow or a movement, but you cannot identify the object, it is near the outer limit of your peripheral vision. The point at which you can first identify the object (a red pencil, a blue car, your mother) marks the beginning, or inner limit, of your peripheral vision. If the object kept moving around your head to the front and then to the other side, it again would reach a point where you could no longer identify it and then a point where it would disappear. These points mark the inner and outer limits of your peripheral vision range. A **range** defines the inner and outer limits of a characteristic such as peripheral vision. Ranges exist in individuals as well as within populations.

Wrap Up

Write the words "Wrap Up" in your notebook. Discuss these questions with your partner and write the answers in your notebook.

1. Describe how well your data table worked for this investigation.
2. Do you and your partner show diversity in your peripheral vision limits?

3. Describe the operational definition your class used to measure peripheral vision. You might also draw a simple sketch or diagram showing the setup that you used to measure peripheral vision.

4. Which variables may have caused a difference in results among teams?

5. Look at the class graph and decide where these points would be:
    a. The smallest value recorded for peripheral vision.
    b. The largest value recorded for peripheral vision.
    c. The top of the curve.

6. Based on the class graph, answer these questions.
    a. What is the range of peripheral vision measurements on the graph?
    b. Do most students have a peripheral vision measurement greater than or less than the measurement at the top of the curve?
    c. Describe the diversity of peripheral vision measurements in your class.

Light, Lenses, and the Eye

What happens to light when it passes through different types of lenses? What can we learn about light and lenses by exploring this question? How are these lenses like the lens in our eyes? Do they work the same way or are they different? What other questions do you have about light, lenses, and your eyes?

For this activity, work in your teams of three. One of you will be the Manager, one of you will be the Tracker, and one of you will be the Communicator. Sit at a table where you can work together. Work on the social skill *Show caring and respect for others and their ideas.*

Topic: Visible Light
Go to www.scilinks.org
Code: physical64

Materials for Each Team of Three:
- 1 set of lenses
- 1 light source
- two 3 × 5-in. index cards, unlined
- a tube of black construction paper with a hole in the side
- small package of modeling clay
- 1 marker

Process and Procedure

Part A—Exploring Light and Lenses

1. Collect the materials that you will need for this activity.
   → This is the Manager's job.

2. Have the Communicator hold the white card so that it faces a distant light source, such as light coming in your classroom window from the opposite side of a room.
   → If you do not have a window in your classroom, your teacher will help you make use of a light source across the room.

 Do not look directly at a bright light source such as the sun or a strong lamp. This could damage your eyes permanently.

3. Describe what you can see on the card.
   → Notebook entry: Record your description in your notebook.

4. Next, the Tracker should position the convex lens so that the curved surface faces the same distant light source. (A **convex** lens is thicker at the center and thinner along the edges.)
   → Be careful to keep the lenses as clean as possible.

CHAPTER 3 Identifying Limits and Diversity

5. The Communicator should hold the white card behind the lens so that the light source goes through the lens and shines on the white card.

6. While the Communicator is doing this, the Tracker should move the lens to vary the distance between the lens and the card (between 5 and 50 cm) until he or she finds the place where the light comes together in the brightest spot or forms a sharp image of a distant object.
   → The light travels more slowly through glass than it does through the air and the shape of the lens causes the light rays to converge (come together) in a focus point behind the lens.

7. The Manager should measure the distance between the lens and the card and record this distance in his or her notebook.
   → This distance between the lens and the card where the image of the distant object is the sharpest is the **focal length** for this lens.

8. Record in your notebook your description of what you see on the white card.

9. Now repeat Steps 4 and 5 using a lens that is thin at the center and thick at the edges—a **concave** lens. What happens?
   → Move the lens back and forth to see if this makes any difference.
   → This type of lens causes the light rays to diverge (break apart) so that no real image is formed.

Part B—Creating and Exploring a Model of the Eye

1. What do light and lenses have to do with your eye? Use the materials that you have to build a model of a specific eye as shown in Figure 3.4. You will be using one of two types of lenses that your teacher provides. Each represents a specific eye.

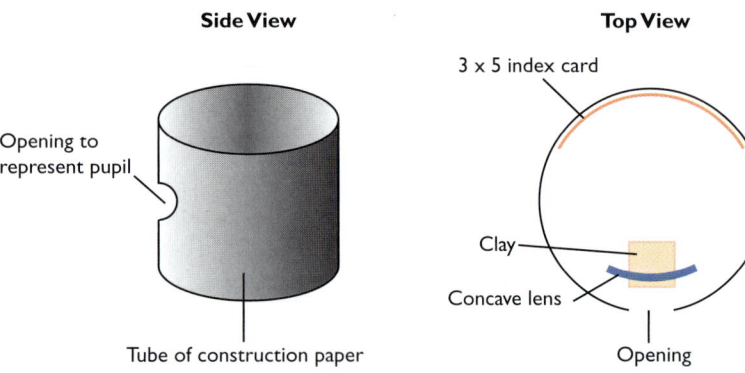

FIGURE 3.4 The hole in the black paper represents the pupil of the eye. The pupil is the small opening in the iris that expands and contracts to let in more or less light. The 3 × 5-in. card represents the retina, which is a place at the back of the eye that receives the image. The lens mounted on a small block of clay and placed directly behind the pupil represents the lens of the eye. The 6-V lightbulb represents the image that our model of the eye will "see." Notice that the light, the center of the pupil, the center of the lens, and the retina all need to be lined up exactly. We suggest that you adjust these so that the center of each is 5 cm above the table.

2. Find the focal length of this lens as you did in Part A.
   ⟶ Your teacher will suggest a starting range for the lens-to-"retina" distance.

3. The Tracker should move the light toward and away from the "eye" while the Communicator watches the image on the "retina."
   ⟶ Your teacher will suggest a starting range.

4. The Communicator should indicate whether the image is more in focus or less in focus as the Tracker moves the light back and forth. Stop when the image is the sharpest.

5. The Manager should record the following information in his or her notebook:
   a. Focal length of the lens (cm),
   b. Distance from the lens to the retina (cm), and
   c. Distance that the object is from the lens when the image on the retina is the sharpest (cm).

6. As a team, discuss other aspects of the image that appears on the "retina."

7. Discuss your results with the rest of your classmates. Did everyone have the same result for Step 5c? If not, what do you think could account for these differences?

    → The Background Information may help you develop an explanation.

8. Read the Background Information that follows the Process and Procedure and discuss this information with your teammates. Share your ideas with other teams.

9. Use what you have learned so far in this investigation to attempt to "correct" the lens in your model eye with another lens.

    → Your teacher will suggest a lens to begin with. Leave the current lens in place and place another lens in front of this one to see what difference it might make.

    → If your model is of a farsighted eye, you will need an additional lens that will allow the eye to focus on an object that is close up (20 cm or less).

    → If your model is of a nearsighted eye, you will need an additional lens that will allow the eye to focus on an object that is far away (2 m or more). If you need help, have the Communicator ask another team for some ideas.

CHAPTER 3 Identifying Limits and Diversity

10. Describe the lens that corrected the sight of your model eye. What are these corrective lenses often called? In what varieties do they come?

11. Explore more about this imaging process by using a model of the eye that can focus on objects that are near.

    → You can use a model of a nearsighted eye or a model of a farsighted eye with a corrective lens.

    a. Draw a stick figure on one-half of a 3 × 5-in. card and mount this drawing on a small block of clay so that the center of the figure is 5 cm above the surface of the table, in line with the center of the lens. Use a desk lamp to illuminate the drawing. Describe what you see on the "retina."

    → Notebook entry: Record your descriptions in your notebook.

    b. Remove the drawing of the stick figure and position your hand or face in this area. Describe what you see on the "retina." What do you see when you position an open book in the same spot?

    → Take turns so that everyone on the team has a chance to explore.

12. If time permits, explore questions that you have about other combinations of lenses and the behavior of light that results.

Background Information

One of the truly marvelous devices that uses light was not invented by scientists. It is the human eye that has developed through millions of years of evolution. Our eyes tell us an immense amount about the world around us. As you know, people's eyes are not identical. Some are green, some are brown, and others are blue, for example, and they come in a variety of shapes. Other properties of the eye vary from person to person as well.

If you could remove your eyeball from its socket, it would feel like a ball of jelly enclosed in a tough case. This case is opaque and white except at the front where the clear cornea is located. If you cut the eyeball in half with a vertical slice through the middle, you would see something like the picture in Figure 3.5.

Behind the cornea is the iris, or the colored part of the eye. The **iris** is a muscle that opens and closes to let more or less light pass through the **pupil**, which is the opening

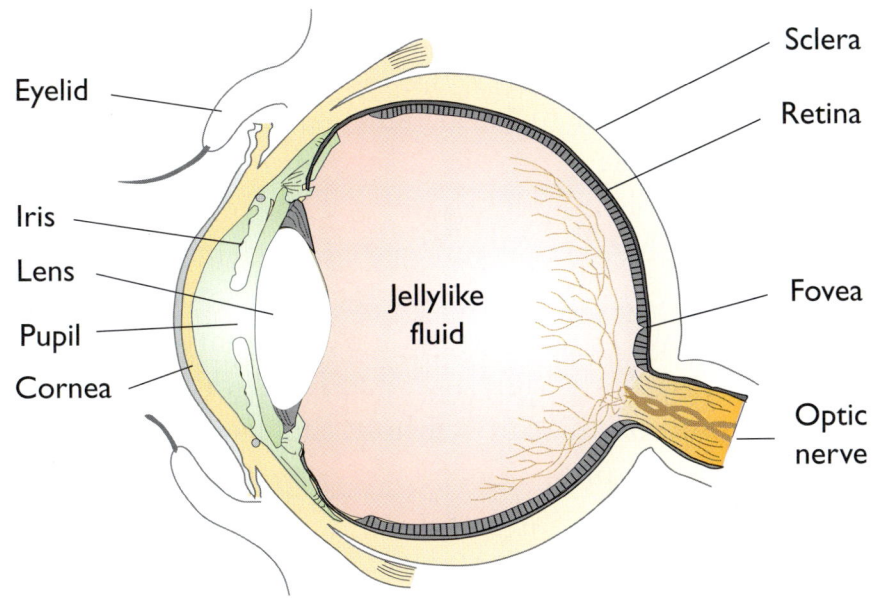

FIGURE 3.5 The eye is composed of three layers that form a fluid-filled sphere. The front part of the sclera is called the cornea, and it is transparent. The colored part of the eye, the iris, is a muscle that opens or closes to control the amount of light coming into the eye. The pupil is a clear opening in the center of the iris that lets light into the eye. It looks dark because most of the light entering the eye does not escape. The transparent lens behind the iris helps the eye focus on images. The innermost layer, the retina, contains the light-sensitive cells—the rods and cones.

CHAPTER 3 Identifying Limits and Diversity

in the iris. It is easy to see the iris in motion as it opens and closes. Take turns looking out the window or past a light source while the rest of your teammates watch your pupil contract. Remember not to look directly at the light source. Then look away from the light source and toward a darker area while the rest of your teammates watch your pupil expand.

Directly behind the iris is the **lens**, which focuses images. The rest of the eye behind the lens is filled with a transparent, jellylike material. Around the edge of this jellylike material is an area called the retina. The **retina** is composed of an amazing network of cells shaped like either cones or rods. These **cones** and **rods** act like detectors that tell the brain that a light signal has arrived. Among other functions, the cone-shaped cells help you see colors, and the rod-shaped cells help you see in dim light. The rods and cones are most sensitive in yellow-green light. This happens to be where the sun emits its maximum energy.

Light that is reflected off an object reaches the retina, where it falls onto the cones and rods. But how does the detailed information about the size, shape, and color of an object make an image on the surface of the detectors? The critical part of this imaging process is the lens. The lens is transparent with spherical surfaces. It is also **convex**—thicker at the center than at the edges. As you experimented with the lenses in Part A, what effect did the lenses have on the light that passed through them? What appeared on the card without the use of a lens or with the use of a concave lens?

An area near the center of the retina is called the **fovea**. Here, the detectors are packed together most tightly, and details of the image are distinguished most easily.

Light entering your eye from the side does not fall on the fovea, but on the part of the retina where there are fewer detectors. You are less able to identify images that result from light coming into your eye from the sides. See Figure 3.6. This explains why peripheral vision is limited. The placement and numbers of cones in your retina limit how well you see colors in your peripheral vision. The placement and numbers of rods and cones differ among people. This explains why we have a diversity of limits in our peripheral vision.

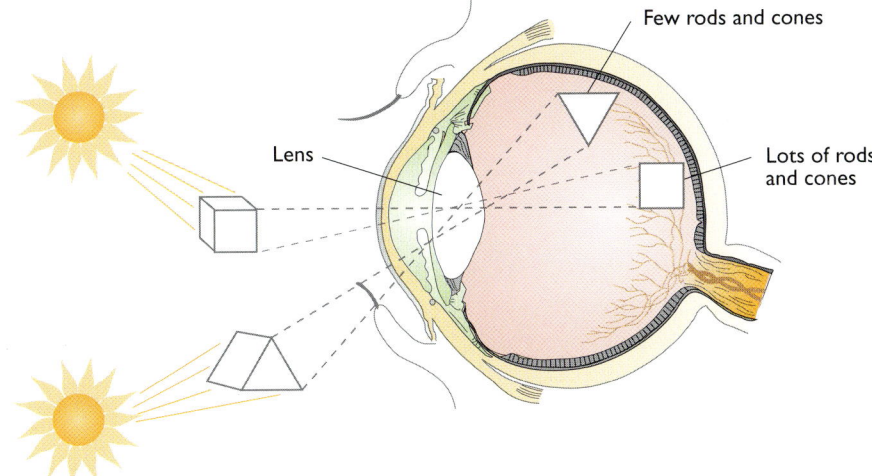

FIGURE 3.6 When you see an object directly in front of you, the light reflecting off that object enters the eye and falls on the fovea, which consists of detectors that are packed together most tightly. This results in a sharp, crisp image. When you seen an object off to the side, the light falls on the part of your retina that has fewer rods and cones. The images that result are not as clear as the images of objects directly in front of you.

The human eye is capable of automatically changing the shape of the lens slightly. In that way, the eye can focus on something in the distance at one minute and something close up the next. The range of change that the lens can accomplish varies from person to person. Some human eyes can focus better on images that are far away but cannot focus well on images close up. Reading a book is difficult. These are called **farsighted eyes**. Some eyes are better at focusing on images that are near, but they cannot focus well enough at a distance to identify the person across the room. These are referred to as **nearsighted eyes**. Is this similar to what your teams experienced with your models of human eyes? Did your team have a model of a nearsighted eye or a farsighted eye?

Do you know anyone who has eyes similar to your model? (Your model may even represent your own eyes.) What do people often do to help their eyes focus on images that are not sharp—either images that are close up or far away?

Wrap Up

Use the drawings and your model to answer these questions. Discuss your answers with your teammates and then write your own answers in your notebook.

1. In what ways is the human eye like this model?
2. In what ways is the eye different from this model?
3. What other items do you know of that make use of lenses? How are these items similar to and different from the human eye?
4. What did you learn about the characteristics of light and lenses?
5. Based on what you have learned, what else would you like to explore about light, lenses, or the eye?
6. How is the human eye limited?
7. How do lenses affect the limits of the human eye?
8. How are human eyes diverse?
9. How do lenses affect the diversity of human eyes?

Limits and Diversity in Animal Senses

Some animals have limits on their senses that are much different from the limits humans have. Some have poor hearing or eyesight, while others have special adaptations that give them extraordinary abilities. How do these animals adjust to these limits?

Although bats have poor eyesight, they are not blind. They "see" in a different, but very effective, way. As a bat flies, it constantly makes high-pitched sounds. As these sounds travel away from the bat, they bounce off objects. The sounds then come back to the bat. When the bat hears sounds coming back, it knows that there is an object ahead. This ability to use sound to locate objects is called **echolocation**. Using echolocation, bats hunt for food and find their way at night. Their echolocation ability is so good that bats can fly higher, faster, and farther at night than most other animals that fly at night, such as owls, moths, and fireflies.

Other animals with different limits to their senses include the following:

- Rattlesnakes and pythons can see the infrared radiation, or "heat," given off by the bodies of living animals.
- Goldfish can see not only infrared but also ultraviolet radiation. (Human eyes cannot detect either of these types of radiation.)
- Frogs have special types of cells in their retinas that trigger an automatic reaction when anything "fly-sized" moves in front of them. Frogs often respond only to things that are of this size. Much of the visual information that people record may never even reach a frog's brain.
- Flies are able to take in many separate images rather than a single image. They see the world as though they are looking through a kaleidoscope. This helps them survive in their high-speed world and track the motions of their predators.
- Ducks have 360 degree peripheral vision. This allows them to see all the way around themselves without moving their heads—a way of detecting hunters and predators nearby.

CHAPTER 3 Identifying Limits and Diversity

Chapter 4: Ranges of Limits and Diversity

In Chapter 3, you and your classmates investigated your abilities to accomplish certain tasks. When you compared your limits, you discovered that your class displayed a wide range of diversity. For example, some people could see farther around their heads than others. Imagine you had a graph of the peripheral vision limits of all the students in your school. Do you think it would show as much diversity as you found in your class, or is your class unusual in its diversity? Take a moment to discuss that question. Then take a moment to think about the ranges of limits and diversity in snowflakes and fingerprints, for example. Now you can investigate diversity in something that is probably very familiar: POPCORN!

ENGAGE / EXPLORE
- An Invasion of the A-maize-ing Popcorans

EXPLAIN
- Diversity Is Part of Our Natural World
- A Diversity of Popcorn
- More on the Meaning of the Bell-Shaped Curve

ELABORATE
- The Value of the Bell-Shaped Curve
- The Bell-Shaped Curve and You

EVALUATE
- A Flag of a Different Color

investigation

An Invasion of the A-maize-ing Popcorans

In this investigation, you will observe imaginary creatures from the planet Popcora and explore the range of diversity in their behavior.

Materials for the Entire Class:
- bowl of creatures from the planet Popcora
- 1 alien life vessel
- small amount of polyunsaturated lipid
- thirty 250 mL beakers

Process and Procedure

1. Listen as your teacher reads the following guided imagery:

   Recently, planet Earth experienced a peaceful invasion of creatures. These creatures landed on the kitchen counter of the White House in a small vessel resembling a salad bowl. Five seconds later, another vessel, more complicated in structure, landed beside the bowl. This vessel resembled a small electric appliance with a base and a lid. Inside the vessel was a message. The message read:

   "We are from the planet Popcora. Our civilization was at war, and just before our planet exploded, we managed to launch ourselves in our life vessel. We were unable, however, to launch our transferring mechanism. We set our destination for your planet, which we have named 'Butter.' Through our studies, we know that you can help us by providing the proper atmosphere in our life vessel. The proper

Work individually as you follow the procedure, then work as a class during the Wrap Up section. Your teacher will let you know where to work.

atmosphere will be achieved with the addition of a small amount of a polyunsaturated lipid substance. Once our life vessel contains the proper atmosphere, transfer us into the life vessel and supply an electric current to provide us with a comfortable temperature. We then will be able to fulfill our prime directive: to shed our protective suits with much jumping, wiggling, and exploding and to be consumed as nourishment by the peoples of many lands and nations."

Scientists around the world have been studying these creatures called "Popcorans." After numerous inconclusive studies, top government scientists have shipped the Popcorans to your scientific community, which has a reputation for conducting the finest scientific investigations in the world. You have, therefore, been chosen by U.S. officials to make the final observations of the Popcorans and to help them fulfill their prime directive.

Chapter 4 Ranges of Limits and Diversity

2. Fill your beaker with creatures from Popcora until the bottom is covered.
   → Do this when your teacher has finished reading and then return to your workstation with your beaker.

3. Observe your Popcoran creatures closely and describe their basic features.
   → Stir up the Popcorans. Let the kernels fall between your fingers and look closely at them.
   → Notebook entry: Be sure you record your observations and answer these questions.
      - How similar are they to one another? How different are they?
      - If each alien had a different name, would you be able to tell one from the other? Why or why not?

4. Provide an atmosphere and some heat to the life vessel and transfer the Popcorans to it.
   → The oldest scientist in the room should do this.

5. Carefully observe what happens to the Popcorans as they shed their protective suits.
   → All scientists in the room should do this.

 The popcorn popper and oil will be very hot. Be careful not to touch either the oil or popper at this point. Unpopped kernels also might be hot.

6. After the Popcorans are finished shedding their suits, the senior scientist will distribute some to each group.

7. Record your observations about the aliens and their changes.
   → Notebook entry: Be sure you include comments about the size, color, and texture.

78 CHAPTER 4 Ranges of Limits and Diversity

Wrap Up

Answer these questions in your notebook. Be ready to share your answers with the rest of the class. Remember to show caring and respect for your classmates during the discussion.

1. Using your recorded observations, rate the characteristics of the unpopped kernels of popcorn from 1 to 5 according to this scale: (1) all are similar, (2) most are similar, (3) some are similar, (4) a few are similar, and (5) none are similar.
    a. color
    b. shape
    c. size
    d. texture
    e. smell
   Record this information in your notebook.
2. Use the same scale and characteristics to rate the popped kernels and the kernels that did not pop.
3. Did all of the kernels pop at once?
4. What do your answers to Questions 1–3 tell you about popcorn?

Diversity Is Part of Our Natural World

In the Star Tracers investigation, some of you were more successful at tracing a star pattern than others. When you tried to thread the eyebolt with a bolt, some of you were more successful than others. Also during the peripheral vision experiment, you discovered that your class had a range of limits in peripheral vision. You plotted your data from the peripheral vision investigation and observed the pattern that your data created. On your graph, you indicated the range of peripheral vision for your class by marking the lowest value, the middle point, and the highest value you measured. Your graph is one way to show the diversity of peripheral vision in your class.

Recall the question from the beginning of this chapter: If you had the peripheral vision test results of all the students in your school, would you expect to find as much diversity as you found in your class, or are your classmates more diverse than any other class in your school? If you tested everyone in your school, you probably would find as much diversity as you did in your class, if not more. A graph of data from your entire school would be similar to the class graph. In fact, if you tested everyone in the world for peripheral vision, you still would come up with a graph that looked something like the graph in Figure 4.1.

Because the graph of human peripheral vision includes more data, the shape of the curve is probably smoother and more bell-shaped than your class graph. Otherwise, the graphs are very likely about the same. The graph of

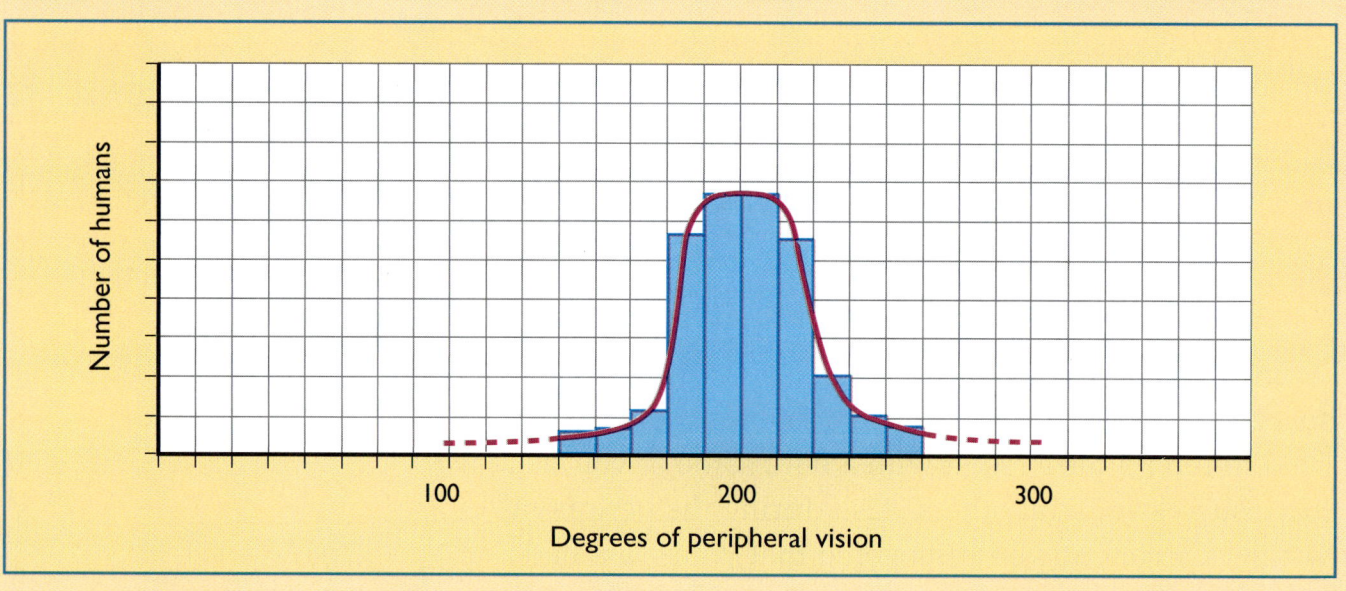

FIGURE 4.1 This is an example of a graph of people's peripheral vision measurements. The graph of your class data might be very similar. Most people probably have peripheral vision close to 200 degrees, but there is a lot of diversity. Notice how the data form a bell-shaped curve.

human peripheral vision shows diversity, just as your class graph does. The end values on the curve of the graph define the range of human peripheral vision. The bulge indicates where the values for most humans' peripheral vision would fall.

The bell-shaped curve that displays the range of human peripheral vision shows us that most people in the world have a peripheral vision range between 180 and 220 degrees. Yet some people have peripheral vision that is above or below that. Where you fall on that curve describes your own personal limit. The entire human peripheral vision curve shows the limits in peripheral vision for the entire human population. It also demonstrates that there is a range of diversity in the entire human population.

By graphing the data, we emphasize the range of diversity. If you included some other animals' peripheral vision curves on the same graph, you would see something like the graph in Figure 4.2.

Other animals exhibit a wide diversity of limits. So diversity is found not just in your class but in entire populations of humans and other animals.

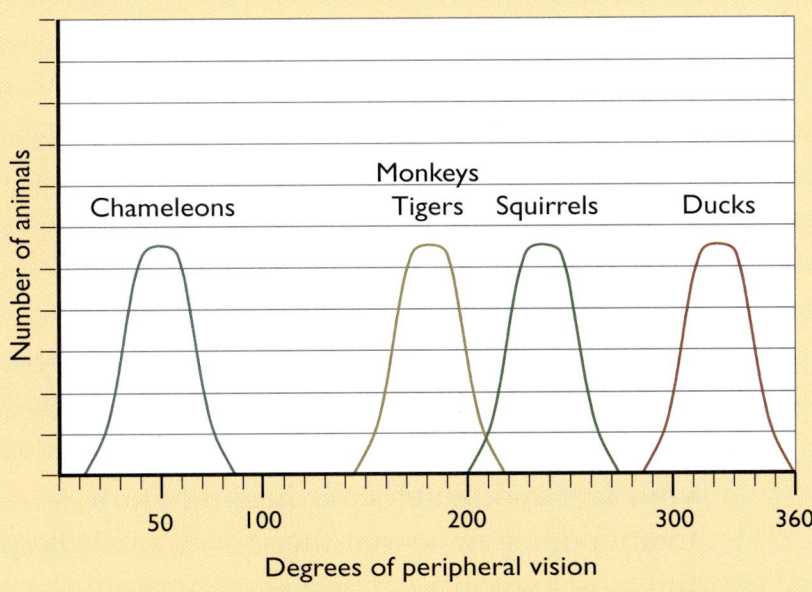

FIGURE 4.2 This graph represents peripheral vision of other animals. Notice that each species shows diversity in its limits. Which animals have peripheral vision most similar to humans? Which animals can see the farthest around their heads while looking straight ahead and not moving either their heads or their eyes?

Chapter 4 Ranges of Limits and Diversity

EXPLAIN investigation

A Diversity of Popcorn

In this investigation, you will use popcorn again to further your understanding of the bell-shaped curve. When you have completed the investigation, see whether you can better describe the bell-shaped curve and what it means.

Materials for Each Student:
- 1 marker, any color
- 1 large desk-sized piece of drawing paper

Materials for the Entire Class:
- enough popcorn and oil to pop 2 batches
- 1 popcorn popper

Process and Procedure

1. Obtain your materials.

    The popcorn popper and oil will be very hot. Be careful not to touch them.

2. Squat as low as you can beside your desk.
   → Your teacher will pop the first batch of popcorn. Think of yourself as a measuring device, something like a thermometer.

3. Indicate the amount of popcorn popping at one time by adjusting your height according to the sounds of the popping.
   → When no popcorn is popping, you should be at the lowest point, which is squatting. As the pops begin, you should slowly rise to your feet and match your height to how many pops you hear at the same time. When the most popcorn is popping, you should be

Work individually. You need a working space large enough for a big piece of drawing paper. Your desktop is fine. If you are working at a table, just leave plenty of room between you and other people. You also need some standing room beside your desk.

at your highest point, or standing on tiptoe. Lower yourself back to a squat as the number of kernels popping diminishes.

4. When you have completed Step 3, sit down and spread your drawing paper across your desk. Hold your marker at the lower left corner of the page and prepare to draw. Close your eyes.
   → Keep your "body gauge" experience in mind (what you did in Step 3). Your teacher will pop the second batch of popcorn.

5. Keep your eyes closed. As the popcorn pops, draw a line that illustrates what you hear.
   → Listen carefully to the sounds of popping. Let the line you draw be the gauge for the popping just as your body was for the last batch. Begin drawing your line at the sound of the first pop and finish at the sound of the last pop.

6. Open your eyes and look at the line you drew. Then look at other students' lines.

Wrap Up

Write answers to these questions in your notebook.

1. Describe the shape of the line you drew.
2. Explain how other students' drawings are similar to or different from yours.
3. Choose the word or phrase that describes the amount of diversity in your classmates' drawings: a lot, some, or none. Why did you choose that word or phrase?

EXPLAIN connections

More on the Meaning of the Bell-Shaped Curve

With your classmates, help your teacher construct a graph on the chalkboard that represents popcorn popping. Label and number the axes and discuss where to place the high and low values. Your class also should decide where to place the center (highest) point of the curve.

Your teacher will make four sections on the graph, labeled A, B, C, and D, and divide them by three lines labeled 1, 2, and 3. Individually study those lines and sections and agree on answers to the following questions. Write your answers in your notebook and prepare to explain your answers to your classmates.

1. Which line drawn through the curve matches where most of the popcorn was popping?
2. In which section(s) of the graph did the most of the popcorn pop?
3. In which section(s) did the least pop?
4. Is there a great difference between the amount of popcorn that popped before the high point on the graph and the amount of popcorn that popped after the high point on the graph?
5. Look again at the line you decided represented where the most popcorn was popping simultaneously. Estimate the percentage of the total amount of popped corn this point represents. Estimate the percentage that sections B and C represent together. How do the percentages compare?

reading ELABORATE

The Value of the Bell-Shaped Curve

As you know by now, bell-shaped curves can show the range of diversity in various limits for a group of organisms. If you looked around your classroom, you would see diversity in the height of the students. If you made a list of every person's height, added the heights together, and divided by the total number of people on your list, you would know the average height of students in your class. You could find the range for your class by determining the shortest and tallest measurements you recorded. The range would include these two points and everything in between. If you were curious, you could even find the average height and the range of heights for all students in your grade level or school using the same method. Or you could use this same method to determine the average height and height range of all students in your grade level in your city, state, or even the entire United States.

FIGURE 4.3 Middle school students display a range of heights.

CHAPTER 4 Ranges of Limits and Diversity

 **1.** You would have a fairly long list of heights just for your class. Discuss how long your list of heights would be if you tried to figure out the average height and the height range for all the students in your grade level or in the United States.

To avoid adding up an incredibly long list and then dividing, you could determine the average height and height range for students in the country by taking samples of heights of students your age from different places in the country. Then you could make a data table, record your information, and plot all the data on a bar graph. The vertical axis would be labeled "Number of students" and the horizontal axis would be labeled "Height in cm." The data would produce a bell-shaped curve. On that curve, you could draw a vertical line down the middle from the high point of the bulge to the horizontal axis. The value on the horizontal axis at that point would tell you the average heights of students in your grade level in the United States without having measured every individual student in the country. The

 end points of the curve would tell you the range. That is a benefit of the bell-shaped curve—it provides a lot of information clearly and quickly.

Breathe a sigh of relief. Now you can determine the average height and the range of heights of students in your grade in this country with less work than if you had used the "add-and-divide" method. Why do you think this is important? Clothing manufacturers, for example, can use data from a bell-shaped curve to decide on clothing sizes. They can construct a curve for the lengths of inseams in jeans that students your age would wear. They then can use the information from this curve to manufacture the appropriate number of jeans in each length. A curve showing the inseam lengths of all 12-year-olds in the country might look like the graph in Figure 4.4.

STOP & THINK

2. What information does the graph show?
3. List two ways you could make use of the information on the graph.
4. How could clothing manufacturers use this graph to help them make clothes?
5. If you were a clothing manufacturer, would you make as many pairs of jeans with a 20-inch inseam as you would with a 30-inch inseam? Why or why not?

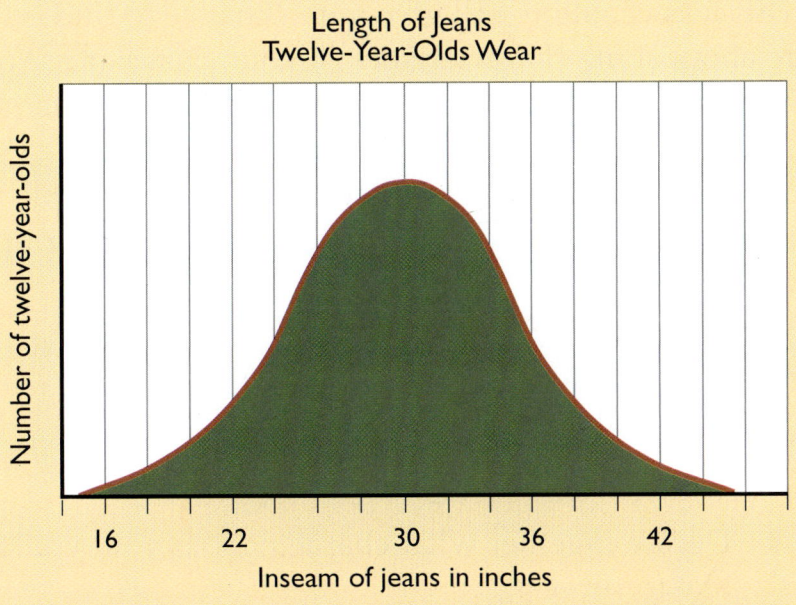

FIGURE 4.4 This bell-shaped curve shows the length of jeans that 12-year-olds wear. If you were a clothing manufacturer, in what lengths would you make the most jeans?

CHAPTER 4 Ranges of Limits and Diversity

Now you have seen how you can apply a bell-shaped curve to many different situations: peripheral vision, popcorn, height, and manufacturing clothes. This type of curve is so useful that you can apply it to everyday life as well as to investigations in science class. Data from many other everyday situations often produce a bell-shaped curve: shoe sizes, weights, number of people in a family, number of hours you sleep at night, number of pets at home, number of hours you watch TV per day, and the list could go on. Using this type of curve can be valuable and fun. You will practice using it in the next investigation.

ELABORATE investigation

The Bell-Shaped Curve and You

This investigation will give you a chance to find out more about some characteristics of your classmates. You will choose a common characteristic and collect data from your classmates. You might choose to investigate something such as foot length or the hour that people have breakfast. Then you will plot these data on a graph.

Materials for Each Team of Two:
- 1 sheet of graph paper
- tape, stapler, or glue

Process and Procedure

1. Decide together what characteristics you would like to survey.
   → Phrase the characteristics as a question—for example, "How tall are students in my

Work cooperatively with your partner. Use the roles of Communicator and Team Member. Push your desks together so you face each other or sit at a table across from each other. Practice the social skill *Move into your groups quickly and quietly.*

FIGURE 4.5 How many items do middle school students have on the floors of their bedrooms? The bell-shaped curve applies to many everyday things.

grade?" You will collect data about this question and use the data to produce a graph.

2. Construct a data table.
    → Notebook entry: Make sure you design your table in a way that will help you record the data you will be collecting. See How To #2, How to Construct a Data Table, for help.

3. Survey the other students in your class and collect data for your question.
    → Be sure to show caring and respect for your classmates as they answer your questions honestly.

4. Prepare the graph paper for graphing the data.
    → While one person is collecting data, the other should do this.

CHAPTER 4 Ranges of Limits and Diversity

5. Plot your data on your graph in the form of a bar graph.

→ Don't forget to label the graph and give it a title. See How To #3, How to Construct a Bar Graph, for more help.

Wrap Up

Prepare to present your graph to the class. In your presentation, you and your partner must answer the following questions:

1. What question did you ask?
2. Why did you choose this question for your investigation?
3. What range does your graph show?
4. How would you describe the diversity your graph shows?
5. Would you say your graph is a bell-shaped curve?

A Flag of a Different Color

You probably can remember several times in your life when you had your photograph taken with a flash camera. For a few minutes after the flash, you had a bright spot in front of your eyes that gradually faded. This occurrence is called an **afterimage**. In this investigation, you will explore afterimages and discover more about limits and diversity among your classmates.

Materials for Each Team of Two:
- 1 clean, unlined sheet of white paper
- 1 stopwatch or clock with a second hand
- 2 sheets of graph paper

Process and Procedure

Part A—The Social Skill

1. Discuss the importance of staying with your group.
   - As you discuss, consider the following questions:
     - How many people would I affect if I did not stay with my group?
     - How does it affect my team when I do not stay with my group?
     - When is it appropriate to leave my group?
2. Discuss and record the strategies you will use if one of you does not stay with your group at the appropriate times.

Part B—The Flag

1. Construct a data table for this investigation.
   - Notebook entry: Construct this in your notebook. Refer to How To #2, How to Construct a Data Table, for help. You will be comparing your data with the data from the rest of the class, so be sure you are doing what you can to promote fair comparisons by having a class discussion to consider the following:
     - What variables will you control to make sure you all are doing the same thing?
     - How can you be sure you are conducting a fair test?
     - How can you be sure you will be able to compare your results with your classmates?

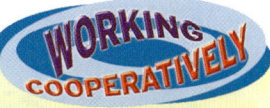

WORKING COOPERATIVELY

Work cooperatively with your partner. One of you should assume the role of Tracker/Communicator, and the other should assume the role of Manager. Practice the social skill *Stay with your group*. Move your desks together so you are facing each other or sit across from each other at a table. Review the role descriptions to be sure you understand your duties.

FIGURE 4.6 Here is a flag of a different color. Use this version of the American flag to test your afterimage. Stare at the bottom right star. Then fix your gaze on a blank sheet of white paper.

2. Obtain the materials you will need.
   → This is the Manager's role.
3. Stare at the bottom right star of the flag in Figure 4.6 for a period of time.
   → You will take turns doing this. The Tracker/Communicator will do this first.
4. After staring at the bottom right star, immediately shift your gaze to a blank sheet of white paper.
   → Have the blank sheet of white paper on your desk next to your book. By immediately fixing your gaze on one spot on the white paper, you should be able to see an afterimage. Try not to blink.
5. Your teammate should take a turn doing Steps 3 and 4.
   → It is the Manager's turn.

6. After both Team Members have observed an afterimage, repeat Steps 3 through 5 and record how long the afterimage lasts.
   → You will need a method for timing how long the afterimage lasts. Let the Tracker time first, then reverse roles.
7. Repeat Step 6 for a total of 5 times per Team Member.
   → Do not take your 5 turns all in a row. Alternate back and forth with your teammate.
   → Notebook entry: Record all times in your data table.
8. Record your data in the class data table.
9. Assist your teacher in constructing a class graph of the afterimage data.
   → Remember to show caring and respect for the ideas of your classmates. Think about the following questions and be prepared to answer them.
      - What labels belong on each axis of the graph?
      - What numbers belong on each axis?
      - What would you title the graph?

Wrap Up

Discuss these questions with your partner and write your own answers in your notebook. Be ready to explain your answers during the class discussion.

1. What variables did you control so that you could compare your results with other groups' results?
2. What operational definition did the class use to measure the length of the afterimage?
3. What is the shape of the graph?
4. Explain the pattern that the graph makes.
5. What is a possible explanation for what you saw on the white piece of paper after staring at the flag?

6. What new questions do you have about your eyes and afterimages? How might you find answers to your new questions? The Sidelight on Nature: The Diversity of Afterimage will provide you with more information about light and afterimages and may help you think of other questions.

7. During this investigation, how much of the time did you and your teammate stay together: more than half the time, about half the time, or less than half the time?

SIDELIGHT ON NATURE

The Diversity of Afterimage

What is afterimage? Why did you see a red (magenta), white, and blue flag on a white sheet of paper after staring at a green, black, and orange flag? To understand this phenomenon, you need to know a few basics about light and color.

White light contains all colors. You can see them if you pass white light through a prism. If you remove one color, let's say green, from white light, then the rest of the colors blend to form another color, which is called its complement. The complementary color of green is magenta. This works for all colors. If you remove orange from white light, what remains is the complementary color of orange—blue. The complementary color of black is white. Thus, the complementary colors of green, black, and orange are magenta, white, and blue. These are the colors of the flag you saw on the white sheet of paper.

How does the eye process color? Recall from the peripheral vision reading in Chapter 3 that the retina of your eye contains cells called cones, which are responsible for your color vision. The longer you look at a color, the less sensitive to that color the cones in your retina become. Eventually, your eyes become so desensitized to that color that you can no longer see it. When you stare at a green, black, and orange flag for a long time, the cones in your retina become desensitized to those three colors. When you shift your gaze to a white sheet of paper, your eyes cannot see the green, black, or orange components of white light. It is as if those colors have been removed from the white light. What you see then are the complementary colors to green, black, and orange: magenta, white, and blue. It takes time for your cones to become sensitive to green, black, and orange light again. When your cones have regained their sensitivity to these colors, the afterimage fades.

You have explored several ways that people have limits: tracing a star in a mirror, threading an eyebolt with a bolt, and seeing peripherally. Now you know that people's limits for seeing an afterimage depend on how long it takes the cones in their eyes to resensitize to a certain color. Some people's cones resensitize very quickly, so their afterimages fade quickly. Other people's cones take longer to resensitize, so their afterimages last longer. People are diverse in the time it takes the cones of their eyes to become sensitive to color again. That is why data on this aspect of afterimage generate a bell-shaped curve.

Chapter 4 Ranges of Limits and Diversity

SIDELIGHT ON NATURE

Blind Spots

Hold your book open at arm's length in front of you and look at the drawing of the plus sign and the dot.

While closing or covering your left eye, focus on the plus sign with your right eye and slowly bring the page closer to you. If you do this slowly enough, you will reach a point where the dot becomes invisible. This point is called your blind spot.

Use this drawing of a plus sign and dot to determine where your blind spot is.

This drawing shows the location of the retina's blind spot in the human eye.

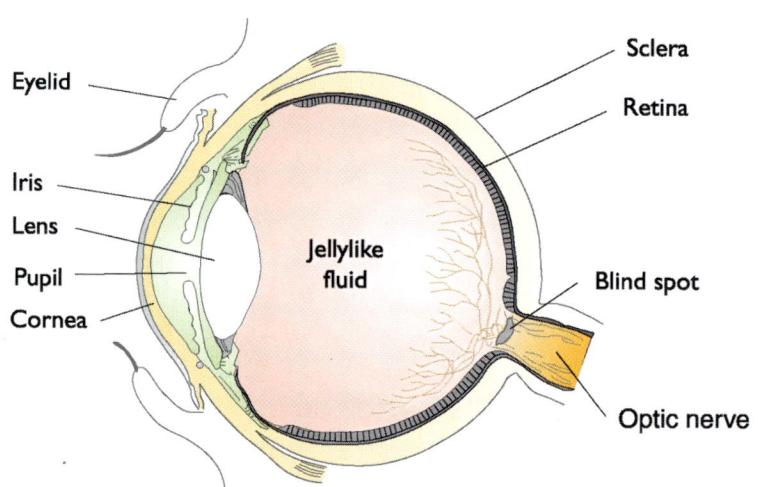

CHAPTER 4 Ranges of Limits and Diversity

Both eyes have a blind spot. The procedure used with this drawing of a plus sign and dot helps you locate the blind spot in your right eye. See whether you can locate the blind spot in your left eye using this same plus sign and dot. How would you change the procedure? Are the blind spots in your right and left eyes located in the same place in each eye?

Just what is a "blind spot"? Inside the back of the eyeball, there is a place in the retina where all the nerves of the eye gather to leave the eye and go to the brain. This collection of nerves is called the optic nerve. That spot does not contain rods and cones, so you are blind to any image that falls on it. People show a diversity in how far away their eye has to be from the drawing before the dot disappears. Try it with your friends and see!

CHAPTER 5

Using Limits to Set Standards

In Chapters 3 and 4, you explored human limits and diversity by investigating vision. When you graphed your class data from these investigations, you saw a bell-shaped curve. You also saw that other data you collected from your classmates formed a graph with a bell-shaped curve. From these experiences, you started learning that diversity is a big part of the natural world.

In this chapter, you will continue to explore the importance of knowing about the diversity of human limits. This time, however, you will focus on gathering the information that you would need to set standards when designing a product—a television for the future.

ENGAGE	■ What Do You Really Know about TV?
EXPLORE	■ Taking a Closer Look at TV Pictures ■ The Ultimate TV ■ TV Pictures and Color TV
EXPLORE **EXPLAIN**	■ A Learning Adventure ■ Your Experiences at the Stations ■ The Optimal TV Viewing Distance
EXPLAIN	■ Setting Standards and Human Factors
ELABORATE **EVALUATE**	■ TV for a New Millennium

ENGAGE investigation

What Do You Really Know about TV?

Although you may spend hours watching TV every week, how much do you know about how a TV works? In this investigation, you will participate in a TV Knowledge Contest. You and your teammates will have a short amount of time to make a list of everything you know about TV.

Process and Procedure

1. When your teacher says "Go," your team will have a 3-minute brainstorming session to create a list of every fact you know about TV.
   → Notebook entry: Record your team's list in the Manager's notebook.
2. When your teacher says "Stop," put your pen or pencil down and hand your team's list to the teacher.

First you will work cooperatively with your partner. Then you will work as a class. As you work with your partner, practice the social skill *Stay with your team.* Your teacher will tell you where your team should sit.

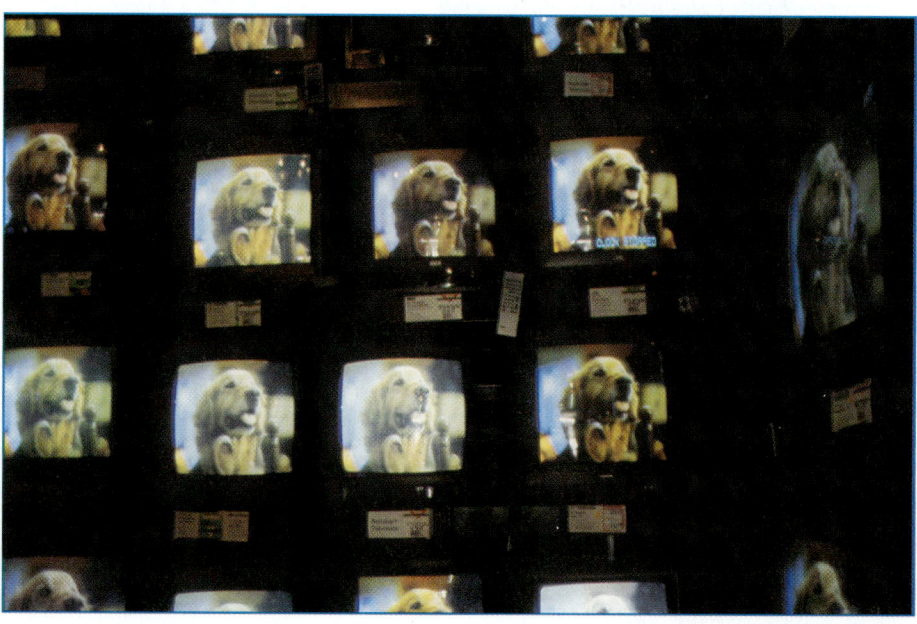

3. Listen and check each item for accuracy as your teacher reads each list to the class.
4. The team listing the most correct TV facts wins!

Wrap Up

Discuss these questions with your partner. Write your answers in your notebook. Be sure you can explain your answers if your teacher calls on you.

1. How similar were the lists the teams generated?
2. What categories or types of information did your classmates list?
3. Is there much diversity in what your classmates know about TV?
4. Do you think your class identified almost everything having to do with TV?
5. Has your team improved its practice of staying together since Chapter 4? If not, discuss strategies that will help you improve.

Taking a Closer Look at TV Pictures

In the last investigation, how many of you listed information about how a television displays pictures? One way to begin exploring the topic of vision and television is to examine TV pictures. Your task will be to observe the television screen carefully and answer some questions about the picture.

Working Cooperatively

Work cooperatively with your partner using the roles of Tracker/Communicator, Manager, and Team Member. Work on the skill *Stay with your group*. You will work at different locations in the room and at your desks for the Wrap Up. Be sure you understand what *Stay with your group* means when you join other teams at a TV or computer.

Materials for the Entire Class:
- 1–6 television sets
- 15 magnifying lenses

Process and Procedure

1. Obtain a magnifying lens for your team.
    → This is the Manager's role.
2. As a team, visit a station with a television.
    → You may work with another team at the station, but the number of students at one station should not exceed 4.
3. Carefully observe the television picture.

 Prolonged exposure to an operating TV at close range may be harmful. Make any close-up observations quickly and do not linger in front of the TV screen.

4. Record your answers to the following questions.
    → Be thorough and use your magnifying lens for close-up observations.
    - What does the TV screen look like when the power is on and you are getting a picture?
    - Step back from the television 3 m. What happens to the quality of the picture as you move toward the set, beginning at a distance of about 3 m?
        → Notebook entry: Record your answers in your notebook. Remember to give details.

Wrap Up

Make sure both members of your team are able to describe your team's observations to the class. After sharing your observations with your partner, write your answers to these questions in your notebook. You will also share your team's observations with your class.

CHAPTER 5 Using Limits to Set Standards

1. What are some of the common observations of a TV screen made by the teams in your class?
2. Did all teams observe the same things?
3. List two observations common to every team in your class.

connections EXPLORE

The Ultimate TV

In your notebook, design the ultimate TV. Use your previous knowledge about TV and your recent observations of a TV screen. You can include sketches as well as a description of the ultimate TV. As you continue through the chapter, you will revise this design as you gather more information about television. Therefore, leave plenty of room in your notebook after this first design so that you can make additional drawings or notes.

FIGURE 5.1 This Ambassador television was a popular model in the late 1940s.

reading EXPLORE

TV Pictures and Color TV

Now that you have looked closely at a TV screen, you are ready to understand more about how a TV works. First you will learn about black-and-white TV; then you will learn about color and digital TVs.

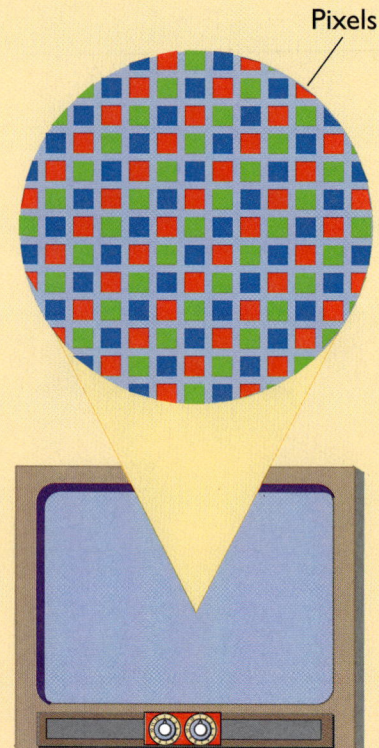

FIGURE 5.2 This drawing represents a close-up view of pixels on a TV screen.

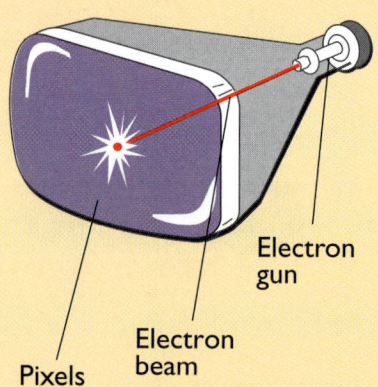

FIGURE 5.3 If you removed the outer cabinet of a black-and-white TV, you would see something like the electron gun in this picture.

When you looked closely at the TV screen, did you notice that it was divided into small sections that appeared as dots? Those dots are called **pixels**, and they make up a grid that covers the entire screen (see Figure 5.2).

Each pixel is made up of a material that glows for a moment whenever it is bombarded with electrons. (Electrons are parts of atoms, and atoms are the tiny particles of which all matter is made.) How do you bombard a pixel with electrons? With an electron gun! Black-and-white TV sets have an electron gun that fires a beam of electrons at the screen.

An electron gun looks something like the illustration in Figure 5.3. The gun contains a heating element that emits a beam of electrons when it is heated. This electron beam moves very quickly across the TV screen from top to bottom, one horizontal row of pixels at a time. The TV camera sends messages to the electron gun that adjust the intensity of the electron beam according to the brightness of the images. As a result, the black-and-white TV picture is composed of many glowing dots of varying shades of gray.

Color TV (also known as analog TV) works the same way, but with two exceptions. First, a color TV screen is coated on the inside with pixels containing a material called phosphor. The word "phosphor" comes from the Greek language and means "light-bearing." Second, color TVs have three electron guns, and each illuminates a different color on the screen. The three colors of pixels on a color screen are red, green, and blue.

When a color TV receives signals from the television station through an antenna, a cable, or a satellite dish, the signal goes to the electron guns. The three electron guns project that signal onto the TV screen by scanning the rows of pixels. Each gun lights up only one color of pixel. For example, when the beam from the electron gun specific for red phosphor hits a red phosphor pixel, it produces the color red. The signals indicate which pixel the electron guns should activate on which part of the screen. Just as you would mix different colors of paint to produce new colors, the pixels are close enough together to blend and make different colors.

What keeps each electron gun from hitting other pixels? For example, the electron beam specific for red phosphor must not strike the pixels coated with blue or green phosphor. To accomplish this, each color TV screen comes with a shadow mask. This mask is full of thousands of holes. Each hole is aligned with a set of red, green, and blue pixels on the screen. The electron beams must pass through the holes in the shadow mask, which focuses the beams on their appropriate pixels, as in Figure 5.4.

Digital TV is the latest technology in televisions. The most obvious difference between digital and analog TV is how clear the picture is, or the **resolution**. Most TV stations and networks have switched to digital cameras, which produce a much higher resolution than video cameras. They also use digital signals to send the images to your house, rather than radio waves from tall towers. Now the pictures they send are much clearer, and if you have digital cable service and a digital TV—called DTV

Topic: Television Technology
Go to www.scilinks.org
Code: physical105

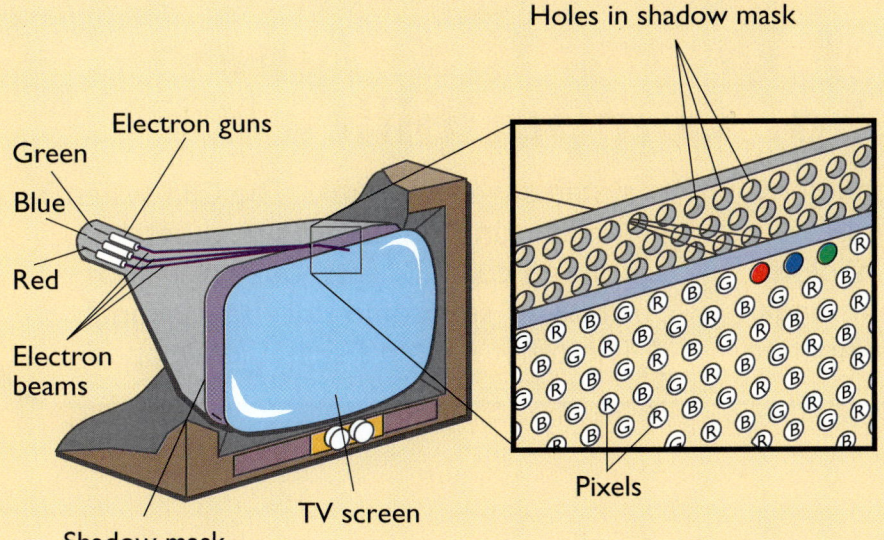

FIGURE 5.4 The holes in the shadow mask behind the TV screen guide each electron beam to its specific color of pixels.

CHAPTER 5 Using Limits to Set Standards 105

or HDTV—the picture you see on your TV is better focused, down to the tiniest detail. For example, if you like watching sitcoms, you can see the brand of cereal the family eats, and the stars painted on the wall that you used to think were just dots. Digital TVs are more expensive than analog TVs, especially because you need to purchase some accessories with a digital TV to take full advantage of this technology. You can't just bring it home, plug it in like an analog TV, and expect to get a better picture. You have to have a satellite dish or digital cable service to receive the signal, and you need a setup box attached to the TV. Even though they are expensive, digital TVs seem to be the "wave" of the future. Someday you may look back and tell stories to your grandchildren about how you had to plug a cable into the back of your TV to watch your favorite shows!

SIDELIGHT ON HISTORY

Color TV and the Problem of Compatibility

A new or improved technology may be difficult to introduce if it is not compatible with the old technology. In the 1950s when companies began producing color TVs, two competing systems were available. The CBS system's transmitting signal produced better color in color TV sets, but black-and-white TV sets could not use its signal. The other system's signal, developed by NBC, produced a picture of lesser quality than CBS's signal, but black-and-white TV sets could use it.

At first, the Federal Communications Commission (FCC) chose the CBS system. Back in those days, however, very few people had color TVs. Few people, then, were receiving the programs broadcast in color. Not surprisingly, CBS color programs were canceled after a few months for lack of viewers.

Eventually the FCC switched to the NBC system. Though the picture quality was not as good as the CBS system, everyone with a TV set could tune into the programs, whether they were broadcast in color or in black-and-white.

A Learning Adventure

In this investigation, you will embark on a journey around your classroom that will take you to five different locations to explore five different activities. We call these locations *learning stations*. At each station, you will investigate something about the limits of human vision.

At Stations 1 through 4, you will explore another limit of human vision. Recall that in Chapter 4, during A Flag of a Different Color, you discovered that people were diverse in their afterimage recovery rates. At these stations, you will explore a phenomenon known as **persistence of vision** in which the human visual system retains the image it sees for a short time after that image is no longer on the retina. At Station 5, you will explore a limit people have for how they perceive something at a distance.

You will move from station to station and perform different tasks that will help you set standards for TV pictures. Your teacher will tell you where to begin. At each station, there will be another team doing the same activity. As you move among stations and work at different stations, practice improving your skill of staying with your group. Because you will generate many different ideas with your partner, be especially aware of showing caring and respect for others and their ideas. While working at the stations, adhere to the rules in the chart.

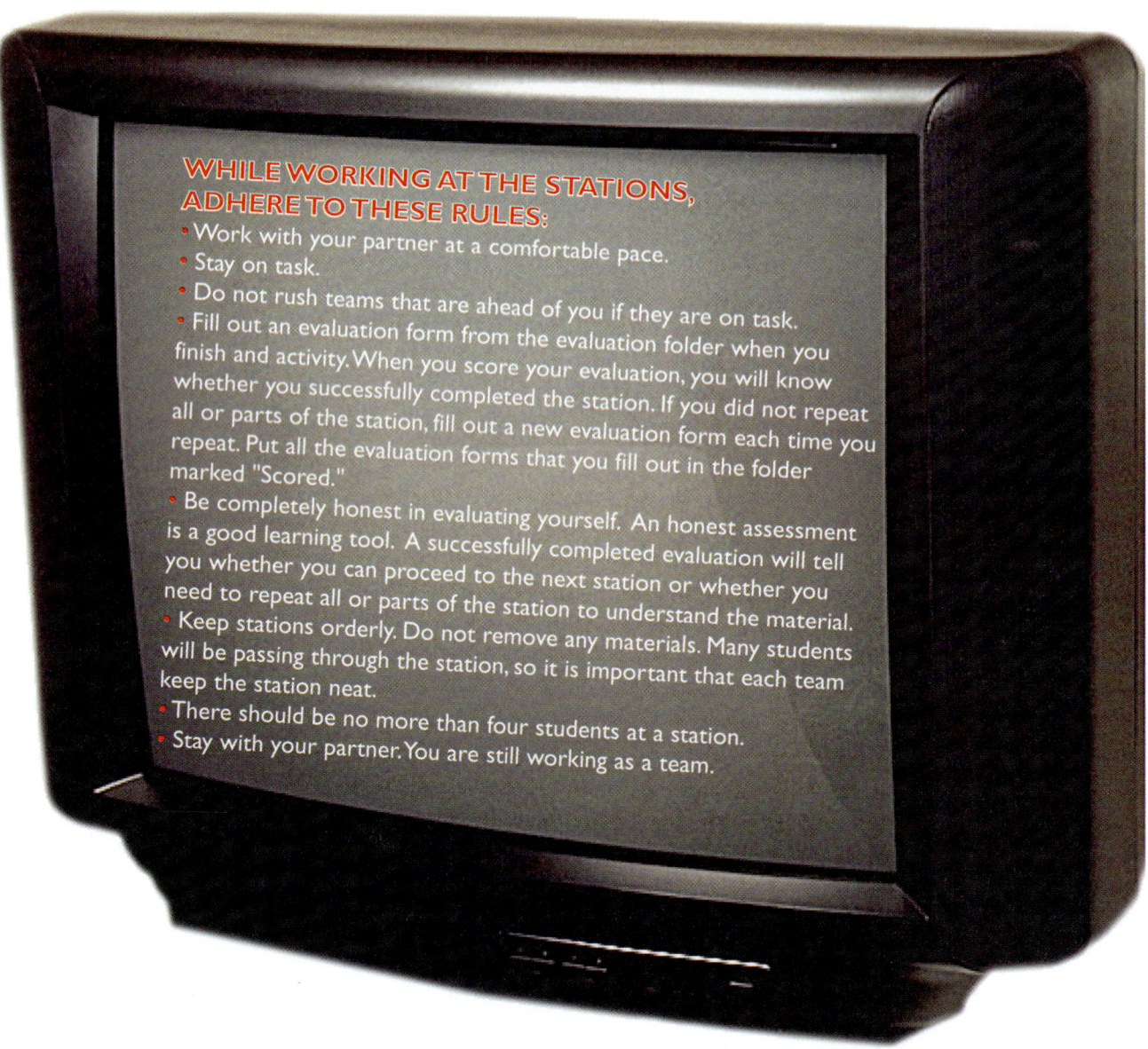

WHILE WORKING AT THE STATIONS, ADHERE TO THESE RULES:
- Work with your partner at a comfortable pace.
- Stay on task.
- Do not rush teams that are ahead of you if they are on task.
- Fill out an evaluation form from the evaluation folder when you finish and activity. When you score your evaluation, you will know whether you successfully completed the station. If you did not repeat all or parts of the station, fill out a new evaluation form each time you repeat. Put all the evaluation forms that you fill out in the folder marked "Scored."
- Be completely honest in evaluating yourself. An honest assessment is a good learning tool. A successfully completed evaluation will tell you whether you can proceed to the next station or whether you need to repeat all or parts of the station to understand the material.
- Keep stations orderly. Do not remove any materials. Many students will be passing through the station, so it is important that each team keep the station neat.
- There should be no more than four students at a station.
- Stay with your partner. You are still working as a team.

Station 1: Constructing a Spinner

This activity will demonstrate persistence of vision. Recall what you read about persistence of vision in the introduction to this activity. Think about the meaning of this activity in terms of television. After the activity, try to answer this question: How does your experience with the spinners help explain why the entire TV screen appears lit up even though each electron beam hits only one pixel at a time?

Materials for Station 1:
- 3 × 5 in. index cards, blank
- pencils (not to write with)
- transparent tape

Process and Procedure
1. You will each need 1 index card and 1 pencil.
   ⟶ The Manager should obtain these.

Work cooperatively in your teams of two. You both will be Team Members. One of you also should be the Tracker/Communicator and the other should be the Manager. The social skill you will be working on is *Share your thoughts and ideas*.

CHAPTER 5 Using Limits to Set Standards

2. Fold the width of the index card in half so you have 2 rectangles measuring $3 \times 2\frac{1}{2}$ in. on each side of the fold.
3. Turn the folded index card so the fold is up and the opening is down.
    → The top of each rectangle is now the fold and the bottom of each rectangle is where the card opens.
4. Draw a fish on the front of 1 of the folded cards and a fish bowl on the back of the same folded card. Draw a bird on the front of the other folded card and a birdcage on the back of the same folded card.
    → Make sure your drawings are centered from side to side and from top to bottom.
5. Tape the edges of the card together.
    → Leave a slit at the bottom.
6. Stick the eraser end of a pencil into the slit at the bottom of the card.
    → Push it up until it meets the fold and can't go in any farther.
7. Center the pencil and tape the slit closed around the pencil.
    → You should not be able to move the pencil easily from side to side or in and out.
8. Spin the bottom of the pencil back and forth between the palms of your hands so that you see both sides of the card flipping back and forth.
    → Spin the pencil slowly at first and gradually increase your speed.
9. Observe what happens as you spin.
10. Trade cards with your teammate and spin each other's spinner.
    → If you have the time and the interest, make and use other drawings to spin and observe.

Wrap Up for Station 1

Each person should obtain an assessment form from the evaluation folder and complete it.

Record your answers to these questions in your notebook.

1. Were you able to see the bird in the cage?
2. Suggest how this experience could explain why the entire TV screen appears lit up even though each electron beam hits one pixel at a time.

Station 2: Continuous Motion through Animation

At this station, you will investigate how persistence of vision can make a series of separate pictures seem like continuous motion. As you complete this station, recall that pixels do not glow continuously, yet the entire picture appears lit up.

Materials for Station 2:
- stacks of flip cards numbered 1–39
- stapler
- paper
- class data table
- clock or stopwatch

Process and Procedure

1. Make sure you each have flip cards numbered 1 through 39 in order in a stack.
    → The stack should have number 1 showing on top and number 39 at the bottom.
2. Place the cards in groups of 8 so that they overlap and staple each group together. One team member should remove every other card in his or her deck and make 2 separate decks so that the team has a total of 3 decks.

CHAPTER 5 Using Limits to Set Standards

FIGURE 5.5 Before you staple the groups of eight, make sure they are in order. Stack #5 will have only seven cards.

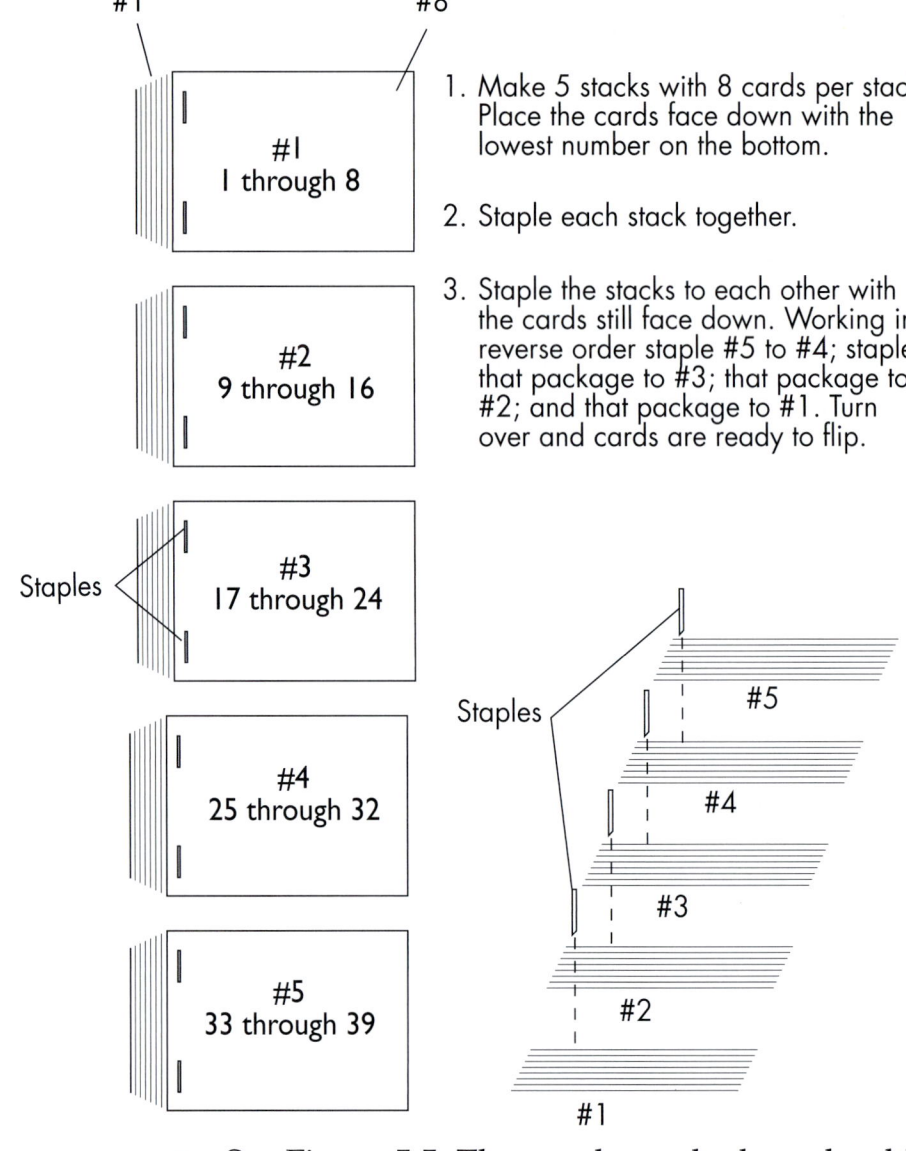

1. Make 5 stacks with 8 cards per stack. Place the cards face down with the lowest number on the bottom.

2. Staple each stack together.

3. Staple the stacks to each other with the cards still face down. Working in reverse order staple #5 to #4; staple that package to #3; that package to #2; and that package to #1. Turn over and cards are ready to flip.

→ See Figure 5.5. The overlapped edges should not be more than 1 mm wide. Stack #5 will have only 7 cards.

3. Staple all the groups together.
   → See Figure 5.6.

4. Flip through the decks to produce a mini-movie.
   → Share the deck or decks you made with your partner. Both Team Members should have a chance to experiment with all 3 decks.

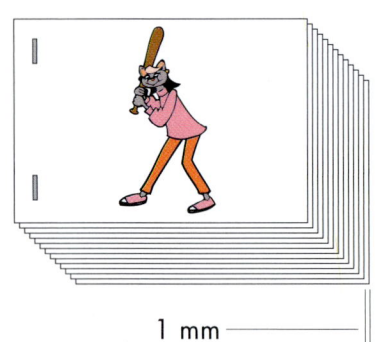

FIGURE 5.6 Your flip book will look like this. You might have to staple the cards from the top and from underneath the deck to hold them all together.

5. Determine how fast you have to flip through 39 flip cards for the appearance of continuous motion.
   → This is called the **continuous motion rate**, and it is measured in flip cards per second. The simplest way to measure the rate is to have someone determine (by trial and error) the minimum rate necessary to achieve continuous motion. Then your partner should time how long it takes to flip through all the cards at that rate. You can calculate the number of flip cards per second as follows:
   Flip cards/sec = total number of flip cards ÷ time in seconds to flip through the cards.
6. Record this value in the class data table located at this station.

Wrap Up for Station 2

Each of you should obtain an evaluation worksheet and complete it. Score yourselves and then place your worksheet in the folder marked "Scored." Then complete these tasks in your notebook.

1. Predict what will happen to the continuous motion if you remove every other frame.
2. Explain your prediction.
3. Test your prediction and describe what happened.

Station 3: Continuous Motion at the Movies

You probably already know that motion pictures, or movies, are a sequence of still pictures. But how fast does the film need to go so that we perceive the separate pictures as continuous motion? You will answer that question as you continue to explore persistence of vision at this station.

Materials for Station 3:

- one 16-mm film projector
- 1 reel of 16-mm film
- masking tape
- rulers
- stopwatch or clock
- scratch paper

Process and Procedure

1. Use a small piece of masking tape to mark where the film exits the film channel, as in Figure 5.7.
   → The projector should be set up so the film is partially wound on the take-up reel.
2. Start the film and note the time or start the stopwatch.
3. Stop the film after 10 sec.
   → The Tracker should say "Stop" after 10 sec. The Manager should stop the film.

FIGURE 5.7 Place your masking tape on the piece of film like this. Make sure you know which edge of the tape is closest to where the film exits the film channel.

Place masking tape here

4. Again mark the new place where the film exits the film channel with a small piece of masking tape.
   → You should now have a strip of film marked on both ends with a small piece of masking tape.

5. Measure the length of film that went through the projector in 10 sec.
   → Measure the length between the edges of the 2 pieces of tape that are closest to where the film exited the film channel. Make your measurements in centimeters.

6. Calculate the length of film that goes through the projector in 1 sec. Find this value in centimeters per second. (# of cm ÷ 10 sec = cm/sec)
   → Record this value in your notebook. Remember to show caring and respect for others and their ideas.

7. Determine how many frames of film move through the projector each second and record this value by counting the number of frames in the length of film that went through the projector in 1 sec (your calculation from Step 6).
   → For example:
   Number of frames of film in the length of film from Step 6 = "x" frames
   Film rate = "x" frames per second

8. Rewind and replace the film on the receiving reel as you found it.
   → Remember to carefully remove the pieces of masking tape from the film.

Wrap Up for Station 3

Each person on your team should obtain an evaluation worksheet and complete it. Score yourselves and put your evaluation sheet in the "Scored" folder.

Station 4: Flicker-Fusion Frequency

Why do we see a whole picture on a TV screen instead of dots glowing one by one? In this station, you will find the answer to that question by exploring a phenomenon called **flicker-fusion**.

Materials for Station 4:
- 1 variable strobe light
- paper
- rulers
- class data table

Process and Procedure

1. Prepare a data table for your team.
2. Start the strobe light at 15 flashes per sec.
   → You will find the strobe light turned off and set at 15 flashes per sec. When you finish using the strobe light, leave it in the same condition as you found it.

Never stare directly at the strobe light. Also it is dangerous for humans to see flashes below 15 flashes per sec.

3. Stand to the side of the strobe light and move your arm up and down slowly.
   → The Tracker should do this first.
4. Observe what your teammate's arm appears to be doing and record your observations in your data table.
   → The Manager should observe and record first. The Tracker should turn off the strobe light when you are not using it. Strobe lights burn out very quickly.
5. Continue to make these observations and record them as you increase the flash rate on the strobe light.

➡ The Tracker should increase the flash rate in increments of 5 (20 flashes per sec, 25 flashes per sec, 30, 35, and so on) up to 60 flashes per sec. The Tracker should continue to move his or her arm in the strobe light path each time. The Manager should record his or her observations each time.

6. Trade places with your teammate and repeat Steps 2 through 5.

7. Determine from your data tables the frequency at which you no longer perceive a flicker.
   ➡ This is called your **flicker-fusion frequency**.

8. Record the flicker-fusion frequency for both members of your team.
   ➡ Record these on the class data table at this station.

FIGURE 5.8 Al and Ros are demonstrating the proper position for watching your partner in this activity. Be sure that you do not look directly at the strobe light. Notice how Al looks at Ros from the side of the strobe light, while Ros keeps her back toward the strobe light.

Wrap Up for Station 4

Obtain evaluation sheets for both Team Members and complete them individually. Score yourselves and place your evaluation sheets in the folder marked "Scored."

Station 5: How Many Lines per TV Picture?

At this station, you will explore another vision factor that influences the design of a television. You know that a TV picture is made of horizontal rows of pixels. But a TV picture does not look like a bunch of dots unless you examine the screen up close. Just how far away from a TV set do you need to be so that you see a complete picture, rather than separate dots? How do different colored pixels blend to produce a screen that displays all colors? Explore and find out.

Materials for Station 5:
- a propped up page of BLM 5.13, Black Lines
- meter sticks or tape measures
- class data table
- masking tape
- rulers
- calculators
- paper

Process and Procedure

1. Stand directly facing the page of black lines.
   → Take turns doing Steps 1 through 3. The Tracker should do this first.

2. Walk backwards slowly until you can no longer see the spaces between the lines and can no longer make out the distinct lines. Stop at this point.
   → As you walk backwards, your teammate should make sure you don't trip or bump into anything.

3. Measure this distance.
   → Have your teammate place a strip of masking tape behind your heel and then measure the distance from the page of black lines to the tape in meters. Measure to the nearest one-tenth (0.1) of a meter.
4. Record this value in your notebook.
5. Trade places with your teammate and repeat Steps 1 through 3.
6. Now it is time to calculate each Team Member's "*D-to-s*" ratio or *D/s* ratio. *D* stands for the distance that you stood from the lined paper and *s* stands for the space between the lines.
   - *First, measure the space between the lines on the lined paper and confirm that the spacing is 2 mm. Because "s" stands for the space between the lines, s = 2 mm.*

CHAPTER 5 Using Limits to Set Standards

- *Next, convert into millimeters (mm) the distance you stood from the lined page when the lines were no longer distinct. Do this by multiplying the distance (which should be in meters) by 1000. "D" stands for the distance you stood from the lined page when the lines first disappeared. If a person were 3.6 m away from the lined paper when the lines were no longer distinct, then D = 3.6 × 1000 = 3,600 mm.*
- *Now, calculate the ratio of D to s. Do this by dividing D by s.*
  $$D/s = \frac{3{,}600 \text{ mm}}{2 \text{ mm}} = 3{,}600 \text{ mm} \div 2 \text{ mm} = 1{,}800$$

7. Record each person's *D/s* ratio in the class data table at this station.

Wrap Up for Station 5

Answer the following questions as a team. Record one set of answers for your team using either the paper at the station or your notebooks.

1. If someone could see small details at a great distance, would his or her *D/s* ratio be small or large? Why?
2. Suppose you had a *D/s* ratio of 100. If a pair of lines were drawn 2 millimeters apart, how far away would you have to be before you could no longer see separate lines?
3. Assume the lines you observed on the lined paper were rows of pixels on a TV screen. If you had a TV screen that was 30 centimeters (300 millimeters) high, how many lines would there be if the lines were separated by 1 millimeter?

4. About how far away from the TV in Question 3 would you have to sit to see a clear picture? (Use your own *D/s* ratio to calculate this.)

5. By now you should have noticed that the number of lines of pixels you need to have a clear TV picture depends on how far away you plan to sit. How does the distance you sit from the TV set affect the number of lines that you need?

After you have finished the Wrap-Up questions, each of you should obtain and complete an evaluation worksheet. Score it against the key and place in the folder marked "Scored."

Wrap Up for All Stations

Part A

When every team has completed all five stations, work with your partner at your desks to complete this Wrap Up. Record answers to questions in your notebooks and be prepared to explain your answers in a class discussion.

Your teacher will assign each team a data table to graph from either Station 2, Station 4, or Station 5.

1. With your partner, use your assigned data table to construct one graph of data from your entire class. If you encounter any difficulties constructing your graph, have the Communicator get help from other teams that are using the same data table.

2. When you have completed your graph, consult the other teams that constructed the same graphs to see whether the graphs are similar.

3. Save your class graphs in the Tracker/Communicator's notebook until you have completed the connections activity called Your Experiences at the Stations.

Part B

Compare your graphs with other teams' graphs. Make sure you see what shape of graphs the other data produced.

1. At which stations did you see the most class diversity?

2. At which stations was it difficult to measure diversity? Why?

3. Which of the class graphs is the best representation of a bell-shaped curve in your class population?

4. Draw a bell-shaped curve on a set of axes. The range of the bell-shaped curve is "poor" to "very good." Put an X on the curve where your team would fall in your use of the unit skill *Show caring and respect for others and their ideas*. Put an O where your team would fall in the activity skill *Stay with your team*.

connections   EXPLORE EXPLAIN

Your Experiences at the Stations

Work individually to answer the following. Write your solutions in your notebook and be ready to share your answers in a class discussion.

1. If pixels on an operating TV screen do not glow continuously, how does the entire picture appear constantly lit up?
2. Turn to the place in your notebook where you drew your ultimate TV design in the connections activity The Ultimate TV. Review your design and make a list of the things that you now would consider when designing your own TV. Record your list in your notebook.
3. Share your list during the class discussion and add to your list any new ideas you hear.

investigation   EXPLORE EXPLAIN

The Optimal TV Viewing Distance

As you moved through the stations, you learned about a variety of diverse human limits that affect how TVs are designed—such as persistence of vision, flicker-fusion frequency, and viewing distance ratios (D/s). If you were going to design your own TV screen, you would have to consider the diversity of people's preferred viewing distances. That way you could manipulate factors such as lines per screen to make the best possible TV picture.

But what viewing distance should you assume? Should you use yourself as the example? Should you use the viewing distance of your family room or living room? Using one person's experience is not the most scientific way to decide on a design. A better way is to collect data—something that you are getting good at! In this investigation, you will collect data that will help you determine the average TV viewing distance of a group of people who may be the best TV critics in the world: you and your classmates.

Materials for Each Team of Two:
- 1 or more pieces of graph paper
- 1 calculator
- scratch paper
- 1 meter stick or tape measure

Process and Procedure

1. Construct a data table for this investigation.
   → You will need 1 per team. The Tracker should do this. Remember to read the entire procedure first before you attempt to construct your data table.
2. Conduct a survey of all the students in class to find out how far they like to sit from the TV while viewing a program.
   → Recall how teams went about conducting the last survey in Chapter 4.
3. Record your data in your data table.
4. Use any, all, or none of the materials your teacher provides to determine the average TV viewing distance of the students in your class.

Work cooperatively as a Team Member with your partner. You will also need the roles of Communicator and Tracker for this investigation. Push your desks together or sit side by side at a table. Your teacher will tell you how to go about conducting a survey for this investigation. Work on the social skill *Move into your groups quickly and quietly.*

Wrap Up

With your partner, complete the following tasks in your notebook.

1. Prepare a summary of your method and results to present to the rest of the class. Include in your summary
   - your survey method,
   - your range from highest to lowest value,
   - your average value, and
   - a comment on the amount of diversity you observed among your classmates.
2. Rate your team on its ability to move into groups. Choose either (a) couldn't be better, (b) needs a little improvement, or (c) couldn't be worse. Explain your choice.

Setting Standards and Human Factors

In the previous investigation, you considered the limits of people when determining the optimal TV viewing distance. Why would anyone need that kind of information? Recall from Chapter 4 the jeans company that used information about the heights of 12-year-olds to manufacture jeans that fit.

In setting standards like sizes, we need information about the way people differ. These differences are called **human factors**. Human factors are important to consider when designing any product. Manufacturers want to be sure that their products fit the people who will be using them.

FIGURE 5.9 What information about humans do the designers of bicycles need to consider? What about when designers design bikes for children or for professional racers?

If you are designing clothes, it is easy to see what we mean by having the product "fit the user." But what does it mean to have a skateboard, a bicycle, a book bag, a calculator, or even a TV "fit" the person who uses it? Take a calculator, for example. A calculator has to be the right size for people to hold in their hands, but not all people have the same-sized hands. When designing a calculator, one also has to make sure that the keys are far enough apart so that most people can touch them without hitting two keys at once. The display needs to be small enough to fit on the calculator but large enough for people to see when it is held at a reasonable distance from their eyes.

Things like hand size, finger size, and limits of vision are human factors. You must consider human factors when you are designing a product that humans will use so that you come up with a standard product or one that fits the most people. It is important to make sure that products are safe, comfortable, and easy to use.

You have spent a good part of this chapter discovering some human factors that influence the way people design TVs.

1. List four of these factors.
2. Describe any limits you think these human factors will have on the design of your ultimate TV.
3. Can you design a standard TV that suits most people?

CHAPTER 5 Using Limits to Set Standards

connections

ELABORATE
EVALUATE

TV for a New Millennium

Work with your partner to read the following discussion and then plan and present the activity. Be sure to show caring and respect for each other's ideas.

Imagine that you and your partner began work at an innovative company called New Millennium Television and Electronics. At New Millennium, the employees not only design and manufacture TVs, but also create advertisements to sell their TVs. You and your partner have been asked to work together to design a state-of-the-art TV for the future. Before you began your task, you had several meetings with the board of directors and the senior partners of New Millennium TV. At each meeting, the board made a decision that you and your partner must adhere to. The decisions are as follows:

Meeting 1: You and your partner will work cooperatively with one of you acting as Manager and the other acting as Communicator.

Meeting 2: You will record in your notebook anything you do during your task.

Meeting 3: You will set standards for the design of your TV before you send it to the assembly line for manufacturing. The senior engineer described a standard this way:

A **standard** is a guideline that industries use to maintain product consistency. Standards help us make products that we can depend on. Imagine, for example, if we had no standards for the sizes of lightbulb sockets and lightbulb bases. Standards are also guidelines we set for the safety and well-being of society. Imagine if there were no standards for how fast cars could travel.

The bottom line for these two new employees is that standards are guidelines we use to design the best product. We need you to set standards for the best TV picture.

Meeting 4: The standards you set will be the same for all TVs regardless of screen size.

Meeting 5: You will concentrate on setting standards for the following three television components as you design your TV:

1. How many pictures per second?

    To decide on an answer to this question, you might review your findings on continuous motion from Stations 1, 2, and 3 and on flicker-fusion frequency from Station 4. Also review the information about these two **phenomena**. (A phenomenon—the plural is phenomena—is an event or occurrence that you experience by using your senses.) The goal is to produce smooth, continuous motion that is pleasing to the viewer.

After you arrive at an answer, justify it with a written list of reasons.

2. How many horizontal lines of pixels?

   You might review your findings from Station 5 that dealt with lines per screen. How will your understanding of D/s ratios influence your answer to this question? You also might consider the results of the survey about people's TV viewing distances.

3. How many pixels per horizontal line?

   Your answer to this question probably will depend on whether the shape of your TV is rectangular or square.

Meeting 6: You are free to be creative in producing your TV commercial as long as you create a new company slogan.

You and your partner are now ready to begin your task: Design a TV with a clear picture by setting the standards described in Meeting 5. Then produce and present a TV commercial to sell your TV. The chair of the board, affectionately nicknamed "Teacher," will tell you how much time you will have and what materials you may use to design your TV. This person also will tell you when you will present your commercial. After you have presented your commercial and watched everyone else's, read the section that follows.

FIGURE 5.10 Computers have video cards that enable you to watch "TV" on the monitor. How are computer monitors similar to TVs? What human factors are involved in designing computer monitors?

The National Television Systems Committee Uses Human Factors

In the previous connections activity, you considered the way a TV works and certain limits to human vision that influence the design of TVs. The National Television Systems Committee (NTSC) is the organization that considers those same factors in setting the standards we currently use to produce TV pictures in the United States and Japan. The standards are called the NTSC standards. It is amazing that standards established more than 60 years ago are still in use today. The connections activity you just completed is similar to what the NTSC did in 1941.

One of the reasons the standards have stayed the same for so long is that the NTSC based them on human factors. The NTSC had to answer the same questions that you answered. To set the standard for how many pictures per second should appear on a TV screen, they used what they knew about motion pictures and the research on persistence of vision and flicker-fusion frequency to arrive at a standard of 30 pictures per second. To get rid of the flicker (remember, humans stop seeing flicker at about 40 to 50 flashes per second), they used a clever technique called **interlacing**. See the Sidelight on Technology: Interlacing for more information.

To set standards for the number of lines per picture, they used information about how well people see at a distance. They used a *D/s* ratio of 2,000 and assumed that people would sit about 1.5 meters (4 feet) away from a 30 centimeter (1 foot) high TV screen. This information helped the committee decide on a standard of 525 lines per screen.

TV screens can be shaped as circles, squares, or rectangles. So the committee had a difficult time deciding what shape the screen should be. Because most of the early programs on TV were movies, they finally decided on a rectangular screen that was about the same shape as a movie screen.

Using a rectangular screen, the NTSC determined how many pixels per line were necessary. They used a ratio of 4 to 3 for the height-to-width ratio of a TV screen, which gave 700 pixels per line (or $\frac{4}{3}$ of 525). After much discussion and investigation, they decided that 700 pixels per line was more than necessary. To keep the TV

CHAPTER 5 Using Limits to Set Standards

equipment as simple as possible, the committee decided that 630 pixels per line was enough.

Your team probably developed standards that are different from those developed by the NTSC, and that is okay. The important thing is that you researched human factors, including human limits, and explored how you can use them to set standards. This is an important skill to carry with you into the next chapter.

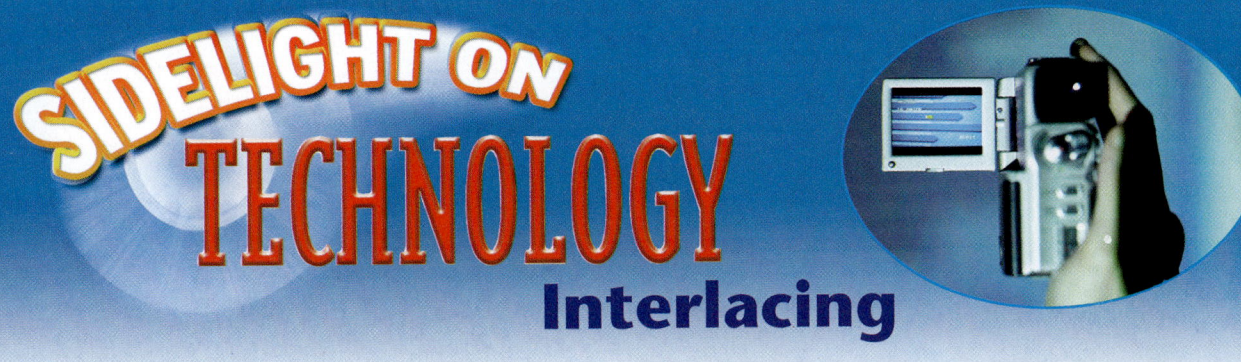

SIDELIGHT ON TECHNOLOGY
Interlacing

In 1940, the National Television Systems Committee (NTSC) decided that one of our TV standards would be that 30 pictures per second should appear on the screen to provide the appearance of continuous motion. Unfortunately, at this rate, the light did not look continuous, and people saw an annoying flicker. Researchers found that the light had to flash about 45 times per second for the flicker to disappear.

The NTSC had a problem. They wanted to limit the number of pictures per second to 30 and still eliminate the flicker. Fortunately, engineers came up with a technique to avoid sending more pictures than necessary. This ingenious technique is called "interlacing." In interlaced television, the electron gun first traces or activates every other line of pixels from the top of the screen to the bottom. It then returns to the top and traces all the remaining lines of pixels. In other words, the electron gun traces half the picture every $\frac{1}{60}$ of a second or a complete picture every $\frac{1}{30}$ of a second. A complete picture, then, requires two tracings over the picture tube from left to right and top to bottom. There is no flicker because the eye cannot detect flicker in such a small area (another human limit). Even though only 30 complete pictures appear every second, the brain perceives 60.

Chapter 6
Using Diversity to Set Standards

Think about the diversity of people who ride bicycles. Some are quite young and others are elderly; some are tall and others are short; some ride for pleasure, others as a way of commuting; and still others ride competitively. When setting standards for the design of various bicycles, designers need to take into account this range of differences in people and how they use their bicycles. Those same issues apply to most products that are designed for humans to use. Automobiles are another example of such a product.

In this chapter, you will consider how our knowledge of human diversity can be useful in setting standards for driving automobiles.

ENGAGE **EXPLORE**	▪ Watching *The Final Factor*
EXPLORE	▪ The Three Phases of Stopping ▪ Your Personal Reaction Time ▪ Determining Reaction Distances and Perception Distances
EXPLAIN	▪ Total Stopping Distances ▪ Life in the Fast Lane
ELABORATE	▪ Setting Speed Limits ▪ The Bell-Shaped Curve and Setting Standards
ELABORATE **EVALUATE**	▪ Don't Drink and Drive

ENGAGE EXPLORE — investigation

Watching *The Final Factor*

You may have a brother or sister who recently earned a driver's license. You also may be anticipating the day when you will be eligible for your driver's license. What does it take to learn how to drive safely? This investigation will introduce you to some of the important factors involved in driving.

Materials for the Entire Class:
- 1 copy of the video *The Final Factor*
- 1 television set with VCR

Process and Procedure

Watch the video *The Final Factor*. Pay close attention to the actions and reactions of the drivers. You may want to take notes to help you remember what you see.

Wrap Up

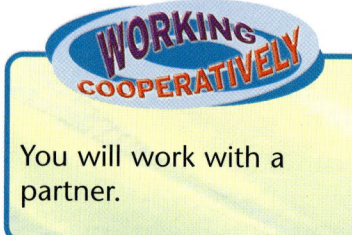

WORKING COOPERATIVELY
You will work with a partner.

With your partner, discuss these questions and write the answers in your notebook. Prepare to share your answers with the class.

1. Describe whether you think the word "factor" is used the same way in the video as it was in Chapter 5 in the phrase "human factor."
2. What are some factors that influence how fast people should drive when operating an automobile?
3. Which of these are human factors?
4. Decide and record what the speed limits for automobiles should be in the following areas.

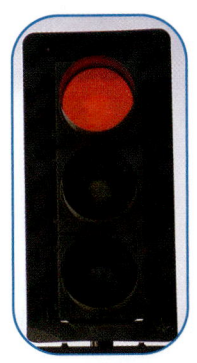

Your limits do not have to agree with the legal speed limits for these areas, but you must be able to explain your reasons for your choices.

- School zone
- Your neighborhood
- Divided limited-access highway (such as an interstate)
- Rural highway

The Three Phases of Stopping

Why do different speed limits exist for different parts of the city and country? For example, why aren't people allowed to drive as fast on a residential street as they are on the open highway? The answers to these questions are related to the limits of humans and their abilities to stop their vehicles during a sudden emergency. Stopping is not an instantaneous event. Stopping takes time, no matter how quickly it seems to occur.

There are three phases in the stopping process. When you were younger, you might have played a game called Green Light—Red Light. Remember that when someone

CHAPTER 6 Using Diversity to Set Standards

PERCEPTION DISTANCE + REACTION DISTANCE + SKIDDING DISTANCE = TOTAL STOPPING DISTANCE

The runner hears a whistle that signals her to stop.

The runner reacts to the whistle and begins to stop.

The runner comes to a complete stop.

FIGURE 6.1 The three phases of stopping include perception, reaction, and skidding. When you add up the distance traveled during these phases, you get the total stopping distance.

yelled, "green light," everybody started running across the yard or room toward the other end and kept running until the person yelled, "red light." At this point, the person turned around and tried to spot people who were still moving. If you were caught moving, you were out. Remember that it was tricky to come to a sudden stop. You had a tendency to keep moving even after the person yelled "red light." And the faster you ran, the longer it took to stop. Although you might not realize it, first your ears received the sound and then your brain perceived the words. The time that elapsed between hearing and perceiving is called the **perception time**. During your perception time, you continued to run before you even began to stop. The distance you traveled during your perception time is called the **perception distance**. This is the first phase in the stopping process. Everyone takes some time to notice or perceive an event. This is especially true if the event is a surprise. Not every person's perception time is the same. You might have been slightly faster or slightly slower than your friends at perceiving the "red light" command. Imagine how difficult it would be to determine the average perception time for the more than six billion humans. Instead, scientists have used a bell-shaped curve to determine the average perception time for all humans: 0.75 seconds.

After your brain perceived the words "red light," it took a little more time for you to react to the words and begin to stop yourself. The time it took for you to react is called **reaction time**. Like perception time, reaction time varies from one person to another. The distance you traveled during that time is called the **reaction distance**. This is the second phase of the stopping process.

Once you "applied your brakes," it took you a certain amount of time to come to a stop. The time that passed between when you first tried to stop and when you finally came to a complete stop is called the **skidding time**. During the skidding time, you traveled a distance known as the **skidding distance**. The skidding distance is the third and final phase of the stopping process.

If you add the perception time, the reaction time, and the skidding time, you get the total time it took someone to come to a complete stop. **This is your total stopping**

Topic: Motion–Speed Relationship
Go to www.scilinks.org
Code: physical137

FIGURE 6.2 What human factor might have affected the stopping process that resulted in this multiple car accident?

CHAPTER 6 Using Diversity to Set Standards   137

time. Similarly, if you add the perception distance, the reaction distance, and the skidding distance, you get the total distance someone covered before coming to a complete stop. As you will discover, a number of things influence each phase of the stopping distance—and, therefore, affect the **total stopping distance**.

Now consider the following example of how the phases of stopping might apply to real life.

You are riding down the highway in your family car. As the car roars along at 60 miles per hour (100 kilometers per hour), the driver suddenly sees something in the road ahead. "What's that?" the driver asks. You both realize what it is at the same time. It is a truck stalled in your lane. The driver slams on the brakes, and your car skids to a stop just inches from the back of the truck.

After breathing a sigh of relief (that you didn't hit the truck and you both were wearing your safety belts), you say, "Wow! You barely had enough total stopping distance to avoid hitting that truck! You must have incredible perception and reaction times!"

"Huh?" asks the dazed driver. "What are you talking about?"

Happy to demonstrate your scientific knowledge, you say, "You see, back there when you saw something in the road, the car kept going while you tried to figure out what it was. That was your perception distance. Then, in the time it took you to take your foot off the gas pedal and hit the brake pedal, the car kept moving. This distance is called your reaction distance. Finally, after you slammed on the brake, the car skidded to a stop. The distance the car moved while it skidded was your skidding distance. Your perception distance, reaction distance, and skidding distance all combine to make your total stopping distance. Either you have quick perception and reaction times, or you just barely had enough space to stop for the speed you were traveling."

You turn to look at the driver, who has fainted at the wheel. "Oh well," you say. "I wish I knew all of the distances for the speed we were going. There's no way that I could measure them now. Maybe I'll learn about that in science class tomorrow."

In the following activities, you will explore the relationships among perception distance, reaction distance, and skidding distance at different speeds.

1. What things do you notice drivers doing that adversely affect their reaction times?

2. Describe an example from the video in which someone's skidding distance was increased.

3. Think of the scenarios presented in the video. What things do people do that dangerously limit (or decrease) the distance they have in which to stop?

Your Personal Reaction Time

You have been exploring ideas about the phases of stopping, and you know that these phases are all influenced by speed. Remember that before you can stop, you must first react. Once you know your reaction time, you can determine reaction distances for different speeds. How long does it take you to react to an event? You will know the answer to this question by the time you finish this investigation.

FIGURE 6.3 How do you react to lightning?

Work cooperatively in your team of two. Besides being Team Members, you will need a Communicator and a Manager/Tracker. Work on the social skill *Stay with your group.* Push your desks together to face each other or sit facing each other at a table. Create a space beside your work area where you both can stand.

Materials for Each Team of Two:
- 1 meter stick
- 2 pieces of graph paper
- 1 ruler

Process and Procedure

1. The Communicator should hold the meter stick as shown in Figure 6.4.
    → Hold the meter stick so that the number 1 is at the bottom.

2. The Manager should hold his or her fingers near the meter stick at the bottom, near the number 1.
    → Do this without touching the stick. You will try to catch the meter stick when the Communicator drops it.

3. The Manager should try to catch the meter stick when the Communicator drops it (see Figure 6.5).
    → Communicator: Let go of the meter stick without any warning. Because you want to measure reaction time, it is important that you give no warning.

FIGURE 6.4 Hold the meter stick with the number 1 at the bottom.

CHAPTER 6 Using Diversity to Set Standards

4. Measure the distance between the bottom end of the meter stick and the point where the Manager caught the meter stick.
   → The Communicator should measure this distance in meters as shown in Figure 6.6 (10 cm = 0.10 m, 27 cm = 0.27 m, 42 cm = 0.42 m, and so on). Record the distance in the Manager's notebook.

5. Switch places and repeat Steps 1 through 4 until each of you has had at least 3 tries catching the meter stick.
   → Take turns dropping the meter stick for each other. You are finished when each of you has recorded 3 distances in your notebook.

6. Average your 3 distances.
   → Record this average distance in your notebook. We will call this distance the "falling distance." If you have never determined averages, refer to How to Determine Averages, How To #5.

7. Use the graph in Figure 6.7 to determine your average reaction time.
   → Look at the example on the graph. Notice that if your average falling distance is 0.60 m (remember, that is 60 cm), then your reaction time is 0.35 sec.

8. Record both your Team Member's and your own reaction times on the class data table your teacher has prepared.
   → This is the time that you have determined from the graph in Figure 6.7.

9. As a team, graph the class reaction times.
   → Notebook entry: Each of you should have this graph in your notebook. If you need to review graphing skills, refer to How To #3, How to Construct a Bar Graph.

FIGURE 6.5 The Manager should catch the meter stick when the Communicator drops it.

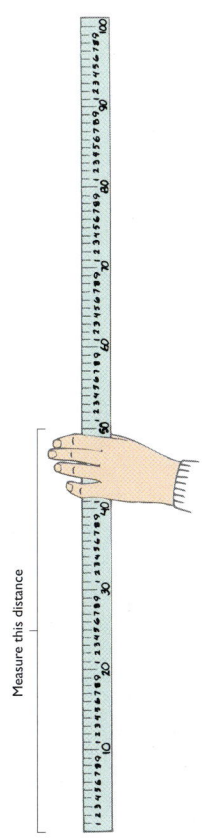

FIGURE 6.6 Measure the distance from the first mark at the bottom to the top finger where you caught the stick.

CHAPTER 6 Using Diversity to Set Standards

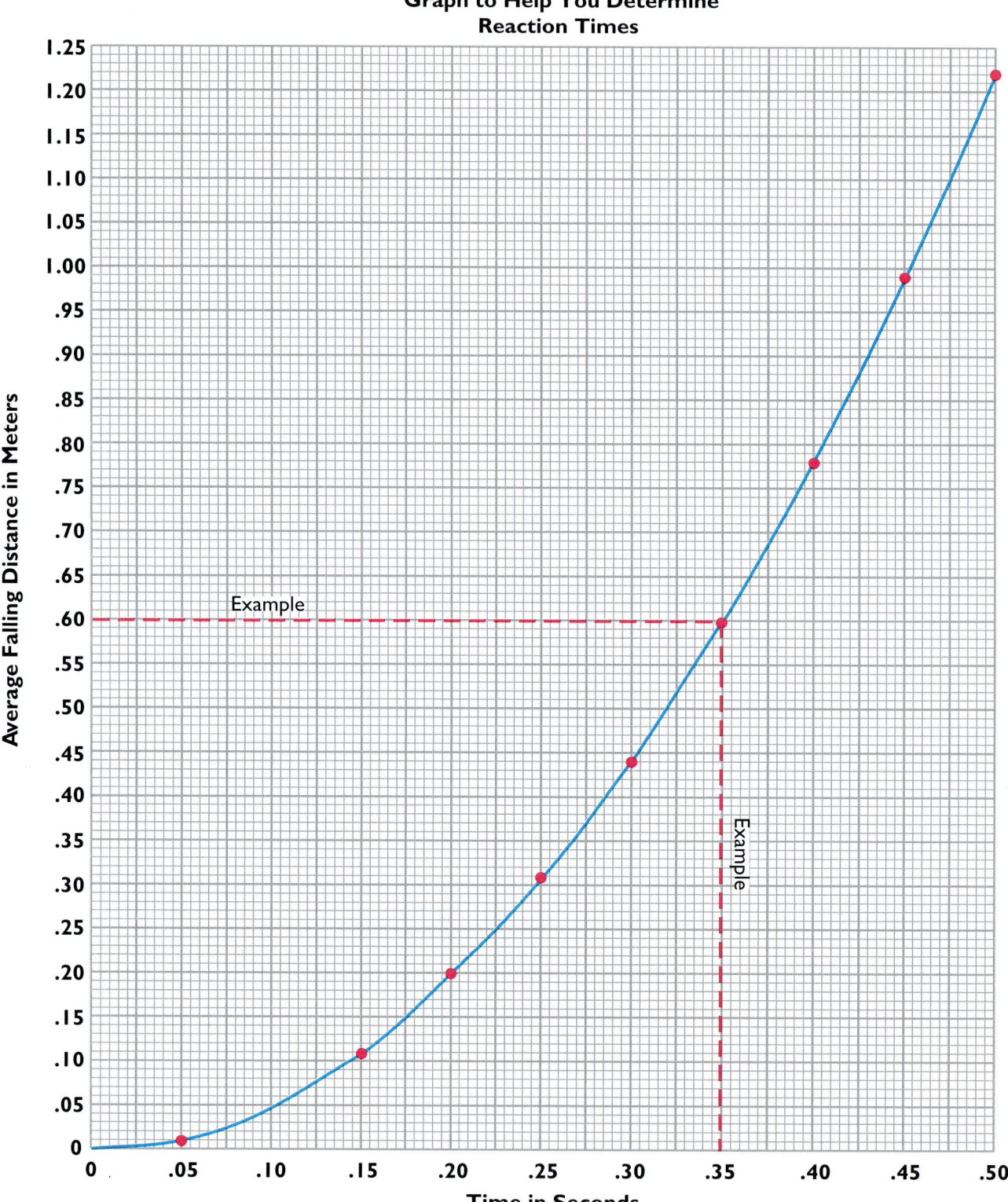

FIGURE 6.7 This graph shows the relationship between time and average falling distance.

Wrap Up

You will need to look at both graphs, the one in Figure 6.7 and the one you made of your classmates' reaction times, to answer these questions. Discuss your answers to these questions with your partner. Record your own answers in your notebook. Be ready to explain your answers when your teacher calls on you.

1. Why was it important to average your three falling distances?
2. Why do we use a graph to determine reaction times? Why not just have someone time how long it takes for you to catch a meter stick?
3. What does the class graph of reaction times resemble?
4. Why does the class graph have the shape it does?
5. What is the range of reaction times for your classmates?
6. Which reaction time is shared by the most people?
7. Based on the class graph, how does your reaction time compare to the reaction times of your classmates?
8. Describe what your class graph demonstrates about reaction times.
9. How far would the meter stick fall if your reaction time were 0.45 seconds?
10. Describe another experiment that you might design to determine reaction times.

CHAPTER 6 Using Diversity to Set Standards

EXPLORE investigation

Determining Reaction Distances and Perception Distances

How far you travel while you react, your reaction distance, depends on how fast you are moving and your reaction time. Before you can react to something, you must first perceive it. How do reaction and perception distances differ when you are moving 60 miles per hour compared to when you are moving only 10 miles per hour? By the end of this investigation, you will know the answers to such questions.

You will be working cooperatively in your teams of two. You will need the roles of Team Members, Communicator, and Tracker. Work on the social skill *Stay with your team*. Move your desks together side by side or sit next to each other at a table.

Materials for Each Team of Two:
- 1 ruler

Process and Procedure

1. Read the Background Information following the procedure.

2. Use the Graph of Reaction and Perception Distances (Figure 6.9) to determine how far you would travel in 1.2 sec if you were moving 40 mph.

   ➞ The distance is 71 ft. This is just practice.

144 CHAPTER 6 Using Diversity to Set Standards

3. Copy the Data Table for Reaction and Perception Distances in your notebook.
   ➡ See Figure 6.8.

FIGURE 6.8 Copy this table into your notebook and fill it in according to Steps 4 through 10 in the procedure.

DATA TABLE FOR REACTION & PERCEPTION DISTANCES

Speed	Distance with Fastest Reaction Time	Distance with Average Reaction Time	Distance with Slowest Reaction Time	Perception Distance
5 miles per hour				
10 miles per hour				
15 miles per hour				
20 miles per hour				
25 miles per hour				
30 miles per hour				
40 miles per hour				
50 miles per hour				
55 miles per hour				
65 miles per hour				
80 miles per hour				
100 miles per hour				

CHAPTER 6 Using Diversity to Set Standards

4. Refer to the class graph of reaction times from the previous investigation. Locate the low, average, and high values.
   → Notebook entry: Record these values in the data table.
5. Use the low value to determine the reaction distances for each of the speeds listed in the data table.
   → Find the reaction time you are working with on the horizontal axis. Then determine the distance for each speed line on the graph. These distances are the reaction distances.
   → Take care not to write in your textbook.
6. Record the reaction distances in the appropriate column for each of the speeds in your data table.
7. Repeat Steps 4 through 6 using the average reaction time.
8. Repeat Steps 4 through 6 using the high reaction time.
9. Determine the perception distance for each of these speeds listed in the data table.
   → Remember that the average human perception time is 0.75 sec. On the horizontal axis, find 0.75 sec and use this time to determine distances for the various speed lines. These distances are now the perception distances.
10. Record the perception distances in the appropriate column for each of the speeds in your data table.

Background Information

In this investigation, you will use the graph shown in Figure 6.9. This graph is different from the graphs you have used in other investigations because it contains more than one line. Each line represents the data for a different

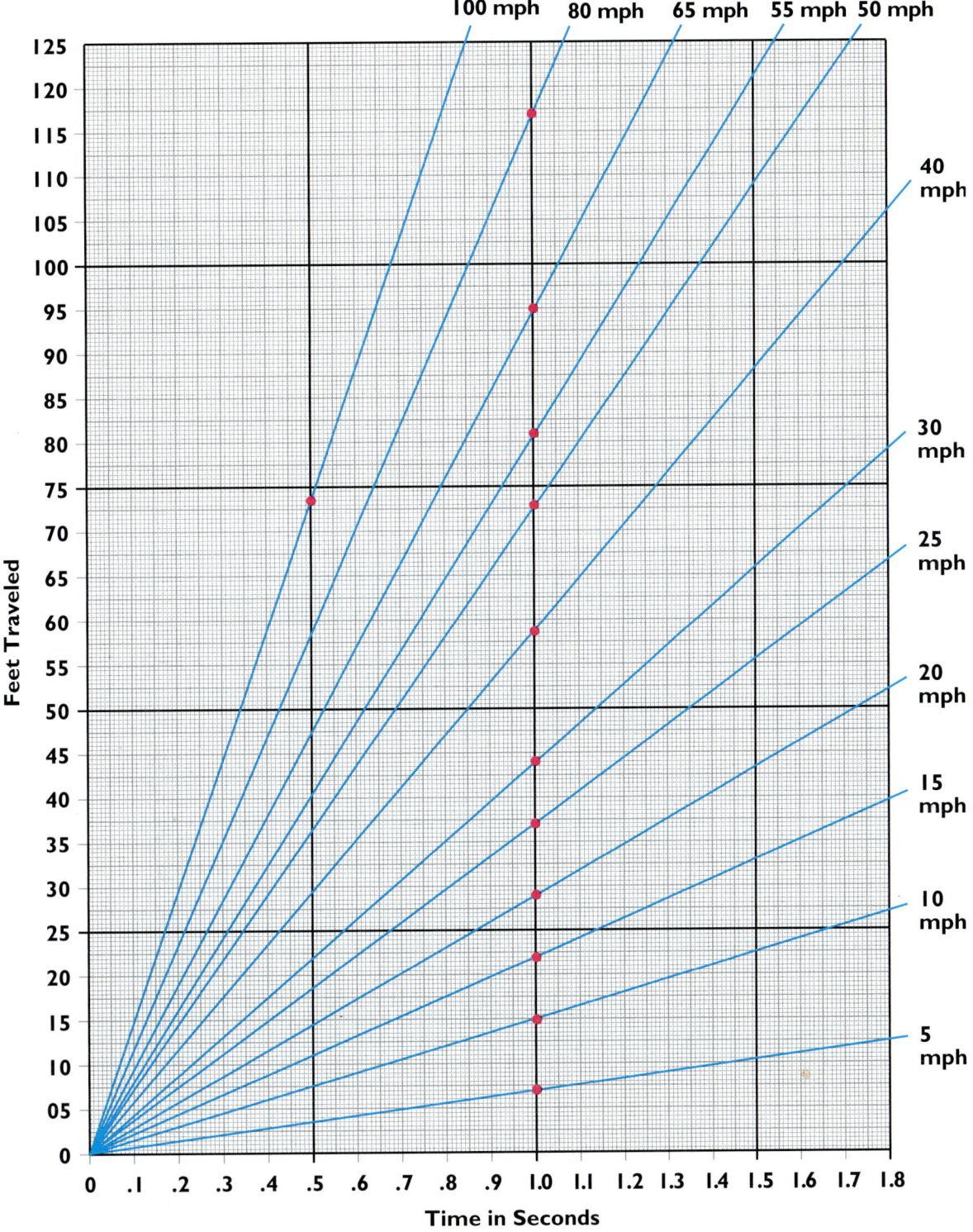

FIGURE 6.9 This graph shows the distance in feet that someone would travel at different speeds and with different reaction and perception times.

CHAPTER 6 Using Diversity to Set Standards 147

speed limit. For example, the line marked "30 miles per hour" shows the number of feet you travel per second if you are in a car going 30 miles per hour. Graphs such as this are an efficient way to show large amounts of information.

For practice, use the graph to determine your reaction distance at 30 miles per hour. First locate your reaction time on the horizontal axis. Next, holding your finger or pencil *above* the page, "trace" the line directly up from that value until you reach the line labeled "30 miles per hour." Then "trace" the line to the left from this point until you come to the vertical axis. The value recorded there is your reaction distance at 30 miles per hour. Be careful not to mark your textbook.

Wrap Up

Discuss these questions with your partner and record your answers in your notebook. Be ready to use your answers in the following connections activity.

1. Describe how your reaction distance changes as your speed increases.

2. Determine the difference in reaction distance between the high and low reaction times in your class for the following speeds:
   a. 10 miles per hour
   b. 30 miles per hour
   c. 65 miles per hour

3. Determine the difference in reaction distance between the high and the average reaction times for the following speeds:
   a. 10 miles per hour
   b. 30 miles per hour
   c. 65 miles per hour

4. How might differences in reaction time influence whether you have enough distance in which to stop to avoid an accident?

Total Stopping Distances

Work individually on this connections activity. The reading The Three Phases of Stopping described how you can add the perception distance, reaction distance, and skidding distance to get the total stopping distance. In the investigation Determining Reaction Distances and Perception Distances, you observed how the distance you travel changes as your speed increases. In this connections activity, you will calculate the total stopping distance for several different speeds.

Copy the Data Table for Total Stopping Distances into your notebook (see Figure 6.10). Transfer the information you need from your Data Table for Reaction and Perception Distances to this data table. Then add the data across each speed limit to determine the total stopping distance for each speed. That means you should choose a reaction distance column from your data table that would result in the safest total stopping distances. Then write answers to these questions in your notebook. Prepare to share your answers with the rest of the class.

1. As speed increases, what happens to total stopping distance?
2. Which reaction time did you use in your calculations: the class average, the high, or the low? Why did you choose that reaction time?

DATA TABLE FOR TOTAL STOPPING DISTANCE

Speed	Reaction Distance (in feet)	Perception Distance (in feet)	Skidding Distance (in feet)	Total Stopping Distance (in feet)
5 miles per hour			1.1	
10 miles per hour			4.7	
15 miles per hour			10.5	
20 miles per hour			19	
25 miles per hour			29	
30 miles per hour			42	
40 miles per hour			75	
50 miles per hour			117	
55 miles per hour			142	
65 miles per hour			198	
80 miles per hour			300	
100 miles per hour			468	

FIGURE 6.10 Copy this table into your notebook. Fill in the columns according to your calculations on your Data Table for Reaction and Perception Distances. You have to decide which reaction distance column you will use.

SIDELIGHT ON TECHNOLOGY
Cellular Phone Safety

Automobile collisions result in about 40,000 fatalities each year in the United States. A majority of states now require that all passengers in an automobile wear seat belts. There also are many penalties for driving while using substances such as alcohol or drugs. These laws are meant to help drivers remember to be safe and cautious while in their cars. There are many factors, however, that contribute to car accidents. Sometimes weather conditions contribute to driving risks. Often, drivers become distracted by things they see or hear. Some drivers may eat snacks, talk to other passengers, or listen to the radio while driving. These activities may lessen their ability to react to emergency situations.

Most states have no restrictions for the use of technologies such as CB radios, stereos, or cellular phones while driving. In February 1997, *The New England Journal of Medicine* published a study that may change the way we think about how safe drivers can be while using cellular phones. Donald A. Redelmeier, M.D., and Robert J. Tibshirani, Ph.D., the authors of this study, observed that using cellular telephones in cars is associated with a quadrupling of the risk of a car crash during the period of a phone call.

Redelmeier and Tibshirani were curious about how the use of cell phones could cause accidents. They found out from other published articles that some countries, like Australia, Brazil, and Israel, were restricting cell phone use based on the belief that cell phones are one cause of automobile accidents. So the authors of this study asked themselves questions like: How does the use of cell phones contribute to motor vehicle collisions? Is there a difference among the age, education, or driving experience of drivers using cell phones who also have been in car accidents? And, what situations are most likely to put a cell phone user at risk for a collision?

A total of 699 individuals participated in Redelmeier and Tibshirani's 14-month study, which was performed in Toronto,

Canada. The authors discovered that a total of 16,870 calls were placed by the participants and 3,643 were received by the participants. The duration of each call averaged 2.3 minutes. Of the 699 participants, 24 percent were on the phone ten minutes before they were in a car crash. When they talked with all the participants, they found that differences in age, education, and driving experience made some difference in the driver's chance of getting in an accident. Those who were younger in age, had not finished high school, and did not have much driving experience were at a slightly higher risk for collision. In no group did a cell phone have a protective effect.

As the authors did more research, they also found that the risk of collision while using a hand-held cell phone—as opposed to a hands-free model—was virtually the same. A study done in relation to Redelmeier and Tibshirani's showed that 72 percent of drivers in collisions tended to be engaged in conversations related to business. Also, high-speed roadways were more likely to be the setting for accidents than low-speed areas. Given the evidence, Redelmeier and Tibshirani think that cellular phones contribute to the risk of collision because they are distracting and interfere with the driver's perceptual process. Cell phones put drivers in the situation of responding to a surplus of unpredictable perceptual tasks.

Answering a phone call or dialing a phone number may increase a driver's reaction and perception time. The driver may not be able to avoid a collision due to his or her longer response time.

The risk of a collision while using a cellular phone is about the same as while driving with a blood-alcohol level of 80 milligrams per deciliter. Although this blood-alcohol level is considered to be at the legal limit, driving with this much alcohol in one's body still increases the risk of collision to four times its normal level. Higher levels of alcohol contribute to even higher accident risks. One difference between cell phone use and driving with alcohol in your blood is that while alcohol circulates through the body for hours, the average cell phone call lasts only minutes. Still, it poses as much danger as driving after drinking.

The authors of this study say that there are many benefits of having a cell phone in your car. For example, you can call for help when an emergency occurs. In fact, hundreds of 911 calls are made from cell phones each day.

The use of cell phones contributes to only some motor vehicle collisions. In what ways do you think drivers could use cellular phones in their cars but decrease their accident risk? Do you think the use of cell phones while driving should be restricted? Why? How do you think using a cell phone while driving interferes with the driver's reaction and perception time?

Life in the Fast Lane

What do you know about acceleration? How do speed and distance relate to acceleration? In this investigation, you will have the opportunity to explore answers to these questions. To do this, you will work in a team to collect information about how far a ball travels down a ramp in specific time intervals. Galileo, an Italian physicist and astronomer who lived during the 16th century, did this same experiment to test what he thought would happen.

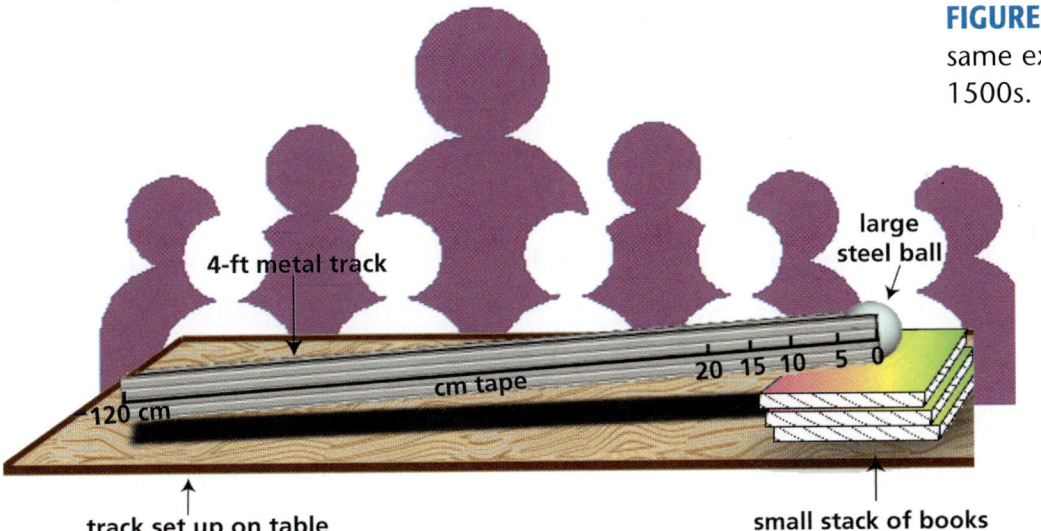

FIGURE 6.11 Galileo did this same experiment in the 1500s.

WORKING COOPERATIVELY

In this activity, you will work in teams of five. One of you will be the Tracker, and four of you will be Team Members. Practice the unit skill *Show caring and respect for others and their ideas.*

Materials for Each Team of Five:
- 1 track set up according to your teacher's directions
- 1 large steel ball
- 1 clock with second hand or stopwatch
- 2 or more pieces of graph paper
- 1 pencil
- 1 calculator

Process and Procedure

1. Read the Background Information that follows this Process and Procedure.

2. Set up the track for this activity as your teacher directs.

3. Read through the investigation and develop a data table in which to record the first part of the data.
   → You might develop a data table that is similar to the one in Figure 6.12.

4. Read the following descriptions of the tasks for this investigation. Be sure you and your teammates understand what the procedure will be.

   The Tracker will keep track of the time. The Tracker also will release the ball. To prepare the other Team Members for their jobs, the Tracker will count down from 3 and then release the ball as he or she says, "GO." At that same moment, the Tracker will also start the stopwatch or begin counting seconds on the clock. The Tracker will announce each second out loud. In other words, "3, 2, 1, GO! 1, 2, 3, 4 …"

   Team Member 1 will read the position of the ball on the track when the Tracker announces 1 second. This position will be the centimeter line closest to the ball at 1 second. Team Member 1 will record this position in his or her data table.

CHAPTER 6 Using Diversity to Set Standards

Data Table #1 – Life in the Fast Lane

	Position 1 1 second	Position 2 2 seconds	Position 3 3 seconds	Position 4 4 seconds
Practice 1				
Practice 2				
Practice 3				
Practice 4				
Event 1				
Event 2				
Event 3				
Event 4				
Average of 4 Events				

Team Member 2 will read the position of the ball on the track when the Tracker announces "Two." This position will be the centimeter line closest to the ball at 2 seconds. Team Member 2 will record this position in his or her data table.

Team Member 3 will read the position of the ball on the track when the Tracker calls out the third second. This position will be the centimeter line closest to the ball at 3 seconds. Team Member 3 will record this position in his or her data table.

FIGURE 6.12 In your notebook, create a data table like this one.

CHAPTER 6 Using Diversity to Set Standards

Team Member 4 will read the position of the ball on the track when the Tracker calls out the fourth second. This position will be the centimeter line closest to the ball at 4 seconds. Team Member 4 then will record this position in his or her data table.

→ Try to be as accurate as possible, but expect some differences in your readings from one event to the next. For example, Team Member 1 may record 7.5 for the first event and 8.3 for the second event.

→ When Galileo did this experiment in the 1500s, he used his pulse to measure the time intervals.

5. Discuss the experiment with your teammates and make sure everyone understands what his or her job is. If you have questions, choose a Communicator to get advice from members of another team.

6. Like many other things that you do, you will perform your tasks better if you practice a few times first. Practice the experiment to see how well you can do it. If you are having trouble after several tries, assign a Communicator to ask your teacher for help.

7. After several practice tries, when you think you have a good feeling for the timing and the tasks involved, complete the experiment 4 times.

→ In your data table, be sure to distinguish between your practice data and your final data.

8. When you have completed the experiment 4 times, find the average position for each second.

→ As you know, people's perception and reaction times vary. Under conditions such as these, it is a good idea to take the average. Why is taking the average a good idea?

→ If you forget how to find averages, ask a teammate for help. Record averages in the appropriate place in your data table.

9. When you have found the average for your position, share your data with your teammates and check each other's work.
    → A team effort is especially important during experiments such as this one. The quality of each person's work is also extremely important. Record each person's average position in your data table.

10. Next you can calculate the average overall speed by taking the average of the total distance traveled and dividing by the total time of 4 seconds.
    → Record this average overall speed in your notebook.

11. Discuss the following questions with your teammates:
    a. Do you think the speed of the ball stayed the same or changed during each event?
    b. If you think it changed, how did it change?
    c. What are some things that you might want to know that the average overall speed does not tell you?
    → Let's look at the experiment in more detail.

12. Using the average position for each second, plot your data on a graph.
    → The x-axis will be "Time" and the y-axis will be "Distance." Zero will be at the origin. See How To # 3 if you need help in creating a graph. Share your graphs with your teammates.
    → What does your graph look like? Does it curve up or down? Is it a straight line?

13. Create another data table like the one in Figure 6.13 in which to record some more information about the experiment.
14. Next, calculate the average speed for each 1-second time interval and record these speeds in the second data table.
    → Calculate the speed by finding the distance in centimeters between each successive position (find the distance between position 1 and position 2, between position 2 and position 3, and so on). Record the **speed** as centimeters per second (cm/sec).

FIGURE 6.13 In your notebook, create another data table like this one.

Data Table #2 - Life in the Fast Lane

Time (Seconds)	Position (cm)	Speed (cm per Second)	Acceleration (cm Divided by Sec²)
0	0	___	
1	___	___	___
2	___	___	___
3	___	___	___
4	___		

15. Create another graph, and for this graph, plot time on the *x*-axis and speed on the *y*-axis.
    → Use the average speed for this time interval to represent the instantaneous speed at the middle of the time interval. (Do you think the speed of the ball changed at all during each interval?)
    → To do this, you will plot the first speed at 0.5 sec, the second speed at 1.5 sec, and so on.
    → What does this graph look like?

16. **Acceleration** is the change in speed divided by the corresponding time interval. Calculate the acceleration for each time interval and record these values in your second data table.
    → For example, take the speed at 1.5 sec, subtract the speed at 0.5 sec, and divide by 1 sec. The units are written as cm/sec^2.

17. Discuss the following questions with your teammates:
    a. Do the numbers that you get when you do the calculations in Step 16 show any pattern?
    b. If so, what type of pattern do you see?

Background Information

People have many different reasons for moving from place to place. Sometimes they need to get to school or to work; other times they travel for fun. The methods that people use to get from place to place vary as well. They might travel by car, boat, bicycle, subway, or even by foot. If people need to travel a long distance in a short time, they may fly in a jet plane or a helicopter. If people need to travel only a short distance and they have a lot of time, they may walk or ride a horse. We see a lot of diversity in the methods and speed of travel.

For each method of travel, there are limits. Some limits are due to the method itself. For example, a bicycle will only go so fast. For others, society places limits for safety reasons. To think about what limits are reasonable for safety, we first must understand the relationships involving distance, time, speed, acceleration, force, and energy. To explore these concepts, we will do a simple exercise.

Wrap Up

Discuss these as a team and record your ideas in your notebook. Be prepared to share your ideas in a class discussion.

1. Review each of the graphs you made and think about the patterns that you see. Hold the graph at a distance so you can see the general pattern of each. Write your ideas in your notebook.

2. Think about the conditions of the experiment. What forces do you think were acting on the ball?
    → A **force** is anything that influences an object or a system, such as a push or a pull.

3. Is there any reason to think that the forces acting on the ball were changing during the experiment?

4. If all of the forces acting on the ball were in balance, then the ball would have remained stationary or would have moved at a constant speed. This is Isaac Newton's First Law of Motion. But in this experiment, the speed did change somewhat, so certain forces must have been acting on the ball. What forces do you think these were and why?
    → Do you remember who Isaac Newton was? If not, review the section in Chapter 1 where Isaac introduces Isaac Newton.

5. Our graph showed the relationship of speed to time. It indicated that the speed increases uniformly with time. In other words, the acceleration of the ball was constant as it moved down the track. This is a special case of Newton's Second Law of Motion.

   When Galileo did this experiment, he also plotted the position of the ball along with the square of the time, or time squared. Now try that same type of graph.

   a. For each time in your data table (these will be seconds 1, 2, 3, and 4), calculate the time squared.
      ➡ For 1, you will get $1^2 = 1$. For 2, you will get $2^2 = 4$, and so on.

   b. Create another graph with the positions on the y-axis and "Time squared" on the x-axis. That is, position 1 should be plotted at $1^2 = 1$ and position 2 should be plotted at time squared $(2^2) = 4$. What does your graph show when you step back from it and view it at a distance? Does it have a constant slope? Galileo expected to get a straight line. This is another way of saying that he predicted that acceleration would be constant.

   c. The accuracy of our experiment was limited by the techniques that we used to measure. For example, we know that we have to allow for people's perception and reaction time. This means that our measurements are not that precise. If you had a research grant to redesign the experiment, what would you do to try to increase the accuracy?

SIDELIGHT ON NATURE

How Is Energy Conserved during an Accident?

Kinetic energy is energy of motion. Although energy may change its form, it is always conserved in some way. Consider an automobile that is traveling 25 miles per hour. If a truck suddenly pulls out in front it, the driver of the automobile must quickly apply the brakes to avoid a collision. At this point, the kinetic energy of the moving car is transferred to heat energy in the braking system of the car. If the kinetic energy of the car is not all transferred to heat energy at this point and the car is still moving forward, the kinetic energy still available will be transferred to all of the objects involved in the collision—to vehicles that smash, to materials that bend and break, and often to people as well. What if the car were going faster? The kinetic energy of the car varies with the mass and increases with the square of the speed. A car traveling at 60 miles per hour will have 16 times as much kinetic energy as a car traveling at 15 miles per hour. (Sixty miles per hour is four times greater than 15 miles per hour, and $4^2 = 16$.)

Because people with a range of skills and a range of perception and reaction times drive cars in a variety of circumstances, it is important to limit many aspects of motorized travel, especially SPEED.

ELABORATE  investigation

Setting Speed Limits

In this investigation, you will use your data from the connections activity Total Stopping Distances to establish your own set of speed limits. Based on what you have just learned about the relationships among time, distance,

162   CHAPTER 6 Using Diversity to Set Standards

speed, and acceleration, how would you set and justify speed limits for different locations?

Process and Procedure

Decide on speed limits for the following locations:

    a. Divided, limited-access highway (interstate)
    b. Rural open highway
    c. School zone
    d. Busy downtown street
    e. Residential neighborhood

Use the information from the Data Table for Total Stopping Distances and consider the following questions for each location:

Work cooperatively as a Team Member with your partner. One of you also will be the Communicator. When your teacher says "Go," try to be the quickest and quietest team to move into its group. Practice this skill at the end of the investigation as well. Place your chairs or desks so that you are sitting beside each other.

- What do you think is an appropriate distance for people to leave between cars in order to be able to stop for an emergency?
- The speed limits will be for all people. Should you reconsider your choice of which reaction time to use?
- What are the pros and cons of using the average perception time of 0.75 sec instead of a slower or faster perception time?
- Do you think you should consider road or visibility conditions when setting speed limits? Stopping on wet asphalt instead of dry asphalt can double the skidding distance. Stopping on an icy road can increase the skidding distance by a factor of 11. Traveling downhill increases the stopping distance, and traveling uphill decreases the stopping distance.
- How many of the cars on the road are new and in good working condition? How many of the cars on the road have bald tires, wipers that don't work, or brakes that need repair? How many of them have anti-lock brakes?

STOP: This is a good time to practice your unit skill. Review the unit skill T-chart if necessary.

Wrap Up

Do the following tasks individually.

1. Record your team's speed limit decisions.
   → Notebook entry: Record these in your notebook beside the speed limits you set in the Wrap Up during the investigation Watching *The Final Factor*.

2. Review this information and be prepared to present and defend your speed limit standards to the rest of the class.

The Bell-Shaped Curve and Setting Standards

Setting standards is not easy. As you learned in Chapter 3, one type of standard is a rule or limit. Such rules often are for the safety and well-being of society. But people in a society are very diverse. When you try to set one standard that accounts for human limits and diversity, you are faced with many difficult decisions.

For example, you just completed an investigation in which you set certain standards, speed limits. You had to decide on one speed that accounted for people's diversity in their ability to perceive and react to emergencies. You might have used 0.75 seconds as the average perception time for humans. You might, however, have adjusted that value to accommodate some people who cannot perceive sudden events as quickly as others.

You discovered that in your class alone there is a wide diversity in reaction times. Imagine how much wider that diversity would be if you considered the entire population of the country. You had to decide which reaction time you would use in setting your speed limits: the highest, the lowest, or an average.

Not only did you have to consider human limits, but you also considered automobile conditions and weather conditions. You might have assumed that all cars on the road were in good condition, or you might have assumed that all cars on the road were in poor condition.

1. Assume that the condition of cars on the road is between average and superior. Would you create potential hazards by setting speed limits too high for the cars that are in the poorest operating condition?
2. You might have set speed limit standards assuming that the roads were dry and visibility was good. Are the roads a safe place to be when the surface is slick and the visibility is poor?
3. What might happen if people on slick roads still drive at the speed limit because they assume it was set to account for bad weather conditions?
4. What might happen if you use the averages or high values of human limits when setting standards? What might happen if you use the low values of human limits?

These are all issues that state and federal governments must keep in mind when setting speed limits. There is a way to minimize the problems of basing speed limits on just the low, high, or average values of human limits. You could use a bell-shaped curve. In using a bell-shaped curve, you could examine the way people's limits are distributed. You could decide to use the areas of the curve that account for the most people without any of the extreme values in limits that account for a minority. Figure 6.14 shows different ways you can divide a curve for setting standards.

In fact, this is the way many organizations set standards. They generate a bell-shaped curve for a set of data. There is no set method of dividing the curve, however. People who are trying to set standards that they hope will be adequate for an entire population of people must make an important decision. They must decide how to section the data on a curve to determine the range of limits for which they want to account.

For example, you probably know that a person has to pass a vision test before obtaining a driver's license. If a person cannot see well enough, that person will not receive a license. What does "well enough" mean? Data from a bell-shaped curve indicate that some people have poor vision (the low values of the curve). When officials set the standards for eyesight, they may have used this bell-shaped curve to section off all the people to the right of the low values and say that they are eligible for a license. There is a point at which officials determine that vision is too poor to be considered acceptable for driving. Drivers with poor eyesight would be a hazard to themselves and to other people on the road, so society places limits on who can drive.

Often society must place limits on behaviors in certain situations. You could use the bell-shaped curve to decide where to section off the low values for reaction time from the rest of the values. For example, you could look at a curve of reaction times and decide what part of the curve represents reaction times that are too long. You could assume that individuals with these long reaction times should not be driving. You then could base your speed limit on the longest acceptable reaction time. (There is, however, no specific law banning people with very long reaction times from driving.) This is one way you could have used your data from your bell-shaped curve to

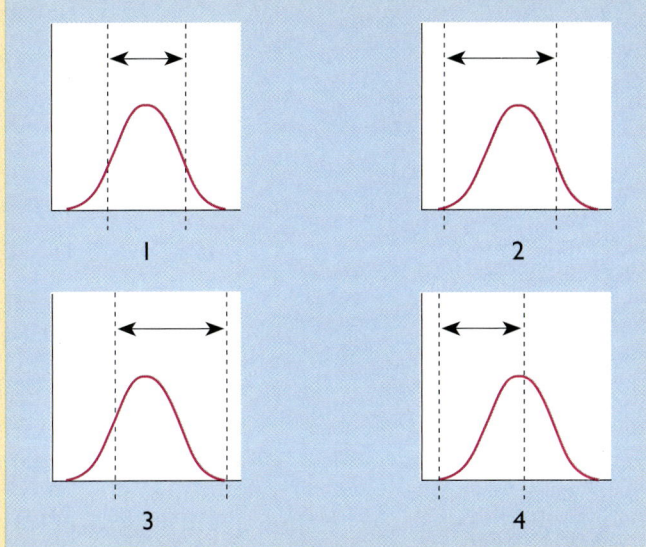

FIGURE 6.14 The peak of the curve shows the value that is measured or noted most commonly. There are many ways you can section off a bell-shaped curve to account for a majority of people. Can you think of other ways to section off a bell-shaped curve?

CHAPTER 6 Using Diversity to Set Standards

determine speed limits. Remember that no one can tell you where to section off your curve or what decisions or assumptions you should make. It is up to the people setting the standards to make their own decisions.

Referring to the bell-shaped curve is one of the methods people use to set standards. Most decisions that officials make based on data from a bell-shaped curve are for the well-being of most people. If you are curious and have some time, you might research how the government makes decisions on speed limit laws and why officials decide what they do. Whether you do that or not, you have experienced how difficult it is to make such decisions. You also have experienced how much people's well-being depends on the decisions of other people.

Don't Drink and Drive

In the investigation Setting Speed Limits, you set speed limits based on the total stopping distance required at a certain speed. You considered diversity in reaction times, inclement weather conditions, road conditions, and car maintenance conditions, among other factors. One factor that you probably have heard a great deal about, but that you might not have considered in setting your speed limits, is **alcohol consumption**. In this investigation, you will learn how alcohol affects people. You then will use this information along with the ideas that you have been exploring in this chapter to set standards for drinking and driving. This activity will help you evaluate your understanding of the ideas in this chapter.

Materials for Each Team of Two:
- 1 sheet of graph paper
- 3 different colors of pencils
- 1 ruler

Process and Procedure

1. Read the Background Information following this procedure.
2. Use the information in Figure 6.15 to answer the following questions.
   → Notebook entry: Write the answers in your notebook.
   a. What do the data suggest about the relationship between drunkenness and Blood Alcohol Content (BAC) levels?
   b. Why do you think these 3 experimenters obtained different results?

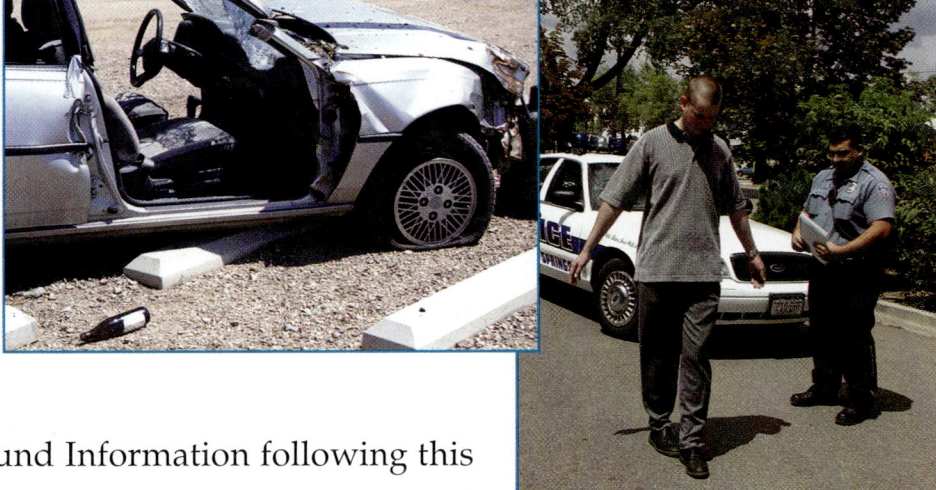

Work cooperatively in your teams of two. You will need a Tracker and a Communicator. Practice the social skill *Stay with your group*. Move your desks together or sit beside each other at a table.

Percentage of Test Subjects (People) Who Became Drunk

BAC	Widmark	Jetter	Hine
0.00 – 0.05	0	10	7
0.051 – 0.10	19	8	18
0.101 – 0.15	31	28	24
0.151 – 0.20	33	34	36
0.201 – 0.25	10	7	8
0.251 – 0.30	5	5	4
0.301 – 0.35	2	1	1
0.351 – 0.40	0	0	2
0.401	0	7	0

Source: Council of Scientific Affairs (1986). Alcohol and the Driver, *Journal of the American Medical Association* 255 (4) 522-527.

FIGURE 6.15 This chart represents the data of three different experimenters (Widmark, Jetter, and Hine) who conducted research on Blood Alcohol Content levels.

CHAPTER 6 Using Diversity to Set Standards

Topic: Drunk Driving
Go to www.scilinks.org
Code: physical170

c. Graph the data in Figure 6.15.
   ➡ Decide what information you should plot on the horizontal axis and what information you should plot on the vertical axis. Be sure to consider how to scale each axis. Use a different colored pencil for each experimenter's data. Smooth out the bars with a curved line using the appropriate color of pencil.

d. Decide on a legal standard for Driving Under the Influence (**DUI**). A person "Driving Under the Influence" is mentally and physically impaired by the effects of alcohol or other drugs.
   ➡ Use the background information on BAC and the graph of the data. Your team should decide at what BAC, and above, it would be illegal for a person to operate a car. You should decide whether you will set just one standard or different standards for males and females, for adults whose stomachs are empty and those whose stomachs are full, and for adults who are small, medium, or large.

3. Share your decision with the rest of the class. Be prepared to justify your decisions.
   ➡ Be sure that both members of your group can explain how you made the decision and why.

Background Information

There are laws against driving under the influence of alcohol. Officials base these laws on current research that tells them how much alcohol a person can tolerate before becoming a dangerous driver. The people who set the standards and make the laws are concerned about the safety of drivers. They usually base Driving Under the Influence (DUI) standards on something known as **Blood Alcohol**

Content or **BAC**. BAC is the ratio of alcohol to total blood volume. This ratio is expressed as a percentage. The range of BAC values is from 0.00 percent, which means there is no alcohol in the blood, to about 0.50 percent, which is usually a fatal amount of alcohol in the blood.

People show diversity in their response to alcohol. Some people act very drunk with a BAC of 0.05 percent, while others do not act very drunk until they have a BAC of 0.20 percent. One thing that happens to people who are drunk is that their reaction times increase. People show a wide range of diversity in how much their reaction times increase at different BACs.

People also show a wide range of diversity in BAC after drinking the same amount of alcohol. A smaller adult who drinks as much as a larger adult will have a higher BAC, because the smaller adult has less blood. One study has shown that men and women differ in how they digest alcohol. According to that study, a woman who drinks the same amount of alcohol as a man will have a higher BAC than the man, even after allowing for differences in size.

Also whether a person has a full or empty stomach affects his or her ability to absorb alcohol into the blood. Because food in the stomach absorbs some of the alcohol before it gets into the blood, the person with a full stomach will have a lower BAC than the person with an empty stomach.

FIGURE 6.16 On the side of the road, police need a convenient way to test for alcohol consumption. They use an instrument called a Breathalyzer, which measures the amount of alcohol in one's breath.

Wrap Up

Discuss these questions with your partner. When you agree on an answer, write the answer in your notebook and prepare to discuss it with the rest of the class.

1. How were the standards other groups set similar to or different from yours?

2. Calculate one DUI standard for the entire class. (You can average the teams' standards, or you can obtain a standard from a graph of the class's standards.)

3. You based your DUI standards on data that told you at what BAC levels people became drunk. What is your operational definition of "drunk"? (Relate your operational definition first to driving, then to walking, then to riding a bike.)

4. Imagine that you were a state legislator. Would you want the DUI laws to be more or less strict than the current DUI laws for your state? Explain how you might want the laws to be different and why. Do you feel safe with the current DUI laws for your state?

5. In your notebook, write a short paragraph that describes how you have improved in your ability to stay with your group.

SIDELIGHT ON TECHNOLOGY
The Hyper Car

Do you ever wonder what cars will look like when you are old enough to drive? Do you think they will look and function the same way as today's cars? Technology, like many things in life, also has patterns of change. Scientists and engineers are developing future cars that may look, feel, and perform in very different ways than current cars.

At the Rocky Mountain Institute in Snowmass, Colorado, scientists and engineers are developing a new type of car, the Hyper Car. It may change the automobile world by being lighter, more fuel efficient, and more environmentally sound than any vehicle presently produced.

Cars built during the 1990s weigh about 3,000 pounds. The bodies of these cars were made mostly of steel, which must be molded, pressed, and pounded numerous times to acquire the desired shape. There is a lot of scrap metal left over, and these cars waste a lot of energy. About 95 percent of a 1990s vehicle's wheelpower hauls the car. Less than 1 percent carries the driver. About 80 percent of the fuel energy is used before it gets to the car's wheels.

The goal of the Hyper Car is to reduce weight and braking energy. The car is designed to be aerodynamically "slippery." In most cars, some kinetic energy is transferred to heat energy each time the driver applies the brakes. Because most cars are so heavy, stopping uses a lot of energy. A Hyper Car would be extremely light because it would be made of composite materials and polymers. Examples of these substances are carbon fiber and titanium alloys already used by designers in race cars and bicycles to make them lighter. Ultralight cars tested in Switzerland weigh 1,000 pounds or less, but they are as crashworthy as heavier cars. Composite materials and polymers take only two pressings to shape, and there is not much scrap material left over. This would make the Hyper Car more environmentally friendly. The sleeker aerodynamic shape of the car would produce less friction and make the car more energy efficient.

The Hyper Car would use a special kind of engine. Engineers are working on a hybrid-electric engine—a cross between an electric battery and an electricity generator. The car would use electric wheel motors, but it would also store fuel to create electricity. This engine would use cleaner, alternative fuels that would allow the car to conserve more energy and travel longer distances than a simple electric car. Solar cells also could recharge and store enough energy for a daily work commute. Scientists predict that even without solar panels, the advanced design would qualify it as a "zero emission vehicle," one that gives off very few air pollutants. The Hyper Car is an example of how technology can change the limits by making use of diverse materials.

CHAPTER 7
Evaluating Your Understanding of Limits and Diversity

Congratulations! You have almost completed the first unit in this book. That is quite an accomplishment! You have explored and learned many things about limits and diversity since the investigation Star Tracers. This chapter should help you put it all in perspective.

Consider the diversity of people that hike and climb mountains. What limits might they encounter? The climber pictured here is blind. What additional limits might he encounter?

ENGAGE	Shhh . . . Hush . . . QUIET!
EXPLORE / **EXPLAIN**	What Have You Learned in Unit 1?
ELABORATE / **EVALUATE**	How Much Noise Is Too Much Noise?

ENGAGE connections

Shhh ... Hush ... QUIET!

By yourself, study the photographs in Figure 7.1 that represent noisy things. Think about which of these you would consider too noisy when you are trying to study at home. Record your decisions in your notebook and share your ideas in a brief, class discussion. How much diversity exists among the students in your classroom when it comes to limits on noise?

FIGURE 7.1 Which things are too noisy when you are trying to study?

176 CHAPTER 7 Evaluating Your Understanding of Limits and Diversity

What Have You Learned in Unit 1?

You began Chapter 3 by conducting investigations in which you measured your classmates' successes at accomplishing certain tasks. You found that people have limits in accomplishing certain tasks and that they show diversity in their limits. Your investigations yielded data, which you learned to organize into data tables and to graph. The shapes of the graphs were similar to one another. After you accomplished all of this, you also learned how to collect data that could be compared to other teams' data by identifying common operational definitions and controlling variables.

FIGURE 7.2 In Chapter 6, you calculated total stopping distances.

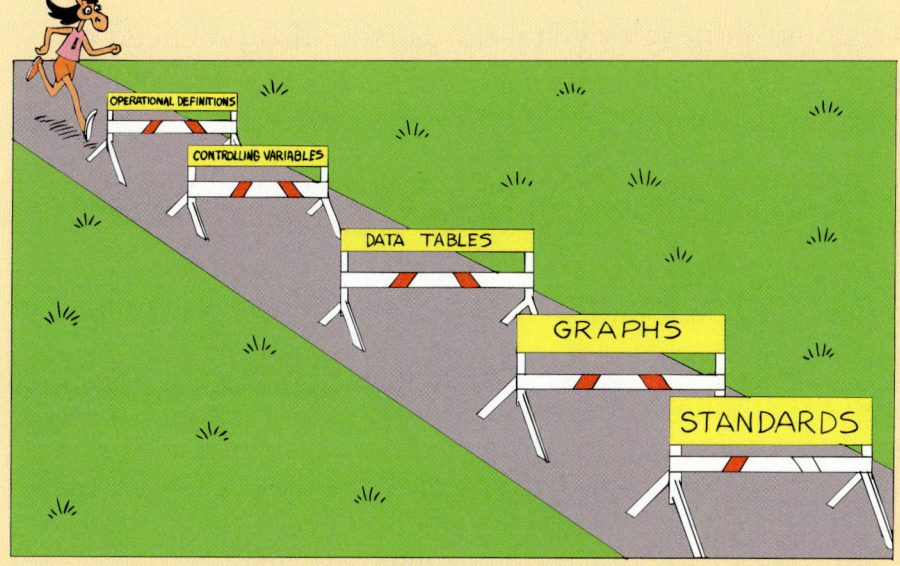

CHAPTER 7 Evaluating Your Understanding of Limits and Diversity

FIGURE 7.3 What is the last phase of stopping?

In Chapter 4, you used popcorn to explore the idea that being different is what is normal. Normal does not mean being able to do things the same as everyone else. You learned this by collecting and studying data that generated a bell-shaped curve and by studying curves of humans' and animals' peripheral vision. You were able to practice your skills in creating operational definitions and controlling variables when you explored the concept of afterimage and discovered more diversity in your class.

In Chapter 5, you had a chance to use TV to identify more limits in humans, and you discovered that humans are diverse in their ability to perceive continuous motion, flicker, and lines at a distance. You also learned a lot about television. You set standards for TV pictures that accounted for human limits.

In Chapter 6, you explored the three phases of stopping and calculated the total stopping distance for people traveling at certain speeds. You learned about the fundamentals of acceleration. You used the diversity in people's ability to react to sudden events and considered those limits and other factors that affect stopping distance. This helped you set standards for speed limits in different areas of the city and country. You learned how using a bell-shaped curve could help you make important decisions about setting standards. You also practiced using a bell-shaped curve to help you set standards for drinking and driving.

In Chapter 7, you are going to use everything you studied in the first four chapters in one investigation in this chapter. You will create operational definitions, control variables, and write a procedure that will help you conduct an experiment and collect data from 25 people. After organizing your data, you will create a graph and use it to set a standard. When you are finished with that investigation, you can compare it with Star Tracers and get a good feeling for what you have accomplished so far.

How Much Noise Is Too Much Noise?

You are competing for the honor of being named to the Noise Patrol Squad at your school. Your school has just lost an entire team from the Noise Patrol Squad because the Thompson triplets moved away. Your team is competing against all the other finalists that you see in your room. (Don't look now; they might think that you are checking out the competition.)

The Noise Patrol Squad has given you a definition of "too noisy." It is when the noise level in the room interferes with a person's comprehension while reading. Your team must design an experiment that will use this definition to find out the diversity of tolerance to noise in a population of 25 people. The procedure section outlines the requirements of this contest. Read the requirements first so that you have a better understanding of what you will do. If you are unsure of what the Noise Patrol Squad wants, have your Communicator ask for help from the judge for the Noise Patrol Squad.

Work cooperatively in your teams of two. You will need a Communicator/Manager and a Tracker. Check the role descriptions to make sure you understand your duties. Move your desks so that you face each other, or sit facing each other at a table. Practice the unit social skill *Show caring and respect for others and their ideas.*

Materials for Each Team of Two:

You will determine what materials you need. (The materials you need may change as you gather results.)

Process and Procedure

1. Have a brainstorming session to create operational definitions. One operational definition should say

Topic: Noise Pollution
Go to www.scilinks.org
Code: physical179

FIGURE 7.4 How much noise is too much when you are studying in a library?

how you will measure a person's comprehension while reading. The other operational definition should say how you will measure the loudness level that first interferes with a person's reading comprehension.

➡ Notebook entry: The Communicator should record all of the ideas in his or her notebook. Be careful not to rule out any ideas yet.

2. As a team, decide which operational definition your team will use while conducting your investigation.

➡ Use the list you generated in Step 1.

3. Create a list of variables that you will need to control while conducting your investigation.

➡ Notebook entry: Record these variables in your notebook.

4. Identify a list of materials and a procedure that your team will use to conduct your investigation.

➡ Notebook entry: Record the materials and procedure in your notebook.

5. Design a data table to help as you collect your data.

6. Conduct your investigation.

➡ Be sure to follow your procedure, use your operational definitions, and control your variables. Collect data from any 25 people you choose. If you find that your operational definitions are not working, you may develop new ones.

7. Construct a graph of your results.

8. Use the pattern that your graph makes to determine the range of tolerance to noise among your 25 test subjects.

9. Set a standard noise level for the classrooms in your school by completing this sentence: "It is too noisy to read when the level of noise in the room exceeds _____."

    ➡ Base this standard on the diversity among your test subjects. Remember, because your team is deciding on the standard level of noise, only your team can determine what, if any, cutoff points there might be on your graph.

Topic: Properties of Sound
Go to www.scilinks.org
Code: physical181

Wrap Up

Prepare a presentation for the Noise Patrol Squad. Begin your presentation with the sentence you completed above: "It is too noisy to read when the level of noise in the room exceeds _____." Also include the following in your presentation:

- your operational definitions and how they worked for your team,
- the variables you had to control and how you controlled them,
- the data you collected,
- the graph that your data generated,
- how you used the graph or graphs to determine a tolerable noise level,
- what a tolerable noise level is for studying in classrooms in your school,
- specific examples of how each member of your team practiced the social skill, and
- your team's overall social skill rating (use a scale of 1 to 10, where 10 is perfect).

Keep your objective in mind: Win slots on your school's Noise Patrol Squad! Because you are part of a team, you may find it easier to make your presentation if each of you prepares two or three of the sections listed above.

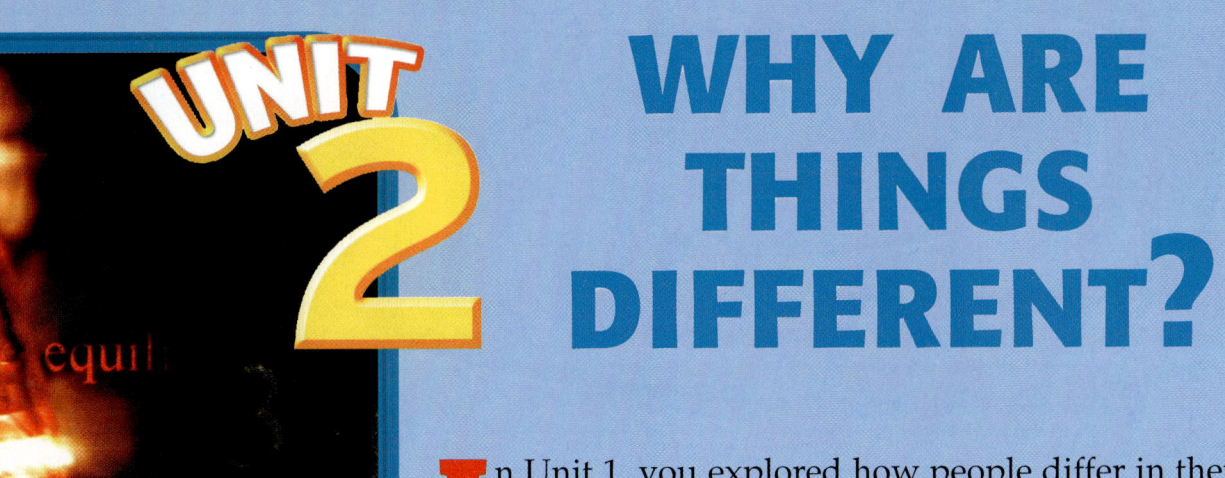

WHY ARE THINGS DIFFERENT?

In Unit 1, you explored how people differ in their limits. You also explored how important these differences are in setting standards for products that people design, such as televisions and automobiles. In this second unit, you will explore how materials are different and how important these differences become when designing products. Why would it matter what materials you use to build a boat, a locker, or a backpack? Why aren't all backpacks made of paper?

In this unit, you will explore materials in much the same way that professional scientists do. Scientists often want to know how and why materials are different. They study and conduct tests on materials, make predictions about the materials, and conduct more tests in an effort to explain what it is about materials that makes them useful for certain types of products.

This unit is about exploring materials and trying to explain their unique properties. As you explore, you will learn how to make scientific explanations and how to test them. By the end of this unit, you should understand more about why things are different and more about using science to help you explain your observations.

Chapter

8 Properties of the Material World

9 Scientific Explanations Are Ancient History

10 Using Scientific Models to Answer Questions

11 Using Models to Test and Predict

Marie is now familiar with working cooperatively in science class. After practicing with social skills and learning to cooperate with others, you too might be feeling more comfortable with the idea of working cooperatively on some science activities.

But what you have learned about working cooperatively is not necessarily confined to your science class. Like Marie in the cartoon, you can apply the skills you have practiced to many other areas of your life. Take a moment now to discuss the characters' comments, and then discuss the following:

1. In what other situations in your life can you apply the cooperative learning model and use of social skills?
2. When does working cooperatively not apply?
3. In what future situations might it be beneficial to know how to work cooperatively?

You will be working cooperatively in teams of three throughout this unit. Make sure that you and your Team Members take turns using each role. Also try to apply all of the social skills you practiced in Unit 1 as you work in your new team. For example, call your new teammates by name and try to be quick and quiet as you form your team. You will be using new social skills for the activities, as well as the new unit skill that Ros and Al mentioned: *Be open to others' ideas.*

The first thing you should do with your teammates is discuss this new unit skill and create a T-chart in your notebooks for it. With a T-chart, you can always go back and modify or add to your ideas.

CHAPTER 8
Properties of the Material World

The title of this unit is Why Are Things Different? Before you can answer the *why* question, you need to explore *how* some of the materials are different. This chapter will give you practice doing just that. You will learn about some of the properties that make materials different from one another, and you will explore and define some common scientific properties of many materials.

The thermometer in this photograph is used to measure air temperature. Designed by Galileo in the 17th century, the small, floating glass balls inside the casing move up or down depending on the temperature. The materials inside the thermometer may look similar, but what makes them different enough to measure temperature accurately?

ENGAGE **EXPLORE**	■ Bounce That Ball
EXPLAIN	■ How Things Are Different
ELABORATE	■ How Can I Learn More about That Property?
EVALUATE	■ Is the Most Always the Best?

ENGAGE EXPLORE

investigation

Bounce That Ball

What is the bounciest material you know? Rubber? Plastic? Silly Putty? What is the least bouncy material you know? Wood? Clay? Stone? In this investigation, you will test different materials and rank them from most bouncy to least bouncy. But first, you will observe a demonstration of "bounceability."

Work cooperatively in your new teams of three. Use the roles of Manager, Tracker, and Communicator. Practice the social skill *Use your teammates' names*.

Materials for Each Student:
- 1 pair of safety goggles

Materials for Each Team of Three:
- 1 wooden ball
- 1 glass marble
- 1 large rubber ball
- 1 steel ball (marble)
- 1 Styrofoam ball
- 1 wooden block (any size)
- 1 steel nail
- 1 wooden splint
- 1 rubber band
- one 8 × 11-in. sheet of aluminum foil
- one 8 × 11-in. sheet of clear plastic wrap
- various measuring tools of your choice such as a pan balance, a metric ruler, a piece of string

Process and Procedure

Part A—Seeing Is Believing

1. Turn your attention to the demonstration that your teacher will conduct.
2. Record your observations in your notebook.

188 CHAPTER 8 Properties of the Material World

3. Use your notes to participate in a class discussion about the demonstration.

Part B—Bounceability Testing

1. Obtain the materials that your team will test for bounceability.
    → You must test all of the items in the list of materials for bounceability except the measuring tools. You may use any or none of the measuring tools as you test the items for bounceability.

2. Decide on an operational definition of bounceability.
    → This means deciding how your team will measure the bounceability of the materials. You may have to try a few ideas before you decide on a definition. It is okay to change your definition if it is not working. You will not need to compare or combine your results with results from other teams.

FIGURE 8.1 What are most golf balls made of? Where would a golf ball fit in your ranking?

Topic: Characteristics of Matter
Go to www.scilinks.org
Code: physical190

→ Notebook entry: Record your operational definition.

3. Construct a data table for this investigation.
   → Read the entire procedure before creating a data table.

 Beyond this step and any time others around you are conducting bounceability tests, wear protective eyewear such as safety goggles.

4. Test the materials for bounceability.
   → Use your operational definition and record a list of the variables that you controlled. Also record the results of your bounceability tests for each material in your data table.

5. Rank the materials from the most bouncy to the least bouncy.
   → Use your results to determine this ranking.
   → Notebook entry: Record your ranking.

Wrap Up

As a team, discuss the results of this investigation. Next, each of you should write a paragraph that summarizes this investigation. You can include anything that you feel is important, but be sure to include the following:

- your operational definition,
- an evaluation of your operational definition and how useful it was in determining bounceability,
- the variables you controlled,
- an evaluation of your data table,
- the results of your tests, and
- how you ranked the materials for bounceability.

CHAPTER 8 Properties of the Material World

SIDELIGHT ON TECHNOLOGY

Does Your Polyvinyl Acetate Lose Its Flavor on the Bedpost Overnight?

Have you ever heard of Santa Anna, the famous Mexican General of the Mexican Revolution?

In addition to being an important historical figure, Santa Anna was responsible for the development of one of America's favorite treats. He liked to chew chunks of chicle, which are pieces of sap from sopadilla trees that grow in the Mexican jungle. Santa Anna discovered that chicle has the amazing property of being chewy for a long time.

Thomas Adams, an American inventor to whom Santa Anna introduced chicle, decided to market it in small, bite-sized balls. People were already chewing wax for pleasure, but they found that the rubbery property of chicle was far more satisfying. Adams's chicle balls became a great success. Later other inventors thought of adding flavor to the chicle. Adams followed suit and introduced chicle that had been flavored with essence of licorice. He called his licorice-flavored chicle Black Jack gum. Black Jack gum, dating back to the 1870s, is still manufactured and enjoyed today and is the oldest brand of gum on the market. The most successful person to

This is a portrait of Antonio Lopez de Santa Anna, the famous Mexican General of the Mexican Revolution. He sat for this portrait in 1858.

CHAPTER 8 Properties of the Material World

jump on the bandwagon was William Wrigley, Jr. The rest is history.

Modern supermarket shelves contain a bewildering variety of gums. The gum you chew today is not Santa Anna's chicle, though. Because scientists study the properties of things, they have discovered an even better chewing material, polyvinyl acetate. This is the marvelous material that you enjoy chewing, snapping, popping, and blowing.

EXPLAIN  reading

How Things Are Different

In the investigation Bounce That Ball, you explored the property of bounceability or how "bouncy" certain materials are.

1. Why do you think you used balls to explore this property?
2. What results in your investigation surprised you?
3. Explain whether items made of the same materials had the same bounceability.

CHAPTER 8 Properties of the Material World

As you might have discovered in Bounce That Ball, exploring properties sometimes can be a complicated process. Many different factors might influence a property. In the case of bounceability, the shape of an object influences the bounceability almost as much as the material composing the object. Scientists have spent a lot of time identifying and defining specific properties of materials. A **property** of a material can be its look, feel, function, or other unique characteristic. One reason scientists have invested so much time in studying properties is that they are curious about the hows and whys of things. Another reason is that understanding properties helps designers construct better products. Consider the following true story.

In 1938, Dr. Roy Plunkett was studying the types of gases used in refrigerators and air conditioners in DuPont's New Jersey laboratory. One evening, he left his experiment, including containers filled with gases, out overnight. The next morning, he found that one of the containers of gases was coated inside with a thin, solid material. He immediately began studying the material to identify its properties. He applied chemicals of all types to see whether any of them would cause the strange material to corrode or

CHAPTER 8 Properties of the Material World

disintegrate. None of the chemicals did. He did other tests to determine the basic texture of the material and found that it was extremely slippery. He named this material for its chemical name, tetrafluoroethylene.

DuPont laboratories thought that it was an amazing material and continued to test it. Eventually they discovered another important property: the material retained its slipperiness even when it was heated. DuPont decided that this material would be very useful in industry to coat certain tools and pieces of machinery.

The *Guinness Book of World Records* listed the material as the slipperiest substance on Earth, comparable with "the slipperiness between two ice cubes rubbing against each other in a warm room." The president of DuPont in the 1950s, Marc Gregoire, decided to have his fishing tackle coated with the material to reduce the incidence of things sticking and tangling on it. Mrs. Gregoire was so impressed by the slippery property of the material that she asked her husband if she could have her pots and pans coated with it. Maybe then food wouldn't stick so much during cooking, and washing dishes would be easier. Mrs. Gregoire's idea worked. Today, pans coated with the strange material are common in many households. If Dr. Plunkett had not decided to examine the strange material and identify its properties, today we might be without Teflon.

4. List as many properties of materials as you can think of.

To get an idea of the types of material properties scientists have identified and how these properties differ from one another, consider three common properties: **translucence**, **hardness**, and **viscosity** (vis KOS ih tee). By yourself, study Figures 8.2, 8.3, and 8.4 and read the property descriptions that go with each one. Record answers to the Stop and Think questions in your notebook. Your teacher may call on any class member to answer the questions in a class discussion.

Translucence

FIGURE 8.2 Will three sheets of paper be more or less translucent than two sheets?

Isaac is right. Scientists measure the property of translucence by how much light passes through a specific material. The more translucent a material is, the better you can see the light passing through it. For example, glass can have varying degrees of translucence. Perhaps you have seen bathroom windows or shower doors that are not very translucent. The glass that makes up the windshield of a car, on the other hand, is very translucent. Even cloth can be translucent. Some curtains let a lot of light in the house. Other curtains are made of cloth that hardly lets any light through. We say that sheer curtains

5. Some materials are translucent only in certain forms. For example, when is wood translucent and when is it not translucent?
6. Identify a product in your home that is made of a translucent material.

7. Think of one reason why the designer of the product you listed in Question 6 might have wanted the product to have the property of translucence.

FIGURE 8.3 How would you define the property of hardness?

are translucent and that nonsheer curtains are less translucent or nontranslucent.

Hardness

Hardness can be a measure of how firm something is. This means that hard materials resist pressure better than soft materials do. That is why Ros and Isaac are complaining about the marshmallows. They can't squeeze them, which means that they are resisting pressure. On the other hand, how would you compare the hardness of an emerald and a diamond, if both resist being squeezed? Scientists consider a diamond harder than any other material because a diamond can scratch any material including itself, but no other material can scratch a diamond.

STOP & THINK

8. Explain which method you could use to measure the hardness of marshmallows.
9. Identify a product in your home that is made of a hard material.
10. Think of one reason why the designer of the product that you listed in Question 9 might have wanted the product to be made of a hard material.

Viscosity

FIGURE 8.4 How would you describe viscosity?

Marie has supplied you with a good description of viscosity. How slowly a liquid pours indicates how viscous a material is. Often we think of viscosity as a measure of the thickness of a substance. This is partly true, but we would be more correct to include a measure of the liquid's "pourability" or "flowability" to define its viscosity accurately. A viscous liquid, such as Marie's paint, does not pour or flow as easily as a less viscous liquid, such as Al's paint.

CHAPTER 8 Properties of the Material World

STOP & THINK

11. Name the most viscous liquid you can think of.
12. Name the least viscous liquid you can think of.
13. Identify a product in your home that is made of a viscous material.
14. Think of one reason why the designer of the product you listed in Question 13 might have wanted it to be viscous.

Now that you have basic definitions for these three properties, you will have a chance to work with different materials to explore these properties more fully. Refer to this reading whenever you are unsure how to investigate one of these properties.

How Can I Learn More about That Property?

In this investigation, think of yourself as Dr. Plunkett. You now have a chance to be a scientist who studies the properties of various materials. In the following investigation, you will study one of the properties that we described in the previous reading. You will have a number of materials to choose from to help you study this property. You and your peer scientists will decide how to study your property, on what materials you will conduct your tests, and finally, how you will rank your materials.

Materials for Each Team of Three:

- any materials you choose to test for your assigned property

Process and Procedure

1. In your notebook, record the property that your team is responsible for investigating.
   → Follow the teacher's directions to determine which property (translucence, hardness, or viscosity) your team will investigate.
2. Obtain the materials that you will test and rank.
3. Decide on an operational definition for how you will measure your assigned property.
   → The Communicator will need to check with other teams, because you will be combining and comparing data with the other teams that are investigating the same property. Remember that to combine and compare data, you need to use a common operational definition. If you have trouble deciding on an operational definition, review the reading for ideas.
4. Construct a data table.
   → This data table should include room for information about what your test is and how each material responds to your test.
5. Conduct your tests.
   → Remember, if your operational definition is not working well, modifying it is acceptable.
   → Notebook entry: Record the results of your tests in your data table.

Work cooperatively in your teams of three. Use the roles of Manager, Communicator, and Tracker. Move your desks together or work together at a table. As you work, practice the social skill *Let others finish without interrupting them.*

Topic: Physical Properties
of Matter
Go to www.scilinks.org
Code: physical200

6. Rank the materials that you tested from the hardest to the softest, most viscous to least viscous, or most translucent to least translucent.
   → Use the results of your test to do this. At this point, do not compare your results with other teams' results.

Wrap Up

As a team, complete the following according to your teacher's directions. Combine your data with data from other teams that investigated the same property and think of some way to represent the data in a graph or picture. You may choose a traditional method such as making a line or bar graph, or you may think of a less traditional way to show the combined rankings in picture form. You will have completed this task successfully as long as you represent your data in a way that explains the following points:

- the property you investigated,
- the ranking of materials for your property,
- your operational definition, and
- the variables you controlled.

FIGURE 8.5 How translucent is this piece of glass?

FIGURE 8.6 How hard is wood? Is oak harder or softer than fir?

Discuss the following questions with your classmates. In your discussion, show how well your team uses the skill *Let others finish without interrupting them.*

1. Think of the property you investigated. Of all the materials in the world, which material is the best example of that property?
2. Which material is the worst example?
3. When you compared data with the teams that investigated the same property that you did, were the rankings identical? Explain your answer.

FIGURE 8.7 How viscous do you want your toothpaste to be?

CHAPTER 8 Properties of the Material World

SIDELIGHT ON HISTORY

Dr. Chien-Shiung Wu

Do you ever dream about discovering new things or exploring the unknown? Often, our dreams seem impossible because we do not know how to go about making them realities. Dreams can be hard to achieve. The story of Dr. Chien-Shiung Wu (Chin-Shing Woo) is about how a young girl had a dream and worked hard to achieve it.

Dr. Wu was born in Liuhe, China, which is a town near Shanghai. At her birth on 29 May 1912, her father named her Chien-Shiung, which means "courageous hero." Chien-Shiung's father was a major guiding figure in her life. Some people think that girls are not as good as boys at things like science, math, and sports. Her father did not believe that. From the day she was born, he wanted her to be a "courageous woman" who could decide on her own what she wanted to do in life.

With encouragement from her family, Chien-Shiung left home when she was nine to study at Soochow Girls School. As a high school student, she enrolled in a teachers' training school but discovered that she was not learning as

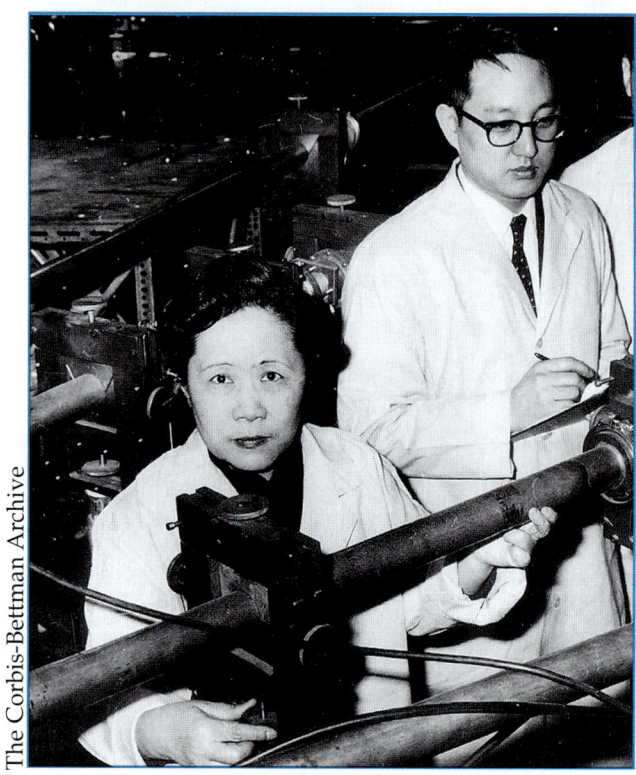

much about math and science as she would have liked. She borrowed textbooks from friends who were taking classes in math and science to teach herself mathematics, physics, and chemistry. She decided that physics was her favorite subject. Chien-Shiung graduated from Soochow Girls School in 1930 with the highest grades in her class.

Although Chien-Shiung loved physics, she thought it would be easier for her to find work in China if she studied education in college. Again her father encouraged her to follow her true interests. He suggested that to prepare herself she should study advanced mathematics and science the summer before she enrolled at the National Central University in Nanjing, China. By the end of the summer, Chien-Shiung was prepared to enter the university and major in physics. When she graduated, she was the university's top student.

In 1936 Chien-Shiung left China to study in the United States. At first she thought she would attend the University of Michigan, but she changed her mind and attended the University of California at Berkeley. At that time, Berkeley was a major center for the study of physics. She had the opportunity to study with a number of notable physicists such as Ernest Lawrence and Emilio Segre who each won a Nobel Prize. Chien-Shiung also met her husband Luke Yuan, who also was a physicist, while attending the University of California at Berkeley.

Dr. Wu is known to many as the "Queen of Nuclear Physics." She was interested in experimental nuclear physics and created experiments to test theories about subatomic particles. Subatomic particles are particles that are smaller than atoms. They were detected with the help of special scientific equipment such as particle accelerators. Dr. Wu's work on subatomic particles led to a major scientific breakthrough that changed the way scientists think about how particles inside the nucleus of an atom behave.

The colleagues who worked with Dr. Wu on this project received the Nobel Prize. Dr. Wu did not receive the Nobel Prize along with them, although many people thought she should have. Dr. Wu had many other successes. She was the first woman to be hired as a professor at Princeton University. She was elected to the National Academy of the Sciences, and she was given the National Medal of Science by President Gerald Ford in 1976.

Dr. Wu lived in New York City and taught at Columbia University from 1944 until her retirement in 1981. Dr. Chien-Shiung Wu's life and work are an example of how creative problem solving and hard work can help people overcome obstacles and achieve their dreams. Dr. Wu achieved many things that other people thought she could not. What do you think some of Dr. Wu's obstacles were? How would you try to overcome such obstacles?

EVALUATE connections

Is the Most Always the Best?

People have a tendency to think that "most" is always "best." For instance, think about the rubber balls from the beginning of this chapter. One bounced very high, and the other hardly bounced at all. You might think that the one that bounced the highest was made out of the best materials. This is partly true, if the designer's idea was to make a bouncy object.

But what if the idea was to make an object out of rubber that had as little bounce as possible? Think about automobile tires, for example. Designers have worked to develop tires that make an automobile ride as smoothly as possible. Tire manufacturers use a special material called

FIGURE 8.8 To make this carnival ride exciting but safe, do you think the designers of these boats wanted the bumpers to be made out of the most bouncy material?

polystyrene-butadiene co-polymer to make tires because of its remarkable property of not bouncing when you drive over bumps in the road. Instead, it absorbs the impact of the bumps. Because of that absorbing property, other manufacturers use the same material to line the large containers in which bomb squads store bombs. Why would this be a good idea?

Can you guess what type of rubber the little ball that didn't bounce was made of? You guessed it: *polystyrene-butadiene co-polymer*! There are other examples in which a material that has the "most" of a property is not the "best" choice when developing specific products.

Write answers to these questions in your notebook. Be prepared to share your answers in a class discussion. Think about your experiences with the properties of viscosity, translucence, and hardness.

1. Describe a situation when the most viscous food (perhaps ketchup or molasses) would not be the best choice.
2. Describe a situation when the most translucent material would not be the best choice.
3. Describe a situation where the hardest material is the best choice.

Consider the property of stickiness. There are times when "the most" is "best," and times when "the least" is "best." Give an example of when it might be better to use a material that was very sticky and an example of when it might be better to use a material that was only slightly sticky.

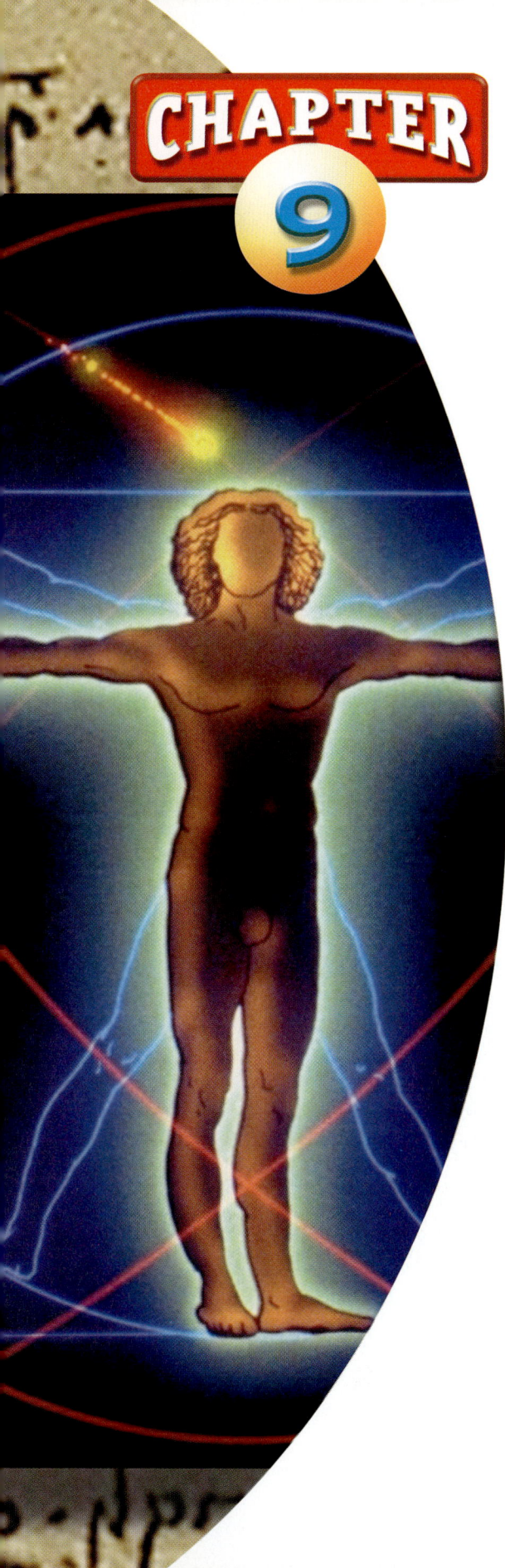

CHAPTER 9
Scientific Explanations Are Ancient History

In the late 15th and early 16th centuries, Italian artist, musician, and mathematician Leonardo da Vinci studied the human body in detail as he worked on his drawings and sculpture. He discovered some interesting things about its balance and symmetry. Some of his ideas are apparent in this drawing in the background. As you can tell by the modern, computer-generated art in the foreground, his ideas are still popular today.

In this chapter, you will learn how scientists who lived hundreds of years ago began to develop explanations for things they observed. You also will learn how they developed models to explain some of the things they could not observe completely, such as the shape of Earth or the structure of materials. Across time, scientists have been able to learn more about how and why materials differ.

ENGAGE — What Happens to Light inside a Prism?

EXPLORE — Strange Phenomena

EXPLAIN — Elements of Explanations

ELABORATE EVALUATE — Thinking like the Ancients

ENGAGE connections

What Happens to Light inside a Prism?

Have you ever used a prism to explore the characteristics of light? Light is a special type of radiant energy that has a very high frequency. Within our eyes are structures that detect this energy. Our eyes interpret a specific frequency as the color red, for example. As we go to higher frequencies, we see yellow, green, blue, and violet in that order. If all the frequencies of light are present together, as they are in sunlight, we see white light. A prism can break apart white light into the range of color that we see, which we call the **spectrum**.

Observe the demonstration that your teacher conducts with light and a prism.

 Do not look directly at the Sun or any other bright light, even after it passes through the prism. This could damage your eyes.

How does a prism do this? What is it about a prism that affects light in this way? Remember from the investigation Light, Lenses, and the Eye that light travels more slowly through glass than it does through air. The shape of the prism also contributes to the creation of the spectrum. When light travels through the thick part of the prism, it spends more time in the glass. This causes the light to bend toward the thick side of the prism. Think about a line of trombone players in a marching band. Imagine that one end of the line is forced to march through some thick mud. There would be a tendency

Topic: Spectrum
Go to www.scilinks.org
Code: physical208

for the line to turn (or bend) in that direction because those trombone players would slow down. Light inside a prism behaves in much the same way. The speed of the violet light is less than the speed of the red light, so the violet light bends more. All of the colors bend a certain amount, but the violet bends more than the red. As a result, when incoming white light passes through a prism, we see the full range of colors from red to violet when it exits. As you remember, the color sensation that you experience depends on the detectors in your eye. Our eyes cannot see beyond the red light (into an area called the infrared) or beyond the violet (into an area called the ultraviolet).

When we understand the characteristics of certain things, such as light and prisms, then we can develop explanations for some of our observations. But how do scientists develop explanations about characteristics in the first place? This chapter will give you an opportunity to explore some of the ways that scientists do this today and ways that they used in the past.

Strange Phenomena

Think of the rubber balls again. The question we want to answer is, Why is one ball bouncy and the other ball not bouncy? We could answer that question if we could answer Marie's question in Figure 9.1.

To find out whether or not materials have insides and what those insides might be, you need to do some more exploring. In this investigation, you will observe three different phenomena. Your careful observations will be important to your understanding of the reading that follows this investigation.

FIGURE 9.1 What does Marie mean about the insides of rubber balls? As Ros said, cut them open and there is rubber inside. Is this what Marie is referring to?

Work cooperatively in your teams of three. Create a work space using your desks or a table that allows each of you equal access to the materials as you use them. Use the roles of Manager, Tracker, and Communicator. Practice the social skill *Let others finish without interrupting them.*

Materials for Each Team of Three:

Part B

- 3 medicine droppers
- 1 small beaker of water
- 1 small beaker of Solution A
- 1 small beaker of Solution B
- 1 lid or 1 base of a petri dish
- one $8 \frac{1}{2} \times 11$-in. sheet of black construction paper
- 3 pairs of safety goggles

Part C

- 1 tablespoon
- 1 resealable plastic bag
- 1 tablespoon of mystery mix
- 1 tablespoon of acetic acid (vinegar)
- 3 pairs of safety goggles

Process and Procedure

Part A—You Can't Touch This

1. Pay attention to the demonstration your teacher will do.
2. Record your observations.

Your teacher will ask you specific questions about your observations, which you should answer in your notebook.

Part B—Puddle Mystery

1. Obtain all of the materials for Part B.

 After this step, you will need to wear eye protection.

2. Place the lid or the bottom of a petri dish on black construction paper.
   → The Manager should do this.

3. With a medicine dropper or thin-stemmed pipet, slowly drop water into the center of the dish until there is a puddle the size of a quarter.
   → The Manager should do this.

4. Carefully fill a second medicine dropper or thin-stemmed pipet with Solution A.
   → The Tracker should do this.

5. Carefully fill the third medicine dropper or thin-stemmed pipet with Solution B.
   → The Communicator should do this.

 Avoid getting any solution on your skin or clothes. If you do, wash the area with cool water immediately. Have a team member notify the teacher of any spills.

6. Hold the filled droppers opposite each other on the outer edges of the puddle.
   → The Tracker and Communicator each hold one dropper. Be careful not to touch the puddle with the tip of the dropper.

CHAPTER 9 Scientific Explanations Are Ancient History   211

FIGURE 9.2 Hold the droppers above the puddle on opposite edges. Remember not to touch the droppers to the puddle.

7. At the word "Go," squeeze 4 drops of liquid from each dropper onto opposite edges of the water puddle.
   → The Manager will say "Go," and the Communicator and the Tracker will squeeze the droppers at the same time.
8. Observe what happens in the puddle of water.
   → Notebook entry: Record your observations.
9. Return the materials for Part B, except for the safety goggles.

Part C—Bubble, Bubble, Toil, and Trouble

1. Obtain the materials for Part C.

    After this step, you will need to wear eye protection.

2. Seal a plastic bag halfway across the top.
3. Through the unsealed half of the opening, place 1 tablespoon of mystery mix into the bag.
4. Add 1 tablespoon of vinegar to the mystery mix and immediately seal the bag completely.

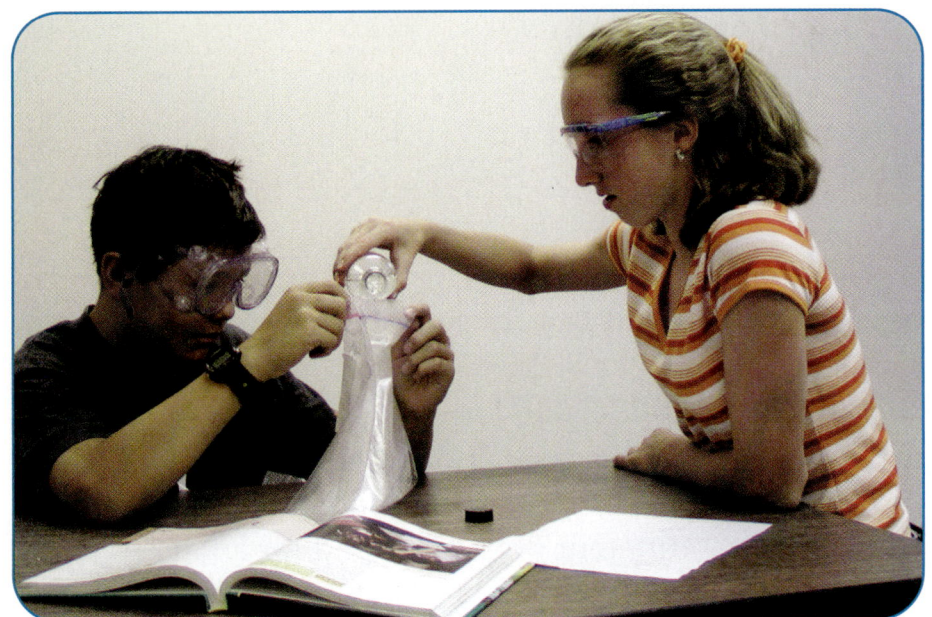

FIGURE 9.3 What happens inside the bag when you add vinegar to the mystery mix?

→ Avoid getting any of these materials on your skin or clothes. If you do, rinse the area with cool water immediately.

5. Gently squeeze the bag to mix the materials.
6. Observe what happens inside the bag and to the bag.
   → Feel the bag with your hands as the experiment progresses.
7. Return the materials for Part C.
   → Dispose of the materials as your teacher directs.

Wrap Up

Complete the following tasks as a team. After you have discussed your answers, record your own in your notebook. Be prepared to share your answers in a class discussion.

For each phenomenon in Strange Phenomena, you recorded some observations. Discuss the observations you recorded and any other observations your teammates can remember. Then try to explain what you think happened inside the materials you used. After each explanation you record, leave space in your notebook because you may want to modify your explanation later.

CHAPTER 9 Scientific Explanations Are Ancient History

EXPLAIN  reading

Elements of Explanations

Now that you have had a chance to explain what you observed, you might wonder how other people might explain why matches burn, why two clear liquid substances can combine to form white particles, or why a mystery mix and vinegar together produce bubbles and heat. For a long time, people have noticed that different materials have different properties, and for a long time they have been asking why.

About 3,000 years ago (1000 B.C.), Chinese philosophers tried to answer *why* questions. They wanted to explain why different materials—such as wood, water, and sand—had different characteristics. Why did things smell differently from one another? Why did wood burn easily, but sand did not?

It seemed possible to these philosophers that something might be going on inside the materials to explain the things that people observe directly. It is much like your teacher demonstrating a match being lit. If you knew what happened with the chemicals, the heat, and the wood, you could explain why you saw what you did.

If you could take the match apart, you could see what is really inside. But there is a catch: you cannot take apart a piece of paper, wood, or rubber to see what is really inside. You can cut these materials apart to see more of the same material, but you cannot see inside the material itself. This story is about figuring out

FIGURE 9.4 The Chinese organize their interior spaces and architecture according to the five elements of fire, earth, metal, water, and wood. This practice, called Feng-Shui, is believed to provide balance and harmony.

what is inside things without being able to see what is really there. Chinese philosophers were among the first people to do this.

The Chinese philosophers decided that everything in the world was made up of just five things. They called these things "elements." The five elements were fire, earth, metal, water, and wood. The Chinese philosophers explained natural occurrences by observing changes in each of these five elements produced by one of the other elements. For example, they would say that wood changes into fire and fire changes into earth (ashes). Earth, they decided, then produces metal. Their evidence for this was that metals such as gold and iron are found in the ground. After observing dew clinging to metallic objects, they concluded that metal must produce water. Finally, they noted that if you watered an acorn in the ground, you eventually get a tree, so they reasoned that water produces wood. If those Chinese philosophers performed the experiments you just did, they would have used these five elements to explain the phenomena you observed.

FIGURE 9.5 This is an artist's rendition of an ancient Chinese philosopher.

About 500 years later (about 500 B.C.), Greek philosophers began developing their own explanations. Unaware of what the Chinese were thinking, they too began to wonder why things had different properties. A Greek philosopher named Thales (THAY leez), who lived between 634 and 546 B.C., thought that water was inside all materials. He thought this because water can take on many shapes; it can be a liquid, a solid, or a gas. Another

CHAPTER 9 Scientific Explanations Are Ancient History

FIGURE 9.6 Empedocles thought fire, water, earth, and air were the four basic elements that made up all materials.

Greek philosopher named Heraclitus (hair uh KLEE tus), who lived between 535 and 475 B.C., thought that fire was the basic substance that made up the world. Why? Because everything in the world was always changing, and to the Greeks, fire represented "that which moves."

A Greek named Empedocles (em PED oh cleez), who lived between 493 and 433 B.C., decided there were four basic things that made up all materials: fire, water, earth, and air. His idea caught on. Using these four elements, the Greeks could explain almost anything. They were able to develop explanations by reasoning that every material was composed of these four things, but not in the same amounts. For example, if they wanted to explain the properties of a piece of cloth, they might have said that it was made of a small amount of fire, and that is why clothing keeps people warm. That same piece of cloth had a little bit more air in it than fire, which gives cloth its characteristic lightweight property. The movement of cloth, the way it clings and flows as it moves, might have suggested that there was quite a bit of water in it. Finally, they might have said that the major element was earth, which gave cloth its solidity. They might have said that cloth was about 10 percent fire, 15 percent air, 25 percent water, and 50 percent earth.

Later the Greek philosopher Aristotle (air ih STAH tul), who lived between 384 and 322 B.C., added to Empedocles's explanation by saying that each of the four materials—fire, water, earth, and air—was a combination of heat, cold, moisture, and dryness. In Aristotle's explanation, fire was hot and dry, water was cold and wet, earth was cold and dry, and air was hot and moist.

The Greeks used Empedocles's and Aristotle's ideas repeatedly in almost any way they wanted to. Their explanations seemed to explain things for them. For about 2,000 years, most people believed that almost everything in the world was composed of fire, earth, water, and air.

FIGURE 9.7 Many of the famous ancient Greeks were the subjects of great works of art, like this drawing of a bust of Aristotle.

connections  ELABORATE EVALUATE

Thinking like the Ancients

Now, imagine that you can go back in time. Suppose that you are either an ancient Chinese or ancient Greek philosopher who has just observed the phenomena you observed in the investigation Strange Phenomena. Explain each of the three phenomena using either the ancient Chinese or ancient Greek ideas. Use words you think the early philosophers might have used.

CHAPTER 9 Scientific Explanations Are Ancient History

CHAPTER 10

Using Scientific Models to Answer Questions

Scientists often use models to help them answer questions about the natural world. Models are especially helpful when scientists are unable to see all aspects of what they are trying to explain. Long ago the sundial was one way that scientists modeled time and the daily cycle of time. Later, precise gears were placed in watches and clocks and used as another way to represent time. Today, many watches use the natural frequency of a quartz crystal to measure the passing of time, one second at a time. Time is an abstract idea, and many models of time only represent certain aspects of this concept. Scientists often must develop models that represent only part of the picture—the part that they currently understand or are able to observe directly. In this chapter, you will work as a scientist to explain the characteristics of materials and why materials are different.

ENGAGE — Mystery Box
EXPLORE

EXPLAIN — Another Explanation

ELABORATE — How Well Does the Particle Model Work?
— Particle Movement—Improving the Model

EVALUATE — Applying the New Model

ENGAGE EXPLORE

investigation

Mystery Box

In this investigation, you will continue answering the question, Why are things different? You will do this by trying to figure out the characteristics of a mystery box, both its outside and its inside. The process you use will help you understand how today's scientists developed their explanations for why things are different.

Materials for Each Team of Three:

- 1 mystery box
- 1 strong magnet
- 1 sheet of light-colored, construction paper
- 2 large marbles
- 2 small marbles
- 1 wooden ruler
- 1 clock or watch
- other materials you choose to use as needed

Work cooperatively in your teams of three. You will need a work space that allows each of you equal access to a large box. Use the roles of Tracker, Manager, and Communicator. Concentrate on the unit skill *Be open to others' ideas*.

Process and Procedure

1. Read the following rules for investigating your mystery box (see Figure 10.1).
    → Make sure that each Team Member understands these rules. It is the Tracker's job to make sure that the team follows these rules.
2. Obtain your team's materials.

CHAPTER 10 Using Scientific Models to Answer Questions

FIGURE 10.1 Refer to these rules as you explore the characteristics of your mystery box.

RULES

To try to determine what is underneath the cardboard for the investigation Mystery Box, you MAY do any of the following things:
1. Shoot marbles or objects at whatever is under the cardboard using a ruler, a pencil, or a pen.
2. Bring the magnet close to the cardboard inside or outside the box or gently touch the magnet on the cardboard.
3. Move the box around and lift the box off the table or desk in order to move the marbles around.
4. Do anything else that does not violate the rules that follow.

You MAY NOT do any of the following things:
1. Touch what is under the cardboard, either directly with your fingers or indirectly using objects other than marbles.
2. Place anything, including the magnet, a ruler, or your fingers under the cardboard. (If a marble gets stuck, ask your teacher to dislodge it.)
3. Lift, push, pull on, or grab the cardboard in any way.
4. Poke holes in the box that will allow you to see under the cardboard.
5. Peek through any exiting holes in the box to see what is under the cardboard.
6. Hit the cardboard with anything. This includes knocking on it and dropping marbles on it.

3. Without violating any of the rules, take a 5-minute turn using the materials to explore the characteristics of your mystery box.
   - You want to learn as much as you can about the mystery box, both what it is like on the outside and what it is like on the inside. Describing the characteristics of the outside will be fairly easy, but what can you describe about the characteristics inside?
   - The Communicator should take the first turn.

4. Observe the actions of the person investigating the box and what happens inside the box.
   → Notebook entry: Record your observations.

5. Stop at the end of 5 minutes.
   → The Tracker is the timer except when he or she is busy with other duties. At that point, he or she should assign the job of timing to another Team Member.

6. Draw in your notebook what you think are some of the characteristics of the mystery box.
   → Each Team Member should do this. It is okay to guess!

7. In your team, take turns showing and describing your work from Step 6.

8. Repeat Steps 3 through 7 until each Team Member has had a 5-minute turn using the materials.
   → During each turn, all Team Members should record their observations and describe some of the characteristics of the mystery box.

222 CHAPTER 10 Using Scientific Models to Answer Questions

Wrap Up

Complete the following tasks and answer the questions before your class discussion.

1. As a team, describe the characteristics of both the outside and the inside of the mystery box based on your explorations. Draw a picture on a sheet of construction paper that depicts these characteristics. Make sure everyone on your team can explain to the class what your team learned about the mystery box. Be ready to explain the evidence you have that your team's drawing is accurate.

2. Present your team's drawing and evidence to the rest of the class. Show the class how well you used the unit skill by indicating how you were open to each other's ideas as you developed your explanation.

3. Answer the following questions in a class discussion:
   a. Did each member of your team have the same evidence with which to determine the characteristics of the mystery box? Why or why not?
   b. Did the members of your team agree each time on the characteristics of the mystery box? Why or why not?
   c. Did all teams agree on the characteristics of the mystery box? If not, explain why not.
   d. What reasons might explain why different teams agreed on the characteristics of the mystery box?
   e. Suppose you could never lift up the cardboard and discover what was really in the mystery box. Could you prove to a friend, without a doubt, that what you thought was under the cardboard was really there? Explain your answer.

EXPLAIN reading

Another Explanation

How Explanations Change

You already have learned how some of the ancient Greeks and Chinese explained what made things different. They decided that a few basic elements such as earth and fire made up the universe. Other philosophers and scientists throughout history developed different explanations for what was inside things. One idea that people kept coming back to was the notion that if you break materials up into smaller and smaller pieces, you eventually would find particles that you could not break apart. One of the first philosophers to develop an explanation like this was Anaxagorus (an ax uh GOR us), who lived around 450 B.C. His explanation went something like this:

All materials are made up of an infinite number of very small "seeds." These seeds are mixed together in a material like a mixture of different colored sand. The properties a material has (such as hardness, color, and smell) depend on how many of each type of seed are in the material.

FIGURE 10.2 As you know, an early explanation for the shape of Earth was that it was flat. When explorers began sailing around the world, a new model developed.

A philosopher named Democritus (de MOK rih tus), who lived about 100 years after Anaxagorus, had a different explanation:

> All materials are made up of very small objects called "atoms." (The Greek word atom means "indivisible.") These atoms are the smallest things that exist. Different materials are made of different types of atoms. Some atoms are rough, and others are smooth. Some are large, and others are small. Atoms don't have color, taste, or smell by themselves. The motion and arrangement of the atoms in a material give it a certain color, taste, or smell.

Scientists' current explanation for what is inside materials is based on Democritus's explanation of the atom, but today it is more sophisticated. An **atom** is a small particle that makes up materials, but it is not the smallest thing, as Democritus proposed. Some of the people throughout history who elaborated on Democritus's ideas include Isaac Newton, John Dalton, J.J. Thompson, E.J. Rutherford, Marie Curie, and Niels Bohr. Perhaps you have heard some of the terms people use when talking about atoms, such as protons, neutrons, electrons, nucleus, neutrinos, or gamma rays. People use different names to describe different types of particles, such as atoms, ions, and molecules. You probably will study types of particles in the future. For now, just focus on the idea that materials are made up of particles.

Topic: Atoms and Elements
Go to www.scilinks.org
Code: physical225a

Two of the first questions you might ask about particles are, How do we know that materials are made up of particles? and, Can we see them? Yes, it is possible to observe larger particles with an electron microscope, which provides evidence that materials are made of particles. We cannot see smaller particles, though, which makes it difficult to gather evidence about them. Thinking about the mystery box investigation might help you realize how scientists gather evidence about things they cannot see.

Topic: Scientists' Biographies
Go to www.scilinks.org
Code: physical225b

Consider the process your team used to try to determine what was under the cardboard during the investigation The Mystery Box. We will now refer to what was under the cardboard as the target. One of the first things you might have done was to pick up the box, shake it, and listen to the sounds it makes.

1. Review your first idea about what the target in the mystery box looked like.
   → You then used the materials to determine whether your first idea was correct.
2. If your first idea was correct, what should have happened when you rolled a marble toward the target? Sketch or describe what your prediction would have been.
3. Sketch or describe what actually happened when you rolled a marble toward the target.
4. Explain how this new evidence changed what you thought the target looked like. Sketch or describe how your idea changed.
5. How many times did your idea change about what the target looked like? (Think about each time a different person conducted tests with the materials and whether or not he or she found new evidence that changed your mind.)

This type of process for determining the characteristics of a target could have gone on and on. You could have decided what the target might be and then tested your idea. Each time you tested your idea, you might have changed your mind about the target. You then tested your new idea to verify it, but you might have discovered new evidence that again changed your mind. Some teams might have changed their ideas repeatedly until the teacher stopped your investigation of the mystery box.

The teams could have continued testing their ideas until they were quite certain what the target looked like without ever seeing it. They then might have concluded something like this: The results of all our tests indicate that the target has a certain shape and is made of certain materials, so we are confident that the target really has that shape and those materials.

Firing marbles at an unseen target is similar to experiments that scientists conduct today to learn about particles. They fire small particles at materials to see what happens. What happens to those particles is consistent with the idea that the material itself is made of small particles. Some of the fired particles pass through the material, which means there might be empty space in the material. Some of the fired particles bounce off at angles (a few even bounce straight back), which means there are solid parts to the material. So without ever being able to see the materials directly, we can conclude something about the characteristics of the materials: that they are made of particles with spaces between them.

Of course, the actual experiments that scientists conduct are not quite as simple as firing marbles at a hidden target. The particles that scientists fire at materials are themselves too small to see. Then how do scientists know that they are really firing particles at the materials? They know because the results of such experiments are consistent with the evidence they have gained from other

FIGURE 10.3 If you could not crack this egg open, how might you investigate what it is made of?

FIGURE 10.4 This experiment showed how light is made up of both waves and particles.

The Corbis-Bettman Archive

experiments. In those other experiments, they have burned, weighed, smashed, and dissolved materials. They have determined how materials respond to electricity and magnets. The one thing they cannot do is look directly inside a material to see whether or not it is really made up of particles. Yet the results of all the experiments scientists can do are consistent with the idea that materials are made up of particles.

Scientific Models

Scientists call the idea that all materials are made up of particles the particle theory. A **theory** is an explanation for a set of observations or a group of phenomena. There is so much evidence in favor of the particle theory that it is the idea most scientists accept as an explanation for the behavior of materials or why things are different. We can use this theory to explain many of the observations that we make about materials. So scientists accept the existence of particles and now spend a great deal of time trying to determine what a typical particle might look like.

In the days when scientists first began firing particles at materials and concluding that materials were made of tiny, unseen particles, they had no idea about the characteristics of particles. From their observations, they could only guess about a particle's characteristics. Today we have powerful microscopes that allow us to study the characteristics of these particles. The picture still is not clear enough to know for sure about all the parts or components of individual particles. Scientists still perform experiments on materials and use their observations to make their best estimate about the characteristics of particles. Scientists call this type of an estimate a **scientific model** because it represents something they cannot observe directly. They refer to the particle theory and their explanation of the characteristics of a particle as the **particle model**.

Topic: Using Models
Go to www.scilinks.org
Code: physical229

You probably were familiar with many scientific models before this reading. If you remember that a scientific model is a representation of something you cannot easily observe directly, you probably can think of some of those models. In fact, you probably have even created a few of your own models. An example was when you drew and described what you thought was under the cardboard in the mystery box. Those drawings would be considered a scientific model because you had to describe something that you could not see, but for which you could collect evidence by conducting experiments. You are probably also aware of other familiar models, such as models of body cells, their tiny parts, and how they function in our bodies. We can see cells and their parts under a microscope, but using a cell model makes it easier for scientists to explain observations about living things. You also may be familiar with models of the universe, which include stars, planets, galaxies, and solar systems. The universe is an example of something so large that we cannot observe all of its components at the same time. Using models of the universe or its parts helps us understand the universe better.

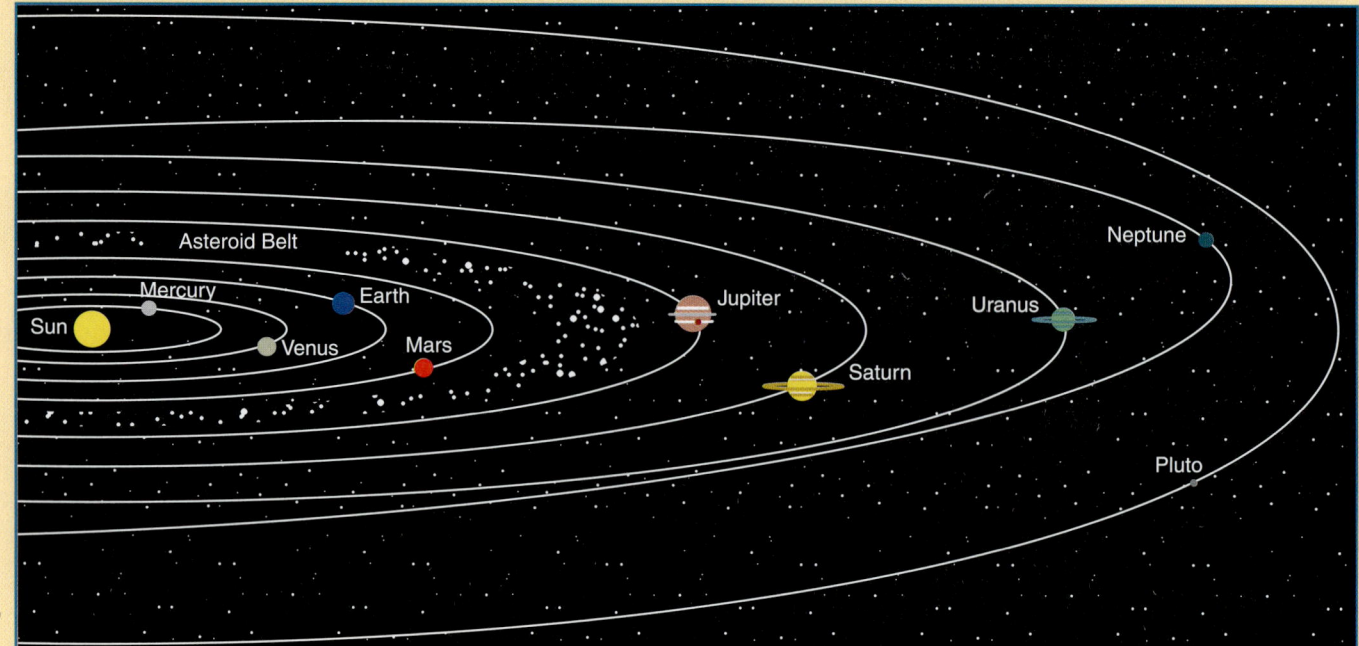

FIGURE 10.5 This is the current model for part of the universe—our solar system.

6. Describe another example of a scientific model.

What Makes a Model Scientific?

Models are not considered scientific models unless they can explain many, if not all, of the observations that scientists make when conducting related investigations. This is one criterion that makes a model scientific. The reason the particle model is used by most scientists is that it does explain many of the observations about the different ways materials act. A model would not be much help if it explained one or two observations but left many others unaccounted for.

7. Based on the criterion that a scientific model explains many, if not all, observations, explain whether you think your mystery box model was scientific.

When you developed your model to explain the mystery box, it probably was not too hard to come up with observations that your model could not explain. So what makes accepted scientific models so thorough? Are scientists smarter or more creative than you? Are there just a few special people who can produce strong scientific models? Some reasons why accepted scientific models seem so strong are as follows:

- There are many different scientists working on any given scientific model at any one time. The models and theories you read about in books usually represent the combined work of hundreds of people, not just one or two.
- Scientists develop scientific models over a long period of time. When scientists encounter observations that their models cannot explain, they often change their models to account for the new observations. However, they make certain they still explain the old observations. When you read about a scientific theory, you rarely are reading people's first ideas on the subject. What you are reading about is a model that has been changed many times.
- When you read about scientific theories, you often do not read about the ones that failed. It is like rehearsing for a play or preparing for an athletic performance. When you are practicing, you do not want the world watching. You try to make your performance as good as possible before you show it to everyone.

Do you think the particle model is a scientific model? Does it explain for you why things are different? The following investigation will give you a chance to decide.

ELABORATE investigation

How Well Does the Particle Model Work?

You will work individually in this investigation at your desk or at a lab station.

Do you think that everything in the world is made of tiny particles? For this investigation, you do not have to be convinced of that. You only have to observe more phenomena and use the particle model to explain your observations. If you can explain all of your observations using the particle model, then you are verifying that the particle model is a valid scientific model.

Materials for Each Student:

Part B—The Balloon
- one 10-in. round balloon
- 1 heat source
- 1 large bowl of crushed ice and water
- 1 metric measuring tape

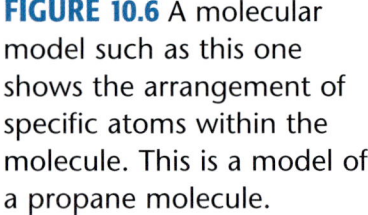

FIGURE 10.6 A molecular model such as this one shows the arrangement of specific atoms within the molecule. This is a model of a propane molecule.

232   CHAPTER 10 Using Scientific Models to Answer Questions

Part C—Red Streamers
- two 600-mL glass beakers
- crushed ice
- cold water
- red food coloring
- hot water
- 1 medicine dropper
- 2 hot pads or 1 set of beaker tongs

Process and Procedure

Part A—Houdini Water
1. Observe the experimental setup that your teacher has assembled.
2. Predict what you think will happen in this experiment.
   ➤ Notebook entry: Record your predictions.
3. Observe what happens in the experiment.
   ➤ Notebook entry: Record your observations.
4. Participate in the class discussion about the experiment.
   ➤ Be prepared to share your ideas with the rest of the class.

Part B—The Balloon
1. Obtain all of the materials for this part.
2. Blow up a balloon as much as you can without popping it. Then let out all of the air.
3. Repeat Step 2.
4. Blow up the balloon a little more than halfway.
   ➤ Do not fill the balloon completely with air.
5. Tie a knot in the end of the balloon to seal it tightly.
6. Measure the balloon with the metric measuring tape around the fattest part of the balloon.

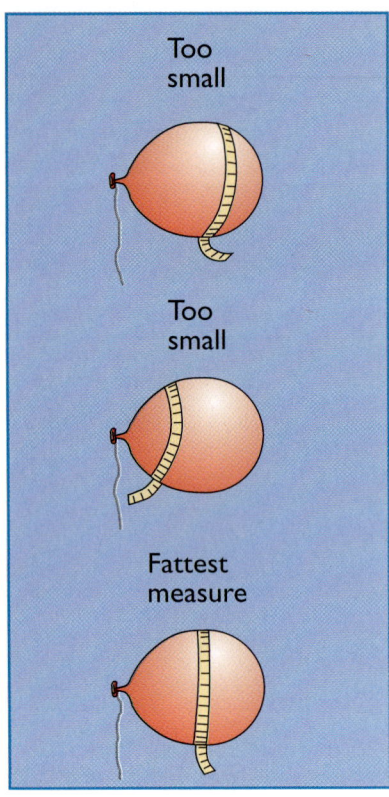

FIGURE 10.7 The fattest part of the balloon is not necessarily the middle part of the balloon as shown here. Where the fattest part of the balloon is depends on its shape. You can determine the fattest part by looking or by taking several measurements of different areas of the balloon.

➡ This measurement is called the circumference of the balloon. Figure 10.7 shows you the right way and the wrong way to measure a circumference.

➡ Notebook entry: Record this measurement.

7. Place the balloon in the bowl of crushed ice and water for 10 minutes.
   ➡ Periodically submerge the balloon into the ice water for 1 minute or so. Do this a total of 4 times during a 10-minute period. Ask your teacher whether it is appropriate to work on Part C while you wait.

8. After 10 minutes, remove the balloon and immediately measure its circumference.
   ➡ Notebook entry: Record this measurement.

9. Place the balloon beside, but not touching, a heat source for 10 minutes.
   ➡ You will have to ask your teacher what heat source you should use. While you are waiting, ask your teacher whether you should proceed to Part C.

10. After 10 minutes, take the balloon away from the heat source and immediately measure its circumference.
    ➡ Notebook entry: Record this measurement.

11. Compare the circumferences of the balloon when it was at room temperature, when it was cold, and when it was hot.
    ➡ Notebook entry: Record your observations.

12. Use the particle model to try to explain your observations.
    ➡ Explain what is happening with the particles of the balloon and with the particles inside the balloon to make it act the way it does at different temperatures.

→ Notebook entry: Record your explanation.

Part C—Red Streamers

1. Obtain the materials for this part.
2. Fill 1 of the beakers with ice and water so the beaker is three-quarters full.
3. Fill the 2nd beaker three-quarters full of hot water.
    → Your teacher will provide this hot water when you are ready.

 Use hot pads to handle the beaker containing hot water.

4. Remove all of the ice from the ice water in the 1st beaker.
    → Use a spoon to remove the ice and dispose of it as your teacher directs.
5. Carefully drop 1 drop of red food coloring onto the surface of the water in the middle of each beaker.
6. Observe the action of each drop of food coloring in each beaker.
    → Notebook entry: Record your observations.
7. Use the particle model to explain your observations.
    → Explain what is different about the particles in hot water from the particles in cold water that results in the different patterns you observed with the red food coloring.
    → Notebook entry: Record your explanation.

FIGURE 10.8 What might be happening inside these balloons?

Wrap Up

Answer these questions in your notebook. Be prepared to present to the rest of the class your explanations for these phenomena using the particle model.

CHAPTER 10 Using Scientific Models to Answer Questions

1. What one characteristic did all of the experiments have in common?
2. Decide how scientific the particle model is. Base your decision on the criterion that a scientific model accounts for many, if not all, observations.

ELABORATE connections

Particle Movement— Improving the Model

How successful were you in using the particle model to explain what happened in the investigation How Well Does the Particle Model Work? Did you have trouble? It is not easy to come up with models to explain your observations. It has taken scientists quite a few years to develop models that explain many observations, and scientists often need to change models as they make new observations.

FIGURE 10.9 Just as models in science change to reflect our better understanding, models involving technology change to reflect our advancements in technology. What are some of the similarities and differences in these spacecrafts? What different purposes did they serve?

236 CHAPTER 10 Using Scientific Models to Answer Questions

To help explain your observations in the previous investigation, you might need some more information about the currently accepted model for what makes things different. First, you can find most materials in one of three forms—liquid, solid, or gas. Water, for example, is most commonly found as a liquid. If you boil water, it will turn into a gas. If you freeze water, it will turn into ice, a solid. What makes these three forms of water different is the way the particles are held together. In all materials, particles attract each other and have "forces" between them. These forces are stronger in solids than they are in liquids. In a gas, the forces are even weaker.

Another important part of the model is that particles in materials are in motion. When particles move faster, the material gets hotter. When particles move slower, the material gets cooler. So particles that make up hot water are, on the average, moving faster than particles that make up cold water. Likewise hot air particles move faster than cold air particles. Particles that make up a warm sheet of steel are moving faster than particles that make up a cold sheet of steel.

Answer these questions in your notebook:

1. How can you now use the particle model to explain your observations in the investigation How Well Does the Particle Model Work? As a class, use this new information about the particle theory to explain what happened in the Red Streamers experiment when you put food coloring into hot and cold water. Does this explanation account for all your observations?

2. Now consider what might be happening with the evaporating liquid in Houdini Water. But before you try to devise an explanation, participate in the class activity that your teacher will conduct.

3. How is what you did in the class activity in Question 2 similar to what happened with the particles in Houdini Water?

4. If a model explained that every once in a while particles in the liquid state are not attracted to each other, would that model explain evaporation? Explain your answer.

5. Use all of the information you have about the particle theory to explain what happened to the green food coloring in Houdini Water.

6. Finally, consider what happened in The Balloon. Sketch a picture of what happens with the particles inside a balloon at different temperatures.

    You have just used three examples to show how a particle model that includes particles in motion can help explain the observations you made in the previous investigation. This does not mean that the particle model is the only model that can explain these observations, though. In other words, just because the particle model is the currently accepted one (and for good reasons), that doesn't mean it explains everything about the world. For the moment, however, it does a good job of explaining why things are different.

    You also should know that we have not given you all the details of the particle model.

7. To see that the model is not complete, try to use it to explain the phenomena you observed in the investigation Strange Phenomena. Use the space you left after the explanations you created in the Strange Phenomena Wrap Up to revise those explanations.

Although you might not be able to explain everything, you can use the model to explain quite a few things, as you will see in the next investigation. You also can use the model as a beginning point for developing your own models for what is inside of things, as you will see in Chapter 11.

Applying the New Model

In this investigation, you will apply your understanding of the particle model to a new situation. You will gather observations and then rate the particle model as to how scientific it is. You will base your ranking on the criterion that a scientific model explains many, if not all, observations.

Materials for Each Team of Three:

Part A—The Amazing Coin Dancers
- two 600-mL glass beakers
- crushed ice
- tap water
- hot water
- one 12-oz glass soft drink bottle
- 1 coin that completely covers the mouth of the soft drink bottle
- 3 pairs of safety goggles
- 1 wax pencil

Work cooperatively in your teams of three with a Manager, Tracker, and Communicator. Practice the unit skill, but also use the activity skill *Let others finish without interrupting them.*

Part B—Rainmakers
- one 250-mL glass flask with a rounded bottom
- 1 rubber stopper to seal the flask
- one 600-mL glass beaker
- 1 hot plate
- tap water
- ice
- blue food coloring
- 1 medicine dropper
- 2 hot pads or beaker tongs
- 3 pairs of safety goggles

Process and Procedure

Part A—The Amazing Coin Dancers

1. Obtain all of the materials necessary for this part.
   → The Tracker should prepare to record the number of interruptions that occur.

 Wear eye protection for this part.

2. Put a soft drink bottle in an empty beaker.
3. Fill the beaker with cold tap water.
   → The water should not overflow.
4. Remove the soft drink bottle.
5. Mark the water level in the beaker with a wax pencil.
   → Mark the level of water in the beaker after you have removed the soft drink bottle.
6. Empty the beaker, and then refill it with hot water up to the line that you marked.
   → Your teacher will provide the hot water when you are ready.

 Use hot pads to handle the beaker containing hot water.

7. Fill a 2nd beaker with cold water and ice to roughly the same level as the level you marked on the 1st beaker of water.
   → This is called an ice bath.
8. Submerge a soft drink bottle in the ice bath for 2 minutes.
   → The bottom of the bottle should touch the bottom of the beaker. The bottle may float, so you may have to hold it down for the entire 2 minutes. Do not let any of the ice water get into the bottle.
9. After 2 minutes, stick a fingertip into the ice water and moisten the rim of the bottle by completely encircling it with your wet fingertip.
10. Remove the bottle from the ice bath.
11. Place the coin on the rim of the bottle to completely seal the bottle.

12. Carefully place the cold bottle sealed with a coin into the hot water.
    → Do not jostle the coin on the bottle during this process. If you do, reseal the coin with more ice water on the rim.
13. Observe what happens.
    → Notebook entry: Record your observations.
14. Use the particle model to explain your observations.
    → Explain what is happening with the particles in the water and the particles in whatever is inside the bottle.
    → Notebook entry: Record your team's explanation.

STOP: Remember to practice your unit skill.

15. Return the materials.
    → Protect your hands with hot pads or beaker tongs when disposing of the hot water.

Part B—Rainmakers

1. Obtain all of the materials necessary for this part.
2. Put about 300 mL of water in a beaker and bring it to a boil on a hot plate.

 Keep long hair and loose clothing away from the hot plate. Do not attempt to handle a hot beaker without hot pads or beaker tongs.

3. Turn off the hot plate, but do not remove the beaker.
4. Fill the empty flask about one-quarter full with very cold water.
    → Add some ice to the flask.
5. Add 5 drops of blue food coloring to the cold water in the flask and put the rubber stopper in the mouth of the flask (see Figure 10.10).

→ If you do not have a rubber stopper, cover the mouth of the flask with a piece of clear plastic wrap.

6. When the boiling in the hot water beaker has stopped, turn the flask on its side and fit the round base over and into the top of the beaker of hot water as shown in Figure 10.11.

7. Observe the water in the beaker for 5 minutes.
   → Notebook entry: Record your observations.

8. Explain your observations by using what you know of the particle model.
   → Explain what is happening with the particles in the water in the beaker and the particles in the water in the flask.
   → Notebook entry: Record your team's explanation.

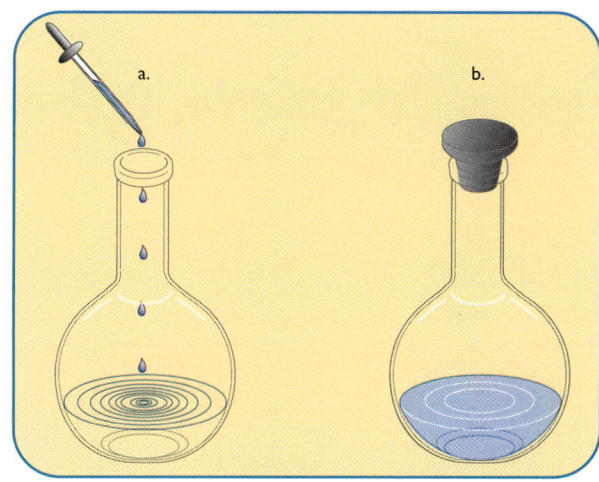

FIGURE 10.10 Add five drops of food coloring and seal tightly with a rubber stopper. The seal must be tight enough so that liquid cannot spill out if you tilt the bottle.

FIGURE 10.11 The bulb of the flask should fit into the top of the beaker. Make sure the water is no longer boiling.

CHAPTER 10 Using Scientific Models to Answer Questions

9. Return the materials.

→ Remember to use hot pads when touching the beaker of hot water.

Wrap Up

Share your results as described in Wrap-Up Question 1, discuss Question 2 with your team, and record your answers in your notebook.

1. Share the models your team created for Parts A, B, and C with the rest of the class. Share with the class how your team was open to each other's ideas.

2. With your teammates, rate the particle model on a scale of 1 to 15—15 being the best—for how strong a scientific model it is. To do this, assign a score between 1 and 5—5 being the highest—for each of these questions.

   - How well does the model help explain some of your observations?
   - How well does the model explain all of your observations?
   - How complete does the model now seem?
   - The overall ranking is _____. (Add up the scores.)

The Tracker should write down the rating you gave the particle model along with the reasons you gave the particle model the rating that you did. Have the Communicator turn in your team's explanations, your particle model rating, and your reasons for assigning the rating you did.

FIGURE 10.12 How would you explain the interaction of particles that caused Al's pain?

Yeow! Ow! That's hot. I guess I really should've used the hot pads!

Be careful, Al! I'll bet it hurts you because the heat from the beaker caused the particles in your skin to start bouncing around like crazy! Ooh! That even sounds painful!

CHAPTER 10 Using Scientific Models to Answer Questions

SIDELIGHT ON HISTORY
Philosophy and Alchemy

People have long tried to explain how and why things are different. Ancient philosophers wrestled with the problems of what different materials are made of and whether something can come from nothing. Recall that Aristotle said that each "element" had distinct properties. Air was warm and moist, and earth was cold and dry. Water was cold and wet, while fire was warm and dry. Fire could become air through the action of heat; air could become water through the action of moistness; water could become earth through the action of coldness; and earth could become fire through the action of dryness. For Aristotle, everything in the world resulted from the combination of these elements and their properties.

Building upon Aristotle's ideas, some people thought that by using the right combination of steps and materials, they could create new materials. These people were known as alchemists, and their art was known as alchemy. Alchemists wanted to find a way to change one type of material (like lead) into another kind of material (like gold). **Alchemy** is said to be a mix of science and magic, because alchemists believed that it was somehow possible to produce a magical substance, which they called the "philosopher's stone," that would change ordinary metal into gold. No one ever produced the philosopher's stone. However, the desire for gold made it possible for many people to be fooled and swindled by alchemists who claimed that they could convert certain materials into gold.

On the other hand, many alchemists were careful scientists. They kept detailed records of their experiments and developed many techniques for separating and preparing materials used today. Alchemists are responsible for inventing many pieces of the equipment that are still in use in chemical laboratories. Through their experiments, alchemists discovered important facts that became the basis for modern chemistry.

CHAPTER 11

Using Models to Test and Predict

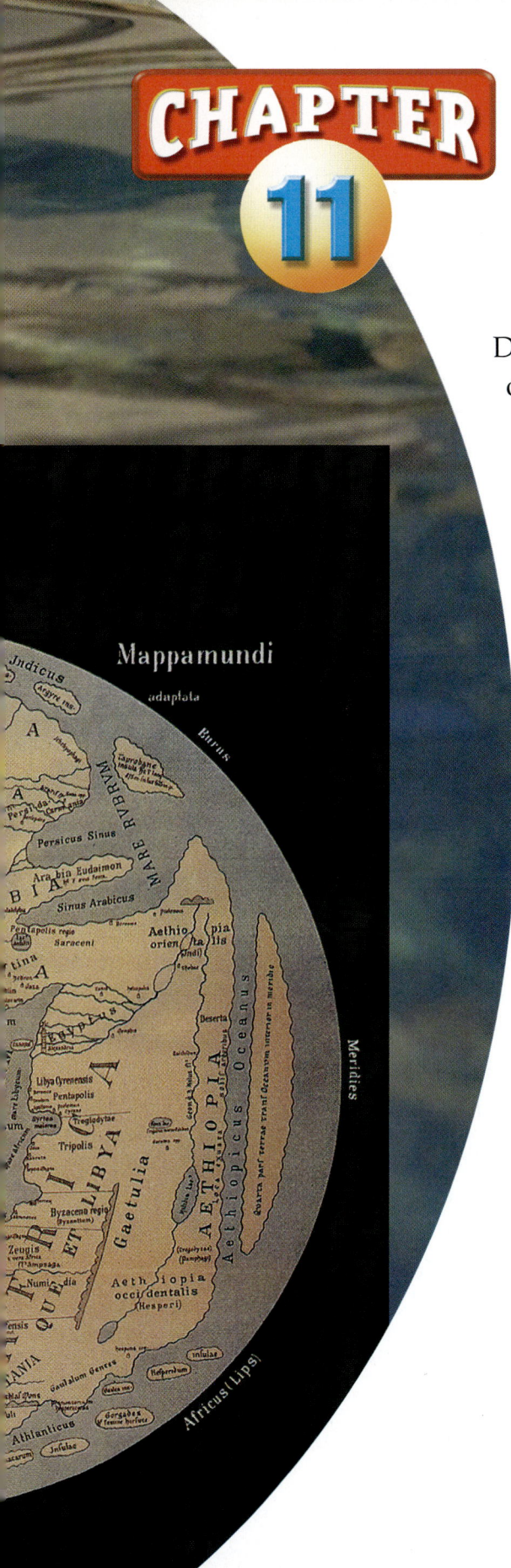

During the Age of Exploration, explorers developed maps of the world that served as models for what they thought the world looked like—how it was shaped and where the land, water, mountains, and deserts were. As the navigators explored, they actually were testing the maps. Often, they found out that their maps were not accurate in places, and they were able to revise the maps to show what they had learned. As you are learning, scientists do much the same thing with the models they develop.

In this chapter, you will have the opportunity to ask more questions about the particle model. As you seek answers to your questions and the questions of others, you will develop new explanations for why things are different.

ENGAGE • Shaping Models

EXPLORE • Gloop

EXPLAIN • More on Models

ELABORATE • Leak-Free Models

ELABORATE EVALUATE • A Penny's Worth of Water

EVALUATE • Properties and Models in Review

ENGAGE connections

Shaping Models

The information you have learned so far about the particle theory is not sufficient to answer all the questions you might have about why materials are different. When scientists have unanswered questions, they often start with an accepted model and then revise it as they test, observe, and learn more. You can use this same process.

The particle model you have used so far states that all materials consist of tiny particles and that the hotter the material, the faster the particles move. You also know that particles in solids and liquids attract one another. But exactly how do the particles attract one another? What is it that connects particles? And how are the particles arranged? Are they close together or far apart? One way for you to improve the particle model is to specify how the particles are arranged and how they interact with one another.

FIGURE 11.1 Before engineers build boats, buildings, or cars, they construct models and use blueprints to test their ideas.

Here is an example. Watch the paper towel demonstration that your teacher will conduct. Then read the following questions and answer them in your notebook.

1. Take five minutes to develop an explanation, based on the particle model, for the paper towel **phenomenon** (or happening) that you just observed.
2. Share your paper towel explanation with the rest of the class.

Al, too, observed the paper towel demonstration. He used the particle model in a different way to explain what makes paper towels absorbent (see Figure 11.2). His model follows:

Al's Model for Sale

Al: I've got it! Here's why paper towels hold water! The towel is made up of tiny particles, and each particle has a little bucket attached to it. When you put the paper towel in water, the buckets fill up. If a towel is really absorbent, that just means the buckets are bigger.

Marie: I don't buy it, Al. Prove it to me.

Al: See? When I squeeze the towel, water comes out. The towel was full of water.

Marie: No, you're showing me the observation that led to your explanation. Now put your explanation to the test.

Ros: I have an idea! Here's a way to prove it, Marie. If the particles are attached to little buckets, we should be able to find which end has the buckets pointing up by filling them and quickly flipping the towel around. The water should spill out of the buckets.

Marie: I'll buy that!

Isaac: Let's try it. I've got a roll of paper towels and some water!

Al: I'm confused again! (The characters try the experiment.)

Isaac: Hmm. Doesn't work. Not a scientific model.

Al: Why not? It explained all my observations!

Ros: Well, now there are more observations!

Al: Okay, then I'll add something to my model. The bucket things are on swivels and never turn upside down.

Marie: Sold!

FIGURE 11.2 Al shows the others his arrangement and the interactions of particles to explain why towels are absorbent. He proposes tiny bucket-like structures attached to each particle.

FIGURE 11.3 This photograph shows droplets of water on a mirror. How might observing water droplets help you better understand the ways in which water is similar to, and different from, other things?

Al used his knowledge that all materials are made of particles and then added his idea that, in paper towels, the particles are attached to bucket-like structures. Al created a new model by using accepted information and his own new ideas.

1. Explain whether you think that Al, or any scientist for that matter, should have to test his or her model in order to convince somebody else.

2. Do you think that Al should have changed his model to fit the new observations that resulted from the test his friends performed? Explain your answer.

Now you know several parts of the particle model: All materials are composed of particles; particles in hot things are moving faster than particles in cold things; particles attract each other; and particles in different materials are uniquely arranged and have unique interactions. The next few investigations will give you the chance to use all of this information to create scientific explanations for more phenomena.

investigation EXPLORE

Gloop

In this investigation, you will determine some of the properties of a very bizarre material. Then you will have a chance to develop your own scientific model by using your own ideas and what you already know of the particle theory. Your model will need to explain the properties of this strange material. You will also see what it is like for someone to ask you to test your model and then have to decide how to change your model to fit the results of the tests.

Materials for Each Team of Three:
- 4 samples of gloop (1 each refrigerated, frozen, dried, and soaked)
- any measuring devices you think you might need

Process and Procedure

Part A—The Social Skill

1. Discuss why the social skill of looking at the person speaking to you is important when working with others.
2. Think of 3 strategies or things you will say to remind each other to use this skill.
   ➡ Notebook entry: Record your ideas.

Work cooperatively in your teams of three with a Manager, a Tracker, and a Communicator. Continue practicing the Unit 2 skill, but also try using the new, activity skill *Look at the person speaking to you.* Work together at your desks or at a table.

CHAPTER 11 Using Models to Test and Predict

Part B—Gloop's Properties

1. Obtain the materials.

2. As a team, explore the gloop.
   → Handle the gloop, look at it, pull it, push it, and manipulate it in as many ways as you can. Spend 10 minutes just exploring the gloop.

3. List all Team Members' observations about the properties of gloop in a list titled "Properties of Gloop."
   → Notebook entry: Make this list in your notebook. Include all of the observations made about how gloop feels, what it looks like, and what it does.

4. Present your team's list of gloop's properties to the rest of the class.

Part C—Explaining the Properties

1. As a team, conduct a brainstorming session to find possible scientific models that would explain all of the properties of gloop.
   → Review How To #4, How to Have a Brainstorming Session.

2. Create one scientific model that explains all of the properties of gloop.
   → Base your explanation on what you know about the particle model and add your own ideas about the arrangement of the particles. Your model should explain why gloop acts, feels, and looks the way it does. Remember you may create only one model that explains all the properties of gloop that you observed. Record your team's gloop model in your notebook.

Part D—Putting Your Model to the Test

1. Read Part D.
2. Construct a data table for recording the information you will obtain as you follow this procedure.
   → Notebook entry: Construct this data table in your notebook. Be sure you make space to record the properties of gloop, experiments on gloop, each prediction you make, results of experiments, and what your revised model states after each experiment.
3. Based on your team's model, predict what will happen if you freeze gloop.
   → This experiment is a way of testing how well your team's model accounts for new observations.
   → Your prediction should answer the following questions:
      - *Will freezing the gloop change the properties that you noted in Part B?*
      - *If so, how will these properties change?*

FIGURE 11.4 Notice how the characters base their predictions directly on the model they created to explain gloop's properties.

4. Tell your teacher that you are ready to conduct an experiment on frozen gloop, and he or she will provide you with a sample.
   → The Communicator should do this.
5. Carefully examine the frozen gloop.
   → Notice any new or different properties that you did not observe during Part B.
   → Notebook entry: Record the properties of frozen gloop in your data table.
6. Decide whether your gloop model accounts for the properties of frozen gloop.
7. If necessary, revise, add to, or change your model so it explains not only the properties you observed in Part B, but also the new properties you observed in Part D, Step 5.
   → Notebook entry: Record any revisions to your model in your data table.

STOP: Remember to include everyone's ideas in your revised model.

8. Based on your new model, predict what will happen if you left the gloop uncovered for 48 hours.
   → Notebook entry: Record your prediction in your data table. Your prediction should answer the same questions as before.

FIGURE 11.5 Do you agree with Marie, or should the characters revise their model?

FIGURE 11.6 Ros wants to get on with the experiment and not make any predictions. What would you tell her?

- *Will leaving the gloop out for 48 hours change the properties of gloop you listed?*
- *If so, how will these properties change?*

9. Tell your teacher that you are ready to conduct the experiment with the gloop that has been left uncovered, and he or she will provide you with a sample.

10. Explore this sample for any new or different properties.
    ➡ Notebook entry: Record in your data table the properties of the gloop that has been left uncovered.

11. Decide whether your revised gloop model accounts for the properties of gloop that has been left uncovered for 48 hours.

12. If necessary, revise, add to, or change your model so it explains all of the properties you observed in gloop, frozen gloop, and gloop that has been left uncovered.
    ➡ Notebook entry: Record your revised model in your data table.

CHAPTER 11 Using Models to Test and Predict

FIGURE 11.7 Make sure, as Ros is doing, that your current model accounts for all of your observations of gloop, not just your recent observations of gloop that has been left uncovered.

13. Predict what would happen to a sample of gloop left soaking in water for 24 hours.
    → Notebook entry: Record your prediction in your data table. Your prediction must answer the same questions as before.
    - *Will soaking the gloop in water for 24 hours change the properties you listed in the original gloop sample?*
    - *If so, how will its properties change?*

14. Tell your teacher that you are ready for a sample of gloop that has been soaking in water.

15. Examine the sample for new or different properties from those you observed in your original gloop sample.
    → Notebook entry: Record your team's new observations in your data table.

16. Decide whether your current gloop model accounts for the properties of the gloop that has been soaking in water for 24 hours.

17. If necessary, revise, add to, or change your current model so it explains all of the properties you observed in gloop, frozen gloop, gloop that has been left uncovered, and gloop that has been soaked in water.
    → Notebook entry: Record your revised model.

→ If you get stuck, remember to make use of the communication system that you have within your group and class as you work cooperatively.

→ Notebook entry: Record your revised model.

18. Return all materials, including all gloop samples, to the appropriate location.

19. Share your team's original gloop model and your team's revised gloop models with the rest of the class.

→ Be sure you tell the class what observations led you to create and revise your model as you did.

Wrap Up

As a team, discuss these questions. Then record answers in your notebooks.

1. Describe how similar and how different the teams' models were to one another.
2. Propose a reason for why the models might have been similar.
3. Propose a reason for why the models might have been different.
4. Pretend that it is 20 years from now. You have just received an International Science and Technology award for your work on a mysterious substance known to the scientific community as "gloop." Your team is at the awards banquet at which you all are expected to give one speech. The prize committee has asked you to explain your modeling process for gloop in your speech. They specifically ask that you tell the assembled scientists how your model changed, how many times you had to revise your model, what caused you to revise your model each time, and how your model includes ideas from each of you. Write your speech in your notebook.

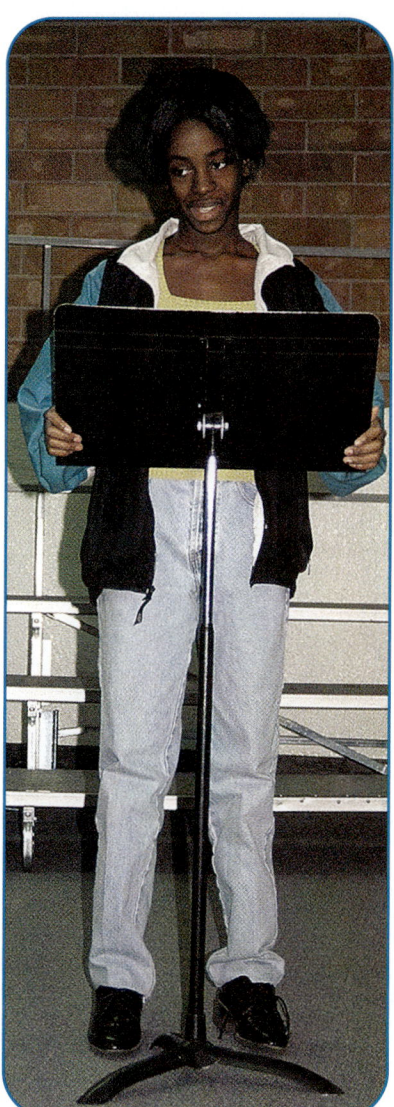

CHAPTER 11 Using Models to Test and Predict

More on Models

Before you began the investigation Gloop, we asked you a very important question: Do you think that Al, or any other scientist for that matter, should have to test his or her model in order to convince somebody else?

1. Now that you have created and tested your own models in the gloop investigation, how would you answer that question?

Creating a model is hard enough. When someone tells you that you have to test your model, the job of modeling becomes even harder. But to be a scientific model, other people should be able to use it to predict future observations. In Chapter 10, you learned that one criterion of a scientific model is that it should explain many, if not all, observations. Another criterion of a scientific model is that we should be able to use it to make predictions.

2. Outline the basic steps you took when you created models for gloop.

When scientists create models, they think of experiments (or tests) that they can do that will result in new observations. Before they conduct their experiments, though, they use their model to try to predict what will happen. Scientists then conduct the experiments to see whether they predicted correctly. If they did make correct predictions, they believe that their model is fairly scientific. If the new observations do not match the

predictions they made, however, the scientists revise their model. This means scientists change parts of their model so that it explains the new as well as the previous observations. In either case, they will think of new experiments and conduct more tests in order to make new predictions. Scientists spend a lot of time trying to improve their model so that it explains many observations.

Making Predictions

Making predictions from a model is more than just thinking about what might happen as the result of a certain test. To make predictions, you need to base your prediction on your model. You also need to keep accurate records of the predictions that you make. One way of accomplishing those goals is by using if–then statements. An **if–then statement** is a logical statement that first states what your model is and then states your prediction, which is based on that model. Let's look at some examples.

Isaac thought that white construction paper was more translucent than black construction paper. His model for

FIGURE 11.8 As you remember, Isaac Newton developed the theory of gravity in the late 1600s. In the early 1900s, Albert Einstein developed a theory about the relationship among various physical principles. Einstein's ideas built on some of Newton's ideas. Today, scientists make and test predictions that are based on the theories of these and other scientists.

Topic: Isaac Newton
Go to www.scilinks.org
Code: physical259

CHAPTER 11 Using Models to Test and Predict   259

what was going on inside the two colors of construction paper is shown in Figure 11.9. Isaac used this model to make a prediction. Notice how he uses an if–then statement:

"If the particles in white construction paper are all lined up and the particles in black construction paper cross over each other, then when I tear the construction paper, the white construction paper should tear as evenly as if I had cut it with scissors, but the black construction paper should be ragged."

Notice that the *if* portion states the model, and the *then* portion states the prediction. In this if–then statement, Isaac predicted the outcome of an experiment by basing his prediction on his model. Isaac now has to perform the experiment to see whether his prediction was correct.

3. Test Isaac's prediction. Is it correct?
4. If Isaac could predict with success how the different papers tear, then he would be on his way to developing a scientific model. Would you say Isaac's model is on its way to being scientific?

FIGURE 11.9 Isaac decided that all construction paper is made of long chains of particles. In the black construction paper (a), the chains cross over one another and don't let much light pass through. But in white construction paper (b), these chains are lined up, and they let light pass through.

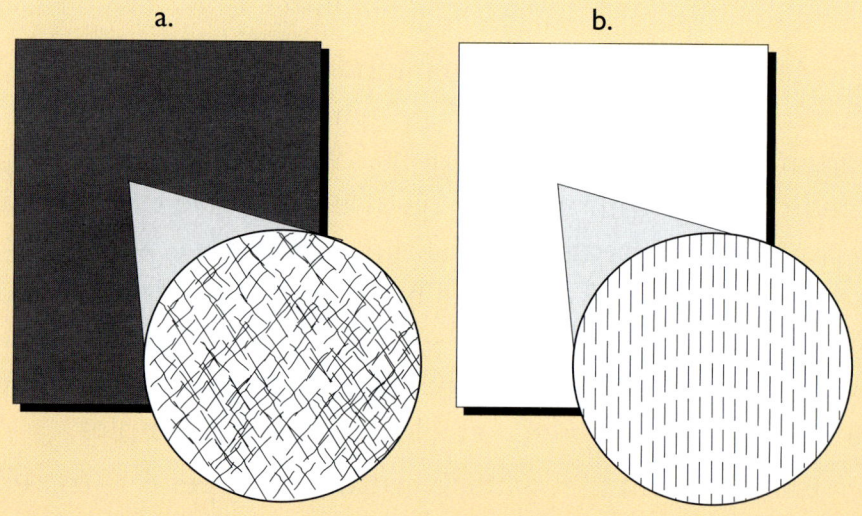

260 CHAPTER 11 Using Models to Test and Predict

Consider again the character scene at the beginning of this chapter. If Ros had used an if–then statement, she might have said:

"If paper towels are made of particles attached to buckets, then I should be able to fill up the buckets with water, turn the towel over, and see streams of water pouring down."

When Rosalind and Isaac conducted the experiment, they found that this prediction was incorrect. Al then appropriately revised his model to account for the new observations. He did not create a new model. He added to the old model.

 5. How did Al revise his model?

We can examine one final example by studying how Marie explains translucence. Her model says that it depends on how close together or how far apart the particles in the materials are. The funny thing is that Al came up with the exact same model to explain the viscosity of corn syrup. He said that viscous materials have particles that are close together, and less viscous materials have particles that are far apart. Al and Marie put their heads together and came up with this if–then statement to make some predictions.

"*If* materials that are not translucent have particles that are close together and *if* viscous materials have particles that are close together, *then* viscous materials should not be translucent."

The characters made a prediction based on both models!

CHAPTER 11 Using Models to Test and Predict

STOP & THINK

6. Do you think Al's or Marie's model needs revising? Why or why not?
7. Rewrite the three predictions that you made for gloop into if–then statements.

Science versus Nonscience

We can make a distinction between what is science and what is not science. One important characteristic for something to be considered science is that it is testable. If you are creating other types of models, you do not have to worry about testability. An ordinary model does not have to be testable. But if you are creating models for science, then you do need to be concerned about testability.

When you create a model, ask yourself these questions:

- Does my model involve elements of magic or superstition?
- Does my model require people to believe something without evidence or require them to "have faith"?
- Is there something about my model, besides lack of technology, that makes it impossible to test?

If your answer to any of these questions is yes, your model is not considered scientific because it does not have the important characteristic of testability. **Testability** is an important characteristic of a scientific model. In part, it separates science from nonscience. If you cannot test a model, it is not scientific. There are lots of models that are not considered scientific because people cannot test them. Here is an example of a model that people cannot test.

Every material on Earth is made of different, tiny, invisible alien creatures from different planets all over the universe. Each type of material has aliens in it that are holding tiny pieces of the ground from their planet. Each piece of ground is one particle of material. The aliens all

FIGURE 11.10 Biosphere II in Arizona is an enclosed system that is modeled after Earth.

stand side by side holding up their pieces of ground from their planet with one hand and holding each other's hands as well. When there are enough of the alien creatures standing side by side and holding hands, you can see a piece of material.

For example, there is a planet one million light years away called Zora. The ground of the planet feels soft, puffy, and hairy. The color of the ground is white. The aliens from this planet, known as Zorites, migrate to Earth on a regular basis and squeeze into plants that are common in the southern United States. Inside the plants, several thousand aliens stand side by side holding up a piece of fluffy, white ground from Zora with one hand. With the other hand, they loosely grasp the free hand of the Zorite beside them. When farmers come around to pick the plant, they see small puffs of white, which they pick and put into the back of a pickup truck. At the end of a long day, farmers are proud of having harvested a big load of what they call cotton.

The farmers ship big groups of Zorites all over the world. Some are processed and stuffed into bottles of vitamins. Some are sent to spinners and weavers. The minute the Zorites feel the spinning process begin, they pop out hook-locks from their hands. These hook-locks interlock and help keep the Zorites from separating so that people can spin and weave them. We will never see these aliens, though, because they set up a force field around their tiny little bodies when they feel that they are being studied. It is impossible to penetrate this force field with anything. That is how it is with all materials, only other materials are made of creatures from planets other than Zora. Creatures from the planet Golee hold up pieces of Golee ground and link hands with springs. Golee ground is hard and bouncy. Goleans make rubber.

This model is not considered scientific because it includes aliens that we have never heard of and planets

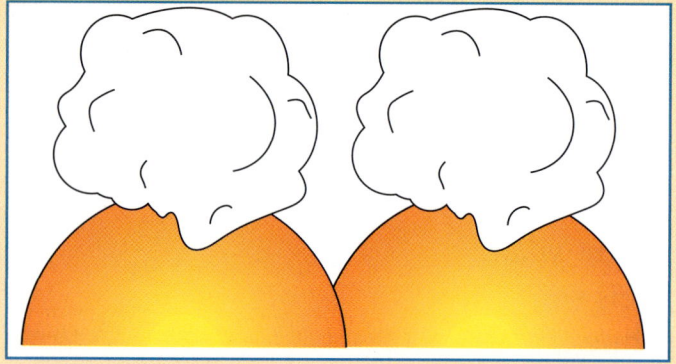

FIGURE 11.11 Would you consider this Zorite model to be scientific? Why or why not?

that we have never discovered. It is not considered to be science because there is no way we could develop an experiment to test the model. Why? Because according to this model, the aliens would "sense" that we were trying to prove their existence, and they would set up force fields that nothing could penetrate.

8. What are other examples of models that would not be considered to be science?

9. Explain whether you would consider your gloop model to be science or to be nonscience.

10. Explain whether you would consider the characters' gloop model to be science or to be nonscience.

11. Complete this statement: I would rate my team's gloop model to be scientific/nonscientific for all of these reasons:

Leak-Free Models

In this investigation, you again will create a scientific model that is based on the particle model. You will use the model you create to make predictions and carry out different experiments. You will state your predictions with if–then statements.

Materials for Each Team of Three:
- 1 plastic vial
- one 2 × 3-in. nylon mesh screen
- 1 resealable plastic bag
- 1 newly sharpened pencil
- a water supply
- 1 sink or tub
- one 2 × 3-in. piece of cheesecloth
- 1 pen
- any other materials needed to test your model

Work cooperatively in your teams of three using the roles of Manager, Tracker, and Communicator. You need a work space by a sink or tub. The unit skill and the activity skill *Look at the person speaking to you* are both important as you create models to explain phenomena.

Process and Procedure

Part A—Look Ma, No Hands!

1. Obtain a plastic vial and a piece of nylon mesh screen.
2. Fill the plastic vial with water.
   → The Communicator should do this.
3. Cover the vial with the piece of mesh screen.
   → The Tracker should do this. See Figure 11.12a.
4. Take the vial of water covered with the screen and place the palm of one hand on top of the mesh screen.
   → The Communicator should do this over a sink or tub. See Figure 11.12b.

5. Quickly turn the vial upside down keeping the palm of one hand on the screen and using the other hand to flip the vial over.

   → The Communicator should do this over a sink or tub as in Figure 11.12c.

6. Gently pull your hand away from the screen without removing the screen.

   → The Communicator should do this over a sink or tub as in Figure 11.12d. You must hold the vial straight down, not tilted to one side.

7. Observe what happens.

   → Notebook entry: Record your observations.

FIGURE 11.12a-d Be sure to perform these steps over a sink or tub.

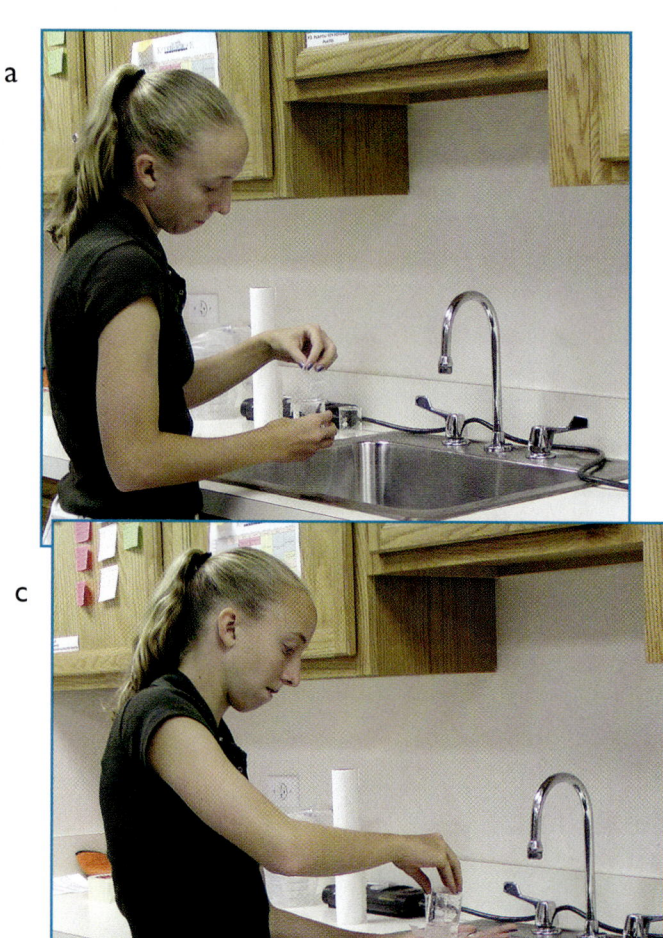

8. Turn the vial back over, and place it on the desk or counter. Create a model that explains what you observed.
   ➡ In your model, include information about how the particles interact inside the water, the screen, and in the air; how they are arranged; and how they might be attracting each other. Also make sure that your model is testable.
   ➡ Notebook entry: Record your model.

9. Read the following test:
   ➡ Replace the screen with a piece of cheesecloth of the same size.

10. Predict what will happen in the test by completing the following statement: *If* _____, *then* when we invert the vial of water using cheesecloth instead of screen, the water in the vial will _____.
    ➡ Notebook entry: Write this statement and fill in the first blank with what your model states. Fill in the last blank with your prediction. (For example, will the water spill out?) Remember that you must base your prediction on the model you created for this phenomenon.

11. Test your model as described in Step 9 and observe what happens to the water.
    ➡ Notebook entry: Record your observations and describe whether your prediction was correct.

12. If you based your prediction on your model and it was correct, you do not need to revise your model, and your team can proceed to Step 13. If your prediction was not correct, discuss as a team how you could change your current model to account for new observations.
    ➡ Notebook entry: Record your revised model.

CHAPTER 11 Using Models to Test and Predict

13. Think of a second test of your model.
    → The second test is up to you. Look over the materials that your teacher has provided and send your Communicator around to get hints from other teams for another test you could perform. As a team, discuss what materials you will use to test your model. Remember to look at the person speaking to you.
    → Notebook entry: Record the test your team designs.

14. Predict what will happen in the test by completing this statement in your notebook;

    If _____, then when we _____, the water will _____.

    → The first blank space is for you to describe your model. The second blank is for you to describe your test. The third blank is for you to tell what will happen as a result of your experiment based on what your model says the particles are doing.
    → Notebook entry: Record your completed statement.

15. Conduct your test and determine whether or not your prediction was correct.
    → Notebook entry: Record your observations and whether or not your prediction was correct.

16. If your prediction was correct and you based it on your model, you do not need to revise your model at this point, and your team can proceed to Part B. If your model did not predict successfully what would happen in your test, discuss with your teammates how you could change your model to account for the new observations.
    → Notebook entry: Record your revised model.

Part B—Porcupine Water Bags

1. Obtain a resealable plastic bag and a newly sharpened pencil.

2. Use a scale of 1 to 10 (10 being the highest) to rate yourselves so far in using the unit skill and in looking at the person who is speaking.
   → Discuss this with your teammates. Record your own score.

3. Fill the plastic bag half full with water.
   → The Communicator should do this.

4. Seal the bag shut.
   → The Communicator should do this.

5. Using the newly sharpened end of the pencil, poke the bag of water below the waterline and continue slowly pushing the pencil through the bag until the point comes out the opposite side.
   → The Tracker should do this holding the bag over a tub or sink. The eraser end of the pencil should be sticking out of one side of the bag, the middle of the pencil should be inside the bag submerged in water, and the sharpened end of the pencil should be sticking out of the other end of the bag as shown in Figure 11.13.

FIGURE 11.13 Make sure that you are holding the bag over a tub or sink for this step. Notice how the pencil goes straight through the water bag, not at a slant.

CHAPTER 11 Using Models to Test and Predict

FIGURE 11.14 This is a photo of Winifred Goldring. She was a paleontologist who specialized in Devonian fossils. What kinds of models might she have tested?

Topic: Paleontology
Go to www.scilinks.org
Code: physical270

6. Observe what happens to the water.
   → Notebook entry: Record your observations.

7. Discuss a possible model that explains your observations.
   → Be sure your model includes how the particles of the bag, pencil, and water could be arranged and what they could be doing. Be sure your model is testable.
   → Notebook entry: Record your model.

8. Think about a test in which you follow the same procedures, except that you use a pen instead of a pencil to puncture the bag.

9. Predict what will happen by completing the following statement:

   If _____, *then* when we substitute a pen for the pencil, the water will _____.
   → Notebook entry: Write this statement in your notebook and fill in the blanks. The first blank is for you to tell what your model says. The second blank is your prediction. (For example, will the water leak out?) You must base your prediction on your model.

10. Test your model as described in Step 8.
    → Notebook entry: Record your observations and determine whether or not your prediction was correct.

11. If you based your prediction on your model and the prediction was correct, you do not need to modify your model at this point, and your team can proceed to Step 12. If your prediction was not correct, modify your model to account for your new observations.

→ Think of other ways the particles could be arranged or other things the particles could be doing that you didn't include in your model.

→ Notebook entry: Record your revised model.

12. Think of another way to test your model.

    → The next test you will perform is up to you. Look over the supplies that your teacher provided. This should help you to think of another test. If you still cannot think of another test, ask your Communicator to get ideas from other teams. As a team, discuss what supplies you will use to test your model.

    → Notebook entry: Record the test you design.

13. Predict what will happen in your next test by completing this statement:

    If _____, *then* when we _____, the water in the bag will _____.

    → The first blank is to record what your model says. The second blank is to describe the experiment you decided to perform. The third blank is to record what you think will happen as a result of your test.

    → Notebook entry: Record your completed statement.

14. Conduct your next test using the materials you selected as a team.

    → Notebook entry: Record your observations and whether or not your prediction was correct.

15. If your prediction was correct, you do not need to modify your model, and your team can proceed to the Wrap Up section. If your prediction was not correct, revise your model to account for any new observations.

    → Notebook entry: Record your revised model.

Wrap Up

Share your results as described in Wrap-Up Question 1, discuss Question 2 with your team, and record your answers to Question 3 in your notebook.

1. Share your models with the class. Explain how you revised your models and why you revised them as you did. When another team is presenting its model, offer constructive criticism. This means telling them what they did that was good and how they could improve their model even more.

2. Describe how the ranking you assigned your team in Part B, Step 2 of the investigation changed.

3. An example of a model for Porcupine Water Bags is that the graphite particles that make up the pencil point do not allow water to seep out. What would you predict would happen, based on this model, if you used a pen instead of a pencil?

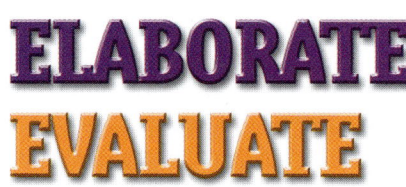

ELABORATE EVALUATE — investigation

A Penny's Worth of Water

A penny is a fairly small thing. So is a drop of water. Have you ever wondered how many drops of water would fit on the surface of a penny? In this investigation, you will have a chance to find out. Then you will have a chance to use all that you know about scientific models to evaluate a model that might explain the phenomenon you will observe.

Materials for Each Team of Three:
- 2 medicine droppers, one for water, one for soap
- one 50-mL beaker filled with water
- one 10-mL beaker of liquid soap
- 1 penny
- 2 sheets of paper towels

Process and Procedure

Part A—The Social Skill

1. Review the social skills that you have used this year.
2. For this investigation, your team will choose a skill to practice. Choose one from the list that you think will benefit your team. You also may develop a skill of your own.
    ⟶ Notebook entry: Record the skill you choose.

Part B—Drops on the Penny

1. Obtain all of the materials.
2. Lay a penny flat on your work area.
3. Have each Team Member guess how many drops of water will fit on the surface of the penny without spilling off.
    ⟶ Notebook entry: Record each teammate's guess.
4. Using the first dropper and the water in the beaker, place drops of water on the surface of the penny 1 drop at a time.
    ⟶ The Manager should do this slowly and patiently without touching the penny or the water on the penny. Be sure to keep the work area stable. Try not to jerk it or bump it suddenly.

Work cooperatively in your teams of three with a Manager, a Tracker, and a Communicator. You will choose your own social skill to practice. Work together at your desks or tables.

5. Help the Manager keep track of how many drops have been placed on the penny by quietly counting aloud each drop as the Manager drops it.
   → The Communicator should do this as the Tracker uses tally marks to record the number of drops the Communicator calls out.

6. Stop placing drops of water on the penny immediately when the first bit of water spills over the edge of the penny.
   → The Communicator should not count the drop that caused the water to spill off the penny.
   → Notebook entry: Record the total number of drops that fit on the penny. Notice how close or how far your guesses were from the actual number.

Part C—Evaluating a Model

1. Read the following model, which explains why so many drops of water can fit on the surface of the penny.

   Water particles are extraordinarily lightweight. They are so light, in fact, that if you slowly release water in very small amounts, gravity does not have any effect on the particles. If you release water in large quantities (from a bucket or a faucet for example) there are so many particles attracting one another that gravity can have an effect on them.

If water particles slowly come out of a dropper, however, the drops are small enough that, even though the drop might contain more than one particle, it does not contain enough to be affected by gravity. Therefore, as you drop water onto a penny, gravity does not pull the water off the penny until there are enough particles of water on the penny to be heavy enough. Then gravity pulls the water off. When you add the drop of water that makes the rest of the water spill off the penny, that last drop added a few too many particles to the water dome. That made the water dome heavy enough for gravity to pull it down and off the penny.

2. Discuss this possible model with your team.

   Decide whether it sounds like a scientific model. As you discuss, consider the following questions:
   - Does the model account for all of the observations you made when you slowly placed drops of water on the penny?
   - Is the model testable? If so, how? What test could you perform on this model? Think carefully! When you have thought of a test, check with your teacher.
   - Try to predict the outcome of your test based on the model. Remember to base your prediction on what the model says should happen, not on what you think will happen.

3. Make an if–then statement about the test you would perform and the prediction you would make based on the model by completing the following statement:

If the model is correct, *if* we _____, *then*, based on this model, the result should be _____.

→ Notebook entry: Record your completed statement.

4. Conduct the test according to your if–then statement.

   → Check your prediction to see whether the model correctly predicted the results.

5. Answer the following questions as a team:
   a. What was the closest guess about how many drops would actually fit on a penny? Were some of you surprised?
   b. Would you consider the model to be science, or is it nonscience?
   c. Did the sample model pass all of the criteria for being a scientific model? In what criteria did it succeed, and in what criteria did it fail?

   → Notebook entry: Record your team's answers.

Part D—The Model Needs Revising

1. Read and consider the revised model below.

   Water is composed of tiny particles that have forces that constantly attract each other. When you drop water on the penny little by little, the particles in one drop attract not only each other, but the particles in the new drop as well. The water particles' attractive force is strong enough to hold each other together in a dome and is strong enough to overcome the force of gravity to a certain point. Although gravity is pulling constantly on the drops of water, it is not until the dome of water bulges out just enough that gravity finally can overcome the attractive forces in the water particles and pull the water down off the penny. It is, therefore, the forces between the water particles that are responsible for this phenomenon.

2. Decide whether or not this revised model is a scientific model.
   → As you decide, consider these questions:
   - Does the model explain all of your observations of the water on the penny?
   - Is the model testable? How would you test such a model? If attractive forces were responsible for this phenomenon, is there some test you could do that would destroy the attractive forces of the water particles?
   - If someone told you about a test that you could do that would break the attractive forces, could you correctly predict the results based on the revised model?

3. Think about the following test to which you will subject this model.
   → To test this model of attractive forces, you will repeat the experiment. But this time, you will place fewer drops of water on the penny. Have the Tracker check in his or her notebook for the number of drops that fit on the penny without any water spilling off. Now subtract 10 from that number. This is the number of drops you will place on the penny this time. Then you will place a drop of liquid soap on top of the dome of water on the penny.

4. Complete the following if–then statement:
   If the model is correct, *then* after we place a drop of soap on the dome of water, based on this model, the results should be _____.
   → Notebook entry: Record the completed statement.

CHAPTER 11 Using Models to Test and Predict

FIGURE 11.15 How is building a house of cards similar to placing drops of water on a penny?

5. Conduct the test described in Step 3.
   → Remember to use the dropper that goes with the beaker of soap, not the dropper you used with the water.
   → Notebook entry: Record your observations.

6. Based on the model, explain what the soap particles did to the water particles that caused them to spill off the penny.
   → Your Communicator can check with other groups for possible answers that you did not think of.
   → Notebook entry: Record your explanation.

7. Share your explanations with the rest of the class.

Wrap Up

Discuss the following Wrap-Up questions with your teammates. Then record your answers in your notebook. Each of you should be prepared to explain your answers in a class discussion.

1. Which of the criteria for scientific models did the second model meet?
2. How would you rate this revised model as a scientific model: strong, medium, or weak?
3. Read the following background information about this model.

Background

The revised model that you just evaluated is actually the current scientific model for the dome of water on the penny that you observed. This model can explain other phenomena as well. Have you ever filled a drinking glass too high and noticed how the liquid bulges over the rim but does not spill over? We also could use the revised model to explain that. The attractive forces between water particles actually have a name. They are called **cohesive forces**. And the phenomenon also has a name. It is called **surface tension**. The cohesive forces between water particles hold water together in this way.

EVALUATE connections

Properties and Models in Review

Work by yourself to see what you have learned about properties and models. Feel free to turn back in your science book or in your notebook to help you remember the ideas you need for answering these questions.

1. If it were up to you to define properties of materials for a new student in class, what would you say?
2. What evidence do scientists have for the particle model?
3. What evidence did the Chinese and Greeks have that led them to their ideas about why things are different?
4. Are today's scientists smarter than the ancient Chinese or Greeks?
5. Explain to the new student in your class what a scientific model is, in general, and what the particle model is, in particular.

In this chapter, you used your knowledge of the particle model to create your own models for why things had certain properties. By then you had a fairly good idea about what things make a model scientific. Two criteria of a scientific model are (1) the model explains many, if not all, observations; and (2) the model can successfully predict new observations. You also learned that in order to be considered science, a model must have the characteristic of testability. Continue your discussion as you answer these questions.

6. Why should a model have to explain as many observations as possible? Why can't a model explain only the first observations you make and then a completely new model explain any new observations?

7. Why do scientists test their models?

8. If you are unsuccessful in accurately predicting new observations based on your model, what should you do and why?

You had many new ideas to learn about and work with in this unit. Those ideas might not be completely clear to you right now, but you should be fairly comfortable with them and what they mean. This is the time to ask your teacher and classmates questions if you do not understand something.

UNIT 3
HOW DOES TECHNOLOGY ADDRESS DIVERSITY AND INFLUENCE LIMITS?

Chapter

12 Consumer Concerns

13 Your Designing Ways

14 Why Are There So Many Products That Do the Same Thing?

15 Masters of Design

In Unit 1, you explored the limits that you and your classmates experienced in accomplishing certain tasks. You found that all humans have a diversity of limits. Sometimes, we are able to change our limits. For example, some people wear glasses or contact lenses. These devices allow people to change the limit of how well they can see close up or at a distance.

In this unit, you will have a chance to explore technological design. This is an aspect of science and engineering that focuses on developing tools and processes that change human limits. First, you will learn some basic concepts about design. Then you will explore some familiar technologies that people use to help them overcome limits. In the end, you will design a newspaper and write articles for it that reflect what you have learned about science and technology so far this year.

283

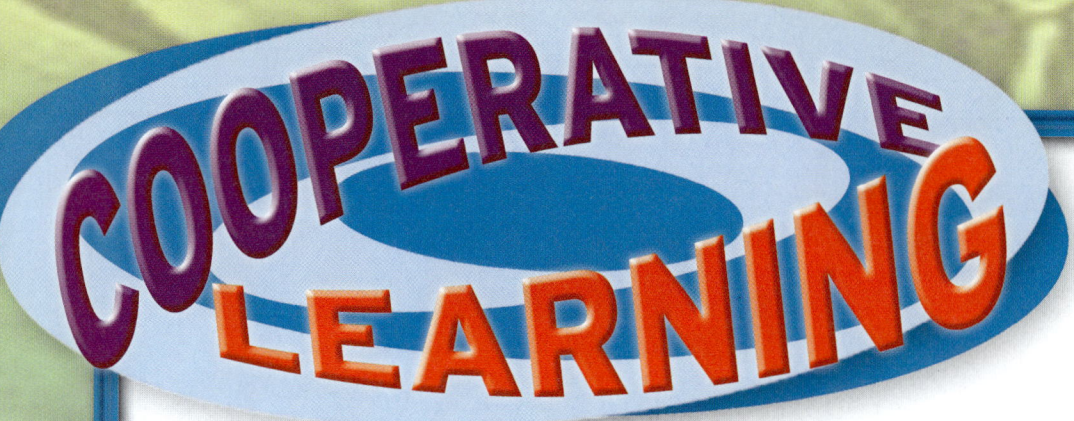

As you complete this unit, keep in mind the social skills from the previous units and practice them when appropriate. Think about how practicing these skills has created a better learning community in your science classroom this year.

Spend a few minutes discussing with your classmates the Unit 3 skill *Disagree with the idea, not the person*. Do this together by constructing a class T-chart for the skill. Make a copy of the T-chart for yourself in your notebook.

CHAPTER 12

Consumer Concerns

If you watch TV, you are familiar with commercials. Manufacturers spend thousands of dollars per minute on television commercials to convince you that their product is the best. But are all products as good as their manufacturers claim? How can each brand of paper towel be the best? How can each breakfast cereal be the tastiest or the healthiest?

Usually, advertisers create catchy slogans or funny commercials to convince consumers to buy their products. These advertising techniques might get consumers to buy the product, but that doesn't necessarily mean that the product is the best. In this chapter, you will investigate paper towels and breakfast cereals and learn to become better consumers. You also will investigate how to determine which products are the best.

ENGAGE
- Putting Paper Towels to the Test

EXPLORE

EXPLORE
- Paper Towel Consumers
- Comparing Ratings

EXPLAIN
- Why Products Fit

ELABORATE
- Do You Understand Criteria and Constraints?
- Part of Your Complete Breakfast

EVALUATE
- Evaluating Your Understanding of Criteria and Constraints

ENGAGE EXPLORE investigation

Putting Paper Towels to the Test

Who makes the best paper towel? How can you find out? This investigation will help you answer these questions. You will test only a few brands of paper towels. Once you figure out how to test a few, you can be a wise consumer and test other brands on your own.

Materials for the Entire Class:
- 1 pan balance
- 1 to 1.5 kg of washers, pennies, or other objects of uniform mass

For Each Team of Three:
- 1 tray
- 1 metric ruler
- 1 beaker or measuring cup

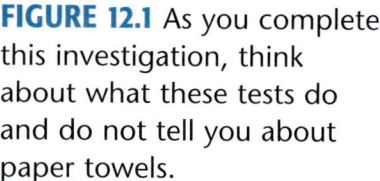

FIGURE 12.1 As you complete this investigation, think about what these tests do and do not tell you about paper towels.

- 1 sample of each brand of paper towels your teacher provides
- a water source
- any other supplies that your teacher provides

Process and Procedure

1. Obtain the materials.
   → The Manager should do this.
2. Label each paper towel sample with its brand name.
   → Write initials or symbols in one corner of each paper towel.
3. As a team, conduct a brainstorming session to create a list of properties that a "good" paper towel has. For example, is a good paper towel strong, absorbent, or soft?
   → Notebook entry: Record this list in your science notebook.
4. Rank the properties on your list from most important to least important.
   → Notebook entry: Record this ranking in your notebook.
5. Pick 1 property from your list that you want to test first in your samples.
   → You can test as many properties as you want, but begin with the property you ranked as most important.
6. As a team, create an operational definition that states how you will measure each property that you chose.
   → Notebook entry: Record your operational definitions. Be creative and try to come up with operational definitions that will not cause you to waste too many paper towels.

WORKING COOPERATIVELY

Work cooperatively in your teams of three. The roles you will need are Manager, Communicator, and Tracker. Practice the social skill *Disagree with the idea, not the person*. Move your desks together or sit in a triangle at a table.

CHAPTER 12 Consumer Concerns

7. For each operational definition, list the variables you will need to control to make your tests fair.
   → Notebook entry: Record the variables. You might not think of all the variables until you are performing your experiment. That is okay. You can record other variables as you work.
8. Construct a data table for your tests.
   → In this data table, you will record all of your tests and the results. The data table also should include a column for ranking the paper towels from best overall to worst overall.
9. Conduct your investigation by testing 1 variable (property) on every paper towel sample. Once you have tested the most important variable, test the second most important variable. Continue testing the variables you listed for as much time as your teacher gives you.
   → In your data table, record any information you need, along with the results you obtain.
10. Rank the paper towels, by brand, from best to worst based on how each brand met the qualifications of your operational definition.
    → Base your ranking on the results of your team's tests.
    → Notebook entry: Record your paper towel ranking.
11. Record your brand rankings on the class data table.
    → This is the Communicator's job. Your teacher will provide this data table on the chalkboard or overhead transparency.

Wrap Up

Discuss these questions with your team. Record your responses in your notebook. Be sure each member of your team can explain the team's operational definitions and the properties that your team decided were important to test.

1. Compare your list of brand rankings with other teams. Did any teams have identical rankings?

2. Describe a reason why there might be a difference in the rankings among teams.

3. Have your Communicator compare the list of properties and operational definitions with those of other teams. How similar was each team's list of properties? Did any teams create the same operational definitions?

4. If some teams used the same operational definitions, did these teams rank the towels identically? Why or why not?

Paper Towel Consumers

The following is an article reprinted from *Consumer Reports*, a magazine that provides information and advice to consumers about purchasing products and services. The magazine's technical staff ranks products just as you did in the previous investigation. We have inserted questions at various places in the article to help you think about specific things. Read each section aloud in your team of three by having the Manager read the section before Questions 1 and 2. The Communicator then should read the section after that, stopping before Question 3.

FIGURE 12.2 Do you think that having more paper towel rolls in one package means that you are saving money?

Then the Tracker should read the section before Questions 4 and 5. Divide the remaining sections into parts and continue to take turns reading in the same order. When you get to Stop and Think questions, discuss the questions with your teammates and write your answers in your notebook. Although this article was written in 1987, the model it provides is still applicable today. Even though some of the brand names and prices have changed, the basic qualities that we look for in a paper towel have not. Also, the process of identifying the criteria that we look for in products and developing ways to test for them is the same.

Paper Towels*

Paper towels are a pretty humdrum product. One supermarket executive told us "they are the closest thing to cordwood my store sells." This is not because of the shape or the way the towels are stacked, but because the retailer treats one roll pretty much the same as another.

The manufacturers, however, go to some lengths to persuade you that paper towels are anything but cordwood. Different colors and decorator designs—even cartoon characters—compete for your attention on the shelves.

Some manufacturers play with the packaging, wrapping two or three rolls together to make you think the bigger pack is a better value than a single roll. Georgia-Pacific's *Mr. Big*, for one, is sold only in a three-roll pack. You can buy most other brands one roll at a time.

 1. Explain the importance of color and decoration to manufacturers and consumers.

Because paper towels are costly to haul long distances, most big paper towel companies make their towels at different mills. Some brands are the same nationwide, but there also are many regional and store brands. In a few cases, a nationally known brand name will vary from region to region. For example, the *Viva* we bought in the Midwest and West is a one-ply towel; the East Coast *Viva* is a two-ply towel. *Hi-Dri* has 100 sheets to the roll on the West Coast, 200 sheets in the East.

Topic: Making of Paper
Go to www.scilinks.org
Code: physical293

 2. Can you think of any reasons for having different packaging in different parts of the country? How do manufacturers come up with these reasons?

FIGURE 12.3 Paper for paper towels is manufactured at large paper mills such as this one.

CHAPTER 12 Consumer Concerns

The manufacturers try to take a bigger share of the market by selling many brands, pitching them to different segments of the market. Scott Paper Co., for example, sells *Job Squad* and *Viva* at a premium price, aiming those brands at consumers who believe that a high price connotes high quality. Scott Paper also sells *ScotTowels*, a moderately priced brand aimed at the consumers who treat towels like cordwood.

Some manufacturers split the market even finer. Proctor & Gamble sells two kinds of *Bounty*, a "regular" towel and one meant for use in a microwave oven. The company wants you to believe that you actually need two kinds of *Bounty* in the kitchen. Scott Paper is trying to segment the market on size; its *ScotTowels Junior*, an $8\frac{1}{4}$-inch wide roll, is pitched as a towel that's "just the right size for saving money." (It is not.)

3. Describe a time when you or someone you know bought something expensive because you or they thought it was better than an inexpensive brand.
4. What problem do you see with the $8\frac{1}{4}$-inch size of *ScotTowels Junior* if all other paper towel rolls are 11 inches wide?

The supermarket executive we spoke with termed the premium-priced towels "over specified," meaning that they are thicker and heavier than they have to be. This term gives the advertiser something to brag about and helps justify the generally higher price for the towel. That in turn pays for both the manufacturing costs and the heavy advertising and promotion expenses.

The Well-Rounded Towel

In our tests of paper towels, we looked for towels that had both good wet strength and good absorbency. Combining those qualities is something of a technical achievement. Most of the papermaking processes that create strength tend to undercut absorbency and vice versa.

To test for absorbency, we weighed dry towels, dipped them in water, skimmed them across the lip of the pan, then weighed them again to see how much water they picked up. In general, the premium-priced towels, both one- and two-ply, were more absorbent than the lower-priced major brands or store brands. The better towels have more of what the trade calls "puff."

Absorbency alone does not help much if it takes the towel a long time to get wet. If you are trying to mop up a large spill, or if you spill something on a carpet, you will want a towel that absorbs liquid fast. To see how quickly the towels could absorb, we ran speed trials using both water and cooking oil.

With the towels clamped in an embroidery hoop, we dripped water and oil on them in separate tests, noting how long it took for the liquid to soak in. Several towels—store brands as well as premium brands—soaked up water almost instantaneously; the slowest took about 20 seconds to absorb the water drops.

All the towels absorbed the oil more slowly. The fastest took two to three seconds to drink a drop of oil and the slowest took a leisurely $2\frac{1}{2}$ minutes.

We measured wet strength two ways. First, to find out how much weight a wet towel could bear, we mounted each towel in an embroidery hoop, wet it, then poured on a steady stream of lead shot. When the towel burst, we stopped pouring and weighed the shot. The weakest towels held only about half a pound of lead; the strongest, more than three pounds. Second, to see how the towels

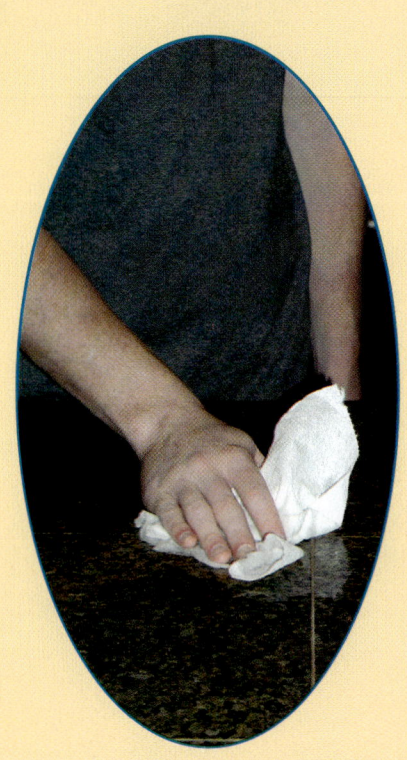

would hold up to scrubbing, we mounted them on a laboratory machine that rubbed them over a textured plastic surface. Some towels disintegrated after about a dozen strokes, while others lasted for a few hundred.

Job Squad proved to be the strongest in these tests. But its high strength—and high price—verge on overkill. The one-ply *Viva* was amply strong, overall, and costs only about half as much as *Job Squad*.

STOP & THINK

5. What properties of paper towels did the *Consumer Reports* investigators test?
6. Why do you suppose the investigators thought these qualities were important to test?
7. From the previous section The Well-Rounded Towel, list the operational definitions the investigators used.
8. List all the variables you can think of that the experimenters had to control for each test they performed.

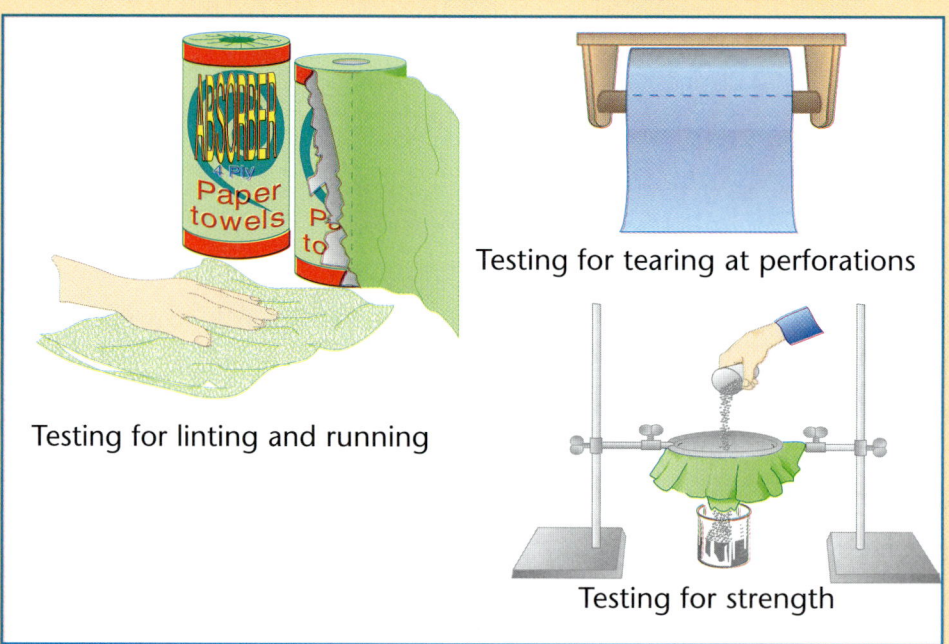

Testing for linting and running

Testing for tearing at perforations

Testing for strength

CHAPTER 12 Consumer Concerns

Linting and Running

Some paper towels shed lint when they are used to wipe a hard surface, a shortcoming that is particularly noticeable if you use paper towels to wash windows or mirrors. The highest-rated towels were about average here.

Towels printed in vivid colors posed another problem: Some of the wet towel's color rubbed off onto white cotton or white-painted panels. Since towel colors and designs change frequently, we cannot say which brands are the most likely to run. But you can always play it safe and use white paper towels.

A towel that does not separate quite right at the perforations can leave you with an avalanche of paper or a mere shred. We tore hundreds of towels off their rolls to see which came off neatly. Nearly all did, provided the towel holder was properly tensioned. Only the *Delta* gave us a ragged portion of towel from time to time.

9. Explain why the following properties might matter to the consumer:
   - linting,
   - colors rubbing off, and
   - towels not tearing easily at the perforations.
10. How does the *Consumer Reports* list of properties compare to the ones you tested?

Recommendations

The strongest, most absorbent towels were the premium-priced brands such as *Job Squad, Viva,* and *Bounty*. Judged strictly on performance, *Job Squad* and the one-ply version of *Viva* earned a check rating. But that does not make those towels the best value. For simple little spills or other small mop-ups, you might want to keep a roll of cheap towels. Look to the Ratings column for unit cost per 100 towels to find the true bargains. For example, store brands such as *Pathmark* (38 cents per 100), Safeway's *Marigold* (51 cents), Kroger's *Cost Cutter* (39 cents), or *A&P* (59 cents) are low in the Ratings yet adequate for undemanding jobs. By contrast, *Job Squad* costs $1.84 per 100, one-ply *Viva* costs 92 cents per 100, and *Bounty* costs $1.09 per 100.

For bigger jobs, such as washing windows or cleaning the cook top, you can always keep a spare roll of towels under the sink or in the broom closet. In this case, though, you should look for a brand that did well in our test for absorbency and wet strength and that is moderately priced by the square foot rather than by the towel. Check the column for unit cost per 100 square feet to find the better values. *Brawny* ($1.05 per 100 square feet) and *ScotTowels* ($.84) are among the cheaper towels that are also good enough to handle most heavy-duty cleaning chores.

A cents-off coupon or a special store sale may make an otherwise expensive brand a good buy. But do not think that towels in two- or three-roll packs give you a price break: most of the multiple-roll packs we looked at were no cheaper per 100 towels or per 100 square feet than single rolls of the same brand.

11. Why would you look for a brand of paper towels priced by the square foot instead of by the towel?

➡ Use the Ratings chart to answer the following questions.

12. Which brand of paper towels is the most expensive per roll?

STOP & THINK

13. Which brand of paper towels is the most expensive per 100 towels?
14. Which brand of paper towels is the most expensive per 100 square feet?
15. How are the *Hi-Dri* paper towels sold on the East Coast different from the *Hi-Dri* paper towels sold on the West Coast?

Ratings

Better ← → Worse

Paper towels

Product	① Plies	② Price per roll	Towels per roll	Sq. ft. per roll	③ Unit cost Per 100 towels	Per 100 sq. ft.	④ Absorption capacity	⑤ Wet-strength	⑥ Absorption rate Water	Oil	⑦ Tearing ease	⑧ Linting
✓ Job Squad	1	$ .92	50	40	$1.84	$2.30	●	●	●	●	●	○
✓ Viva (West, Midwest)	1	.83	90	71	.92	1.17	●	●	●	●	●	○
Bounty	2	.96	88	73	1.09	1.32	●	◐	●	●	●	○
Bounty Microwave	2	1.02	88	73	1.16	1.40	●	◐	●	●	●	◐
Brawny	2	.77	70	73	1.10	1.05	◐	○	◐	○	◐	○
ScotTowels	1	.74	124	88	.60	.84	○	◐	○	◐	◐	○
ScotTowels Junior	1	.66	95	50	.69	1.32	○	◐	○	◐	◐	○
Summit	2	.78	99	70	.79	1.11	◐	◐	●	◐	●	○
Zee	2	.76	102	72	.75	1.06	○	○	●	◐	●	○
Viva (East Coast, South)	2	.82	90	72	.91	1.14	○	○	●	○	●	◐
Truly Fine (Safeway)	2	.88	115	78	.77	1.13	○	○	○	○	●	◐
Mr. Big	1	.52	100	69	.52	.75	○	○	○	●	○	○
Delta	1	.59	110	75	.54	.79	○	○	○	●	◐	○
Coronet	2	.71	115	79	.62	.90	○	◐	○	◐	●	○
Gala	2	.76	110	77	.69	.99	○	◐	◐	○	●	○
Pathmark	2	.65	115	79	.57	.82	○	◐	◐	○	◐	○
Hi-Dri (East Coast)	2	1.19	200	148	.60	.80	○	◐	◐	○	◐	◐
Hi-Dri (West Coast)	2	.67	100	74	.67	.90	○	◐	◐	○	◐	◐
Pathmark	1	.50	130	89	.38	.56	○	○	●	●	○	○
Marigold (Safeway)	1	.59	115	79	.51	.75	○	○	●	●	◐	◐
Cost Cutter (Kroger)	1	.42	108	77	.39	.55	◐	○	●	●	●	○
Fleece (Kroger)	1	.56	108	77	.52	.73	○	○	●	●	○	○
Marcal	2	.59	100	69	.59	.86	○	●	◐	○	●	●
A&P	2	.64	108	77	.59	.83	○	●	◐	○	●	◐
Mardi Gras	2	.69	110	75	.63	.92	○	●	◐	○	●	◐
No Frills (Pathmark)	2	.54	120	83	.45	.65	○	●	◐	○	●	○
Swansoft (Kroger)	2	.72	108	77	.67	.94	○	●	○	○	◐	○

*Copyright 1987 by Consumers Union of United States, Inc., Yonkers, NY 10703. Adapted from *Consumer Reports*, September 1987.

CHAPTER 12 Consumer Concerns

Guide to the Ratings

Listed in order of estimated quality. Differences between closely ranked models were slight.

1. **Piles.** Useful more as an identifying mark than as a sign of quality.

2. **Price per roll.** Averages of the price we paid for single rolls bought in different areas. The price of Mr. Big, available only in a three roll pack, was adjusted accordingly.

3. **Unit cost.** Shown two ways, **per 100 towels** and **per 100 square feet.** If you use paper towels primarily for small jobs, which require one or two towels at a time, the cost per towel is the more significant factor. If you want paper towels for window-washing and big jobs, the cost per square foot is more meaningful.

4. **Absorption capacity.** This was a key factor in determining the ratings order. In our tests, the best towels absorbed about four times as much water as the worst.

5. **Wet-strength.** Another key factor in determining the Ratings order. It tells you which towel should hold up in wet cleaning and scrubbing.

6. **Absorption rate.** No matter how absorbent a towel is, that quality does little good if the towel takes too long to soak up a spill, or the grease from fried chicken. We dripped water and cooking oil on the towels and clocked the rate of absorption. The thirstiest soaked up water almost instantaneously and absorbed the oil in two or three seconds. Some of the worst, needed a couple of minutes to absorb the oil.

7. **Tearing ease.** Towels that don't separate easily at the perforations are annoying. We found few problems, provided the towel holder was properly tensioned.

8. **Linting.** The lower the score, the more tiny flecks a towel left behind. Linting is especially bothersome if you use paper towels to clean windows or mirrors.

STOP & THINK

16. Is *Cost Cutter* a better buy than *Pathmark*?
17. What operational definition did the investigators use to measure linting (category 8)?
18. Which brand performed better in the wet-strength test, *Viva* or *Gala*?
19. Which brand of paper towels did *Consumer Reports* rank as the best towel?
20. How much has the cost of paper towels changed since 1987?

Comparing Ratings

Use your data from Putting Paper Towels to the Test and the *Consumer Reports* article to complete the following. Record your answers in your notebook.

1. Describe whether all, most, or none of your operational definitions match the operational definitions that the *Consumer Reports* investigators used.
2. Describe how similar your rankings of the best to worst brands are to the final rankings in the *Consumer Reports* article.

EXPLAIN reading

Why Products Fit

Recall from Unit 1 that you designed and advertised a television set. Before you actually designed the TV, you explored some of the limits of human vision that related to the design of a TV screen. Later, you learned that these limits are called *human factors* and that people design products with a diversity of human factors in mind. In the process of designing a TV screen, you set standards based on human factors. When manufacturers do not account for human factors, their products might not "fit" people very well. That is not good for consumers. If it is not good for consumers, it is not good for manufacturers either, because then people will not buy the manufacturers' products.

But what does all of this have to do with paper towels? As you read One Morning at Work, imagine that you are a paper towel designer.

One Morning at Work

Scene: *An office in New York City. All around the office are drafting boards, drawing supplies, paper, and samples of paper towels currently on the market. Four young designers are beginning the task of designing a new brand of paper towels. The designers' names are Isaac, Ros, Al, and Marie.*

Ros: If we want the product to sell, I suppose we want our towel to be the best at everything!

Isaac: It can't be the best at everything. We need to narrow it down.

Al: Yeah, if we narrow it down to being the best at some things, we can make a better towel.

Marie: Okay, then let's set some goals for our towel. What do we want our towel to be?

Isaac: The strongest!

Al: The most attractive on the shelf!

Ros: The most absorbent!

Marie: The most convenient!

Al: Hang on, let me write all that down! (He takes time to record all the ideas. He titles the list "Goals.")

Marie: Okay, any others?

Isaac: That seems like the basics.

Marie: Are there any goals on here we don't think are possible?

Ros: They all look good, but we need to know more about each one because that will make a difference in how we reach our goals.

Al: Like what?

Ros: Well, for starters, when people want a strong paper towel, what does that mean? What are they going to use a paper towel for that it needs to be so strong? I mean, who needs a strong paper towel just to dry their hands?

Al: People don't use towels just for drying hands, Ros.

Isaac: Many people use towels for cleaning.

Marie: So? I don't need a paper towel to be strong just to wipe smudges off a mirror.

Al: No, people use them for more than that. These days they expect paper towels to be like sponges. They use all-purpose cleaner and a paper towel to clean their kitchen sinks.

Isaac: Great, we'll make a towel so strong it doesn't get holes or shred when you scrub with all-purpose cleaner.

Ros: Okay, then what about attractive? What do people find attractive?

Marie: I say we stick with colors and patterns that go with things in most kitchens—like blue, green, pink, and brown.

Isaac: And let's stick with flower patterns.

Al: Good idea. Let's make a column there beside my "Goals" column.

Isaac: Call it, "Things that limit goals."

Al: (Beside "Strongest," he writes, "People use for cleaning with all-purpose cleaner." Beside "Most attractive," he writes, "Must match typical kitchens.") What do I put by "Most absorbent?"

FIGURE 12.4 So far, this is what Al has recorded.

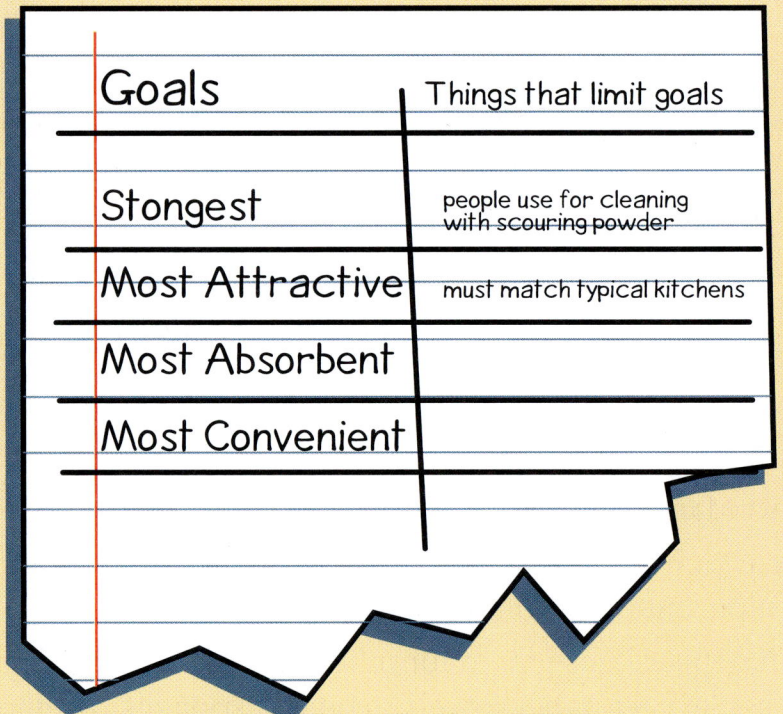

304 CHAPTER 12 Consumer Concerns

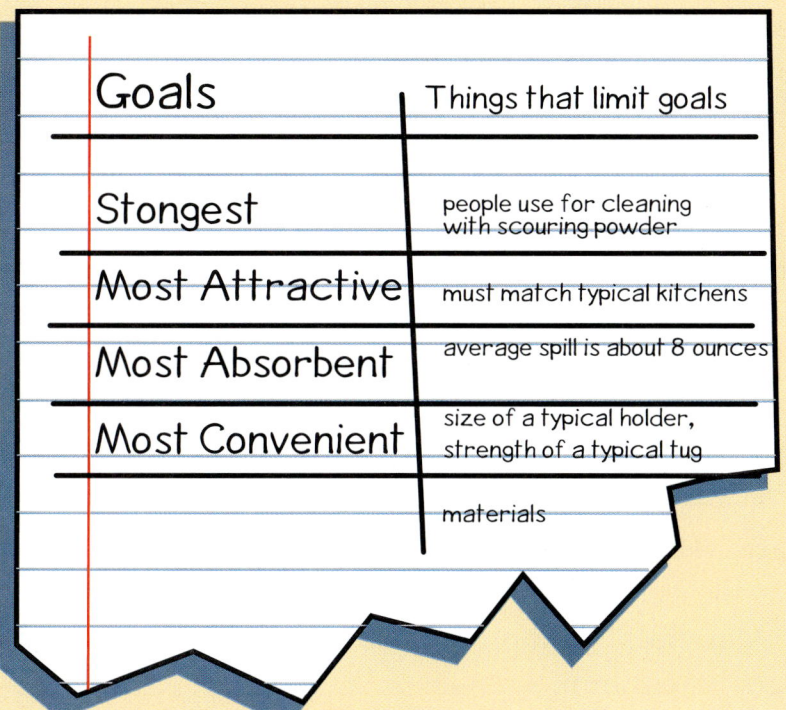

FIGURE 12.5 The two completed columns now include these items.

Ros: At my house, most of the spills happen at the dinner table when someone spills a drink, and then everyone bolts for the paper towels.

Al: Yeah, but you end up using about ten paper towels.

Marie: Then write down, "Average spill is about one full glass."

Isaac: A full glass is about eight ounces. Put "8 ounces."

Ros: What about "Most convenient"?

Marie: Well, we once bought a roll of towels that didn't fit our paper towel holder, so we just left the towels lying around the kitchen. Then we could never find them when we needed them. Finally I knocked them into the sink and ruined half the roll.

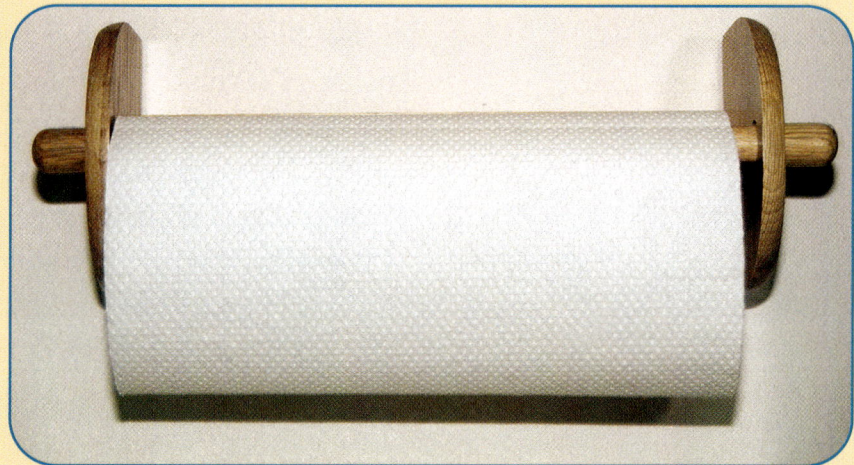

CHAPTER 12 Consumer Concerns

Isaac: Another thing that bothers me is when you're racing to wipe up a spill and you yank at a towel and run and the whole roll comes with you!

Ros: Okay, then put "Size of a typical holder" and also put "Strength of a typical tug."

Marie: You know, you guys, I'm thinking we might get carried away and need too many expensive materials for these great towels we're designing. We can design all we want, but it won't do us any good if the manufacturing department says they can't make them because they don't have enough money!

Isaac: Well, that sort of limits our goals, too. We're limited in the materials we can afford to buy. Al, write "Materials" next on the list in the "Things that limit goals" column.

Ros: So people have a big effect on the goals, but other things do, too. I never would've thought of that.

Al: (Finishes writing. He steps back to look at the two lists side by side.) Hmm . . . something's still missing. We haven't said exactly what our paper towel will do.

Isaac: Sure we have! You just didn't write it down. For "Strongest," we said that when people use it with scouring powder, it won't shred or get holes in it.

Marie: How do we do that?

Isaac: Let's weave nylon thread into each towel!

Ros: Al, I'll make another column. I think a good heading would be "Preliminary decisions." On the "Strongest" line, Ros writes, "Enough nylon thread woven into each towel so it won't tear when you use scouring powder.") And I remember what we said for absorbent. What do you think of this? (She writes, "Make the towel two thicknesses and large enough to wipe up a full glass of milk that spilled.")

Goals	Things that limit goals	Preliminary decisions
Stongest	people use for cleaning with scouring powder	Enough nylon thread woven into each towel so it won't tear when you use scouring powder.
Most Attractive	must match typical kitchens	Blue, green, gold, brown and flower patterns
Most Absorbent	average spill is about 8 ounces	Make the towel two thicknesses and large enough to wipe up a full glass of milk that spilled
Most Convenient	size of a typical holder, strength of a typical tug	Towels will tear easier at the perforated lines and the roll will be 11 inches wide.
	materials	

FIGURE 12.6 The characters end up with a chart composed of three columns like this one.

Marie: Looks good. On the "Attractive" line write, "Blue, green, brown, with flower patterns."

Isaac: And on the "Convenient" line write, "Towels will tear easily at the perforated lines," and "The roll will be 11 inches wide."

Al: How do you know that, Isaac?

Isaac: I just measured this roll of paper towels on my desk. Okay, everyone, that's a wrap. I'll take that down to manufacturing quick so they can start making the perfect towel.

As designers, Al, Marie, Ros, and Isaac are right on target. By getting together and listening to one another, they have pointed out some important ideas about design that you need to understand before you continue this unit.

To design a product, people must first identify the goals for their product. Sometimes, but not always, people phrase the goals they set for their product using general words such as "most," "best," "largest," or "smallest." Or people can state their goals as specific properties, such as "strong enough to pick up average spills." The characters' goals fit the second approach. The characters wanted their towel to be attractive, strong enough to scrub with, absorbent enough to handle a typical spill, and convenient to use. Designers, engineers, scientists, and manufacturers—among other professionals—call goals like these **criteria** ("criteria" is the plural of "criterion").

People judge a good product by whether it meets their criteria. For example, if the characters find that the towel they develop is not attractive to most people, they have failed to meet the "most attractive" criterion. Setting criteria is the first step in deciding on a product design.

In their second column, the characters listed things that will limit or affect how they meet their goals. Often, people's preferences, lifestyles, and physical or mental limits have an effect on criteria. That means human factors play a big role in determining how manufacturers will meet the criteria. Remember, considering human factors helps produce a design that "fits" people. But as Ros pointed out, other things also can affect how designers meet their criteria, and these things in turn affect the final decisions. Most often these other things have to do with the manufacturer's budget, making a profit, and available or affordable materials.

Human factors and other things that affect goals are called **constraints**. If designers tried to make the final design decisions based only on the criteria, they would reach a stopping point. At this point, they would realize that they cannot make just any decision. They are limited by the materials they can use or by certain human factors of the consumer. So determining what constraints there are on decisions is the next step in designing a product.

Only after you have decided on the criteria and determined the constraints can you focus on specific ideas about what the product will do or what the product will look like. The last step in product design is making the final decisions based on criteria and constraints. When you have gone through these steps, as the characters have, you can be fairly certain that your product will fit consumers.

connections ELABORATE

Do You Understand Criteria and Constraints?

When you design something as simple as a trash can, you go through the same steps as when you design something as complex as a computer. Meet with your team of three from the previous investigation to answer the following questions about the classroom trash can. After you agree on answers, record them in your notebook. Be prepared to share your answers in a class discussion.

1. List the criteria the manufacturers might have established for the trash can.
2. List the constraints that might have affected the design of the trash can.
3. List the decisions that the manufacturers made when designing the trash can in your classroom.

ELABORATE investigation

Part of Your Complete Breakfast

MEMO

DATE: Today, This Year
FROM: Your Editor
RE: Your latest writing assignment for
Teen Consumer's Magazine

Dear Pulitzer-Prize winning team:

This memo requires your immediate attention. This month's issue of **Teen Consumer's Magazine** goes to press in three days, and we need one last article. We want to include an article on breakfast cereals for our nutrition column. Please research this idea. You will need to test breakfast cereals and rank them from best to worst for the average teen consumer.

After you have completed your research and performed tests on the cereals, please write an article for our magazine explaining your process. Be sure to tell readers what criteria you considered and to evaluate the products accoeding to how well the manufacturers accounted for constraints. I'd like you to use the same format as you did when you wrote that article for *Consumer Reports* on paper towels. Please e-mail your article back to me and attach all diagrams and artwork.

In case you have lost yours, I have included the step-by-step procedure that all reporters for **Teen Consumer's Magazine** are supposed to follow. I expect your article in three days.

Materials for Each Team of Three:
- 1 measuring spoon
- small cups for the cereal samples
- a supply of water
- 1 tray
- any other materials your teacher provides that you decide you need for your tests
- 1 medicine dropper
- 3 craft sticks, toothpicks, or plastic spoons

Process and Procedure
1. Obtain the materials.
2. Make a list of the criteria that you feel are important for judging the quality of cereals.
    → This should be one column in a data table.
3. Make a list of the constraints that might affect the design of cereals.
    → This should be another column in a data table.
4. Decide on operational definitions to test your criteria.
    → Notebook entry: Record the definitions.
5. Perform your tests based on your operational definitions, and rate the cereals according to each of your criteria.
6. Construct a ratings guide table for the cereals you test.
    → You can refer to the ratings guide in the *Consumer Reports* article for ideas.
    → Notebook entry: Record your ratings guide.

WORKING COOPERATIVELY

Work cooperatively in your teams of three. You will need the roles of Communicator and Manager. Practice the social skill *Choose an explanation that includes the ideas of all of your teammates.* Move your desks together.

Wrap Up
Compose a 200-word article for *Teen Consumer's Magazine*. Include information such as general comparisons of the cereals, what you tested and why, how you tested things, any artwork or diagrams that you

would like to put in the magazine, as well as your test results and cereal rankings. Refer to the article about paper towels as often as you like and have the Communicator get help from other teams. Present your article to the class and describe how successful your team was in using the unit skill and in choosing an explanation that includes the ideas of all your teammates.

EVALUATE connections

Evaluating Your Understanding of Criteria and Constraints

Work cooperatively in your teams of three. The roles you will need are Manager, Communicator, and Tracker. Practice the social skill *Be open to others' ideas*. Move your desks together or sit in a triangle at a table.

Study the character situations on the following pages. Discuss them in your team. Construct a table in your notebook with a column labeled "Criteria" on the left, a column labeled "Constraints" in the middle, and a column labeled "Decisions" on the right. For each product that the characters are discussing, list all of the criteria you think the manufacturers had in mind for that product. Then list the constraints you think limited the manufacturers' decisions and the final decisions the manufacturers made. When you are finished, put your name on your paper and give it to your teacher.

Situation #1. The meal tray

FIGURE 12.7 Study this top view of what their meal trays look like as you determine the criteria, constraints, and final decisions of the designers.

Situation #2. The notebook

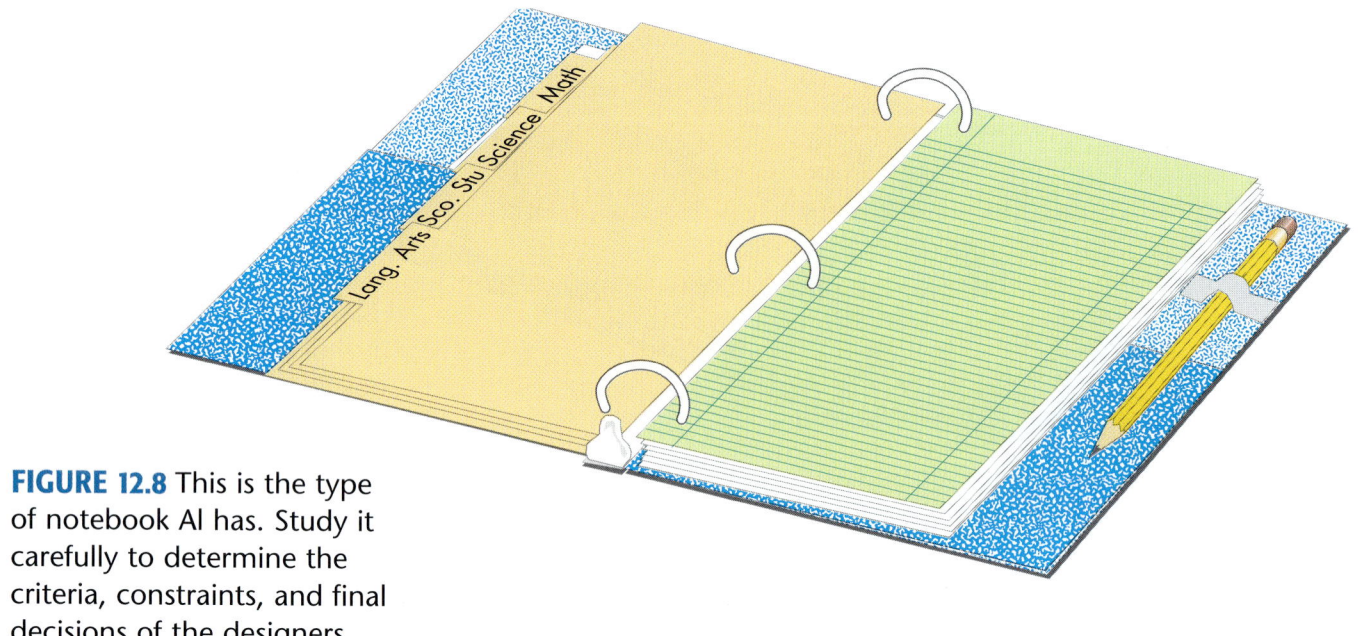

FIGURE 12.8 This is the type of notebook Al has. Study it carefully to determine the criteria, constraints, and final decisions of the designers.

Situation #3. Marie's headache

FIGURE 12.9 The inside of an empty locker at Ros's school looks like this. Study the features carefully to determine the criteria, constraints, and final decisions of the designers.

CHAPTER 12 Consumer Concerns

CHAPTER 13
Your Designing Ways

Now that you have discovered three important parts of designing products (can you name them?), you are ready to explore the design process itself. How do designers make their decisions? What steps do they take to create a product that meets the needs of the people who will use the product? What are their criteria and constraints? In this chapter, you will have an opportunity to design products by using the steps that professional product designers use. Think for a moment about the different types of boats pictured here. What are some of the decisions the designers had to make as they designed each type?

ENGAGE
- Small-Scale Boats

EXPLORE
- Is a Boat a Boat?
- Sails, Propellers, and Gas
- Anchors Away!

EXPLAIN
- Technological Problem Solving

ELABORATE
- Toys for Tots

EVALUATE
- Human Factors as a Design Constraint

ENGAGE investigation

Small-Scale Boats

Work cooperatively in your teams of three. Move your desks together or sit together at a table. Practice the social skill *Share your thoughts and ideas.* Use the roles of Communicator and Manager as well as Team Member roles.

Sometimes as you begin to explore concepts in design, it is helpful to first explore on a small scale. This is definitely the case with boats. In this chapter, you will design and build miniature boats. Before you do that, however, it will help if you know something about how boats move through the water. This investigation and the one that follows will help you explore boat propulsion.

Materials for the Entire Class:
- several tubs or sinks for holding water
- a water source
- "boat fuel" in a small cup
- 10 medicine droppers or thin-stemmed pipets

Materials for Each Team of Three:
- 1 pair of scissors
- 1 sheet of unlined, white paper
- 1 metric ruler
- 1 medicine dropper or thin-stemmed pipet

Process and Procedure

1. Watch closely as your teacher performs a demonstration.
   → Notebook entry: Record any observations you make during the demonstration.
2. Obtain the materials for your team.
   → The Manager should obtain the materials.
3. Design your version of a miniature boat.
   → You have only 5 minutes in which to design your boat. Make your boat as similar to or as

Topic: Design
Go to www.scilinks.org
Code: physical318

318 CHAPTER 13 Your Designing Ways

different from your teacher's as you would like, but you will have to use the same fuel to propel your boat that your teacher used. Your boat also should perform as well as or better than your teacher's boat. Do not put your boat in the water yet.

→ Use this opportunity to be open to the ideas of both your teammates.

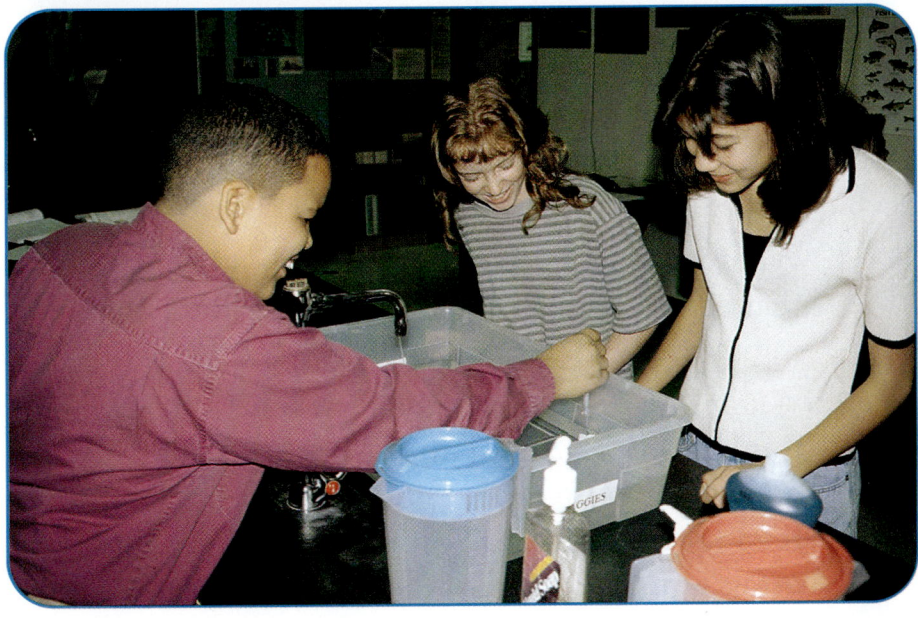

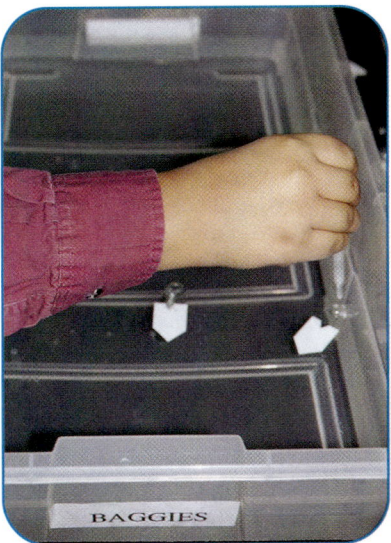

FIGURE 13.1 Remember as you design your boat that you will use the same fuel to propel your boat as your teacher did.

4. Measure your boat.
   → Notebook entry: Record its length and width.

5. Prepare a data table in which you record the length and width of the boat, a description of the shape, a description of the direction traveled, and the speed of each boat in the class. You might also choose to make a miniature drawing of your boat.
   → Ask your teacher how you will determine the speed at which each boat travels.

6. Demonstrate your boat to the rest of the class by putting your boat in the water and adding the fuel.
   → Tell the class your boat's measurements.

7. Observe what happens to your boat, as well as what happens to the other teams' boats.
   → Notebook entry: Record your observations in the appropriate columns in your data table.

8. As a team, decide which boat from all those in the class was the best, based on speed and direction.

9. Compose a statement that tells what type of miniature boat goes the fastest and what type goes the straightest.
   → Notebook entry: Record your statement.

Wrap Up

Discuss these questions as a team. You then will discuss them as a class, so be sure each of you can answer each question.

1. What were some common criteria and constraints that each team had while designing and creating a miniature boat?

2. What do you predict is the boat fuel? Why do you think it propels a miniature boat? It is okay if you do not know for sure. Think about it first and then read the Background Section that follows this Wrap Up. Now discuss the question again.

3. Rate your team on a scale of 1 to 10, 10 being the highest, for how well you used the unit skill. Then rate yourself, using the same scale, for how well you practiced the skill of sharing your thoughts and ideas.

Background

When you placed your small-scale boat on the surface of the water, the water particles below the boat and surrounding the boat were being held together by cohesive forces that resulted in surface tension. (If you do not remember these concepts, go back and review them in Chapter 11.) In addition to those forces, there were also forces between the boat's paper particles and the water particles. Those forces are called **adhesive forces**. Because the cohesive forces among the water particles were stronger than the adhesive forces between the paper particles and the water particles, the boat stayed afloat on top of the surface. Then you added a drop of soap to the back of the boat just as you added a drop of soap to the water on the penny. The addition of the soap caused the boat to suddenly race forward across the water as if it were being pushed. Why?

Think back to A Penny's Worth of Water in Chapter 11. What happened to the water on the penny after you added the drop of soap? The dome fell off the penny, right? If it were the cohesive forces in the water that held the water in a dome shape, then the soap must have done something to break apart the forces so that the water would not hold together anymore. In fact, the soap particles slipped in between the water particles, blocking the cohesive forces between the particles so that they could not hold together any longer.

While your small-scale boat was floating on top of the water, it was surrounded by water that was held together by cohesive forces between the particles. The cohesive forces of the water particles also pulled on the boat, as shown in Figure 13.2a. When you added the drop of soap to the water at the back of the boat, the soap slipped between the water particles in that area and blocked the cohesive forces. Now remember that particles in liquids move around quickly. When the cohesive forces in water particles at the back of the boat were gone, the particles could bounce around more freely, even bouncing against the back of the boat. The rest of the particles to the sides and front of the boat were still held together by cohesive forces, but the particles at the back exerted pressure at the back of the boat, pushing it across the top of the water. The rest of the water particles were still bonded cohesively. The diagram in Figure 13.2b illustrates this particle phenomenon.

We say that soap broke the surface tension at the back of the boat, just as it broke the surface tension of the water on top of the penny. We have solved two mysteries by one model at the same time. Models definitely have their place in science.

FIGURE 13.2a–b (a) First the boat floats on top of the water because the forces between the particles of the paper boat and the forces of the water particles are not as strong as the forces between the water particles themselves. (b) After you add soap, the bouncing movement of particles that are no longer cohesively bonded at the back of the boat overcome the adhesive forces at the sides and front of the boat. The water particles at the back of the boat therefore push the boat along the surface of the other water particles, which are still cohesively bonded.

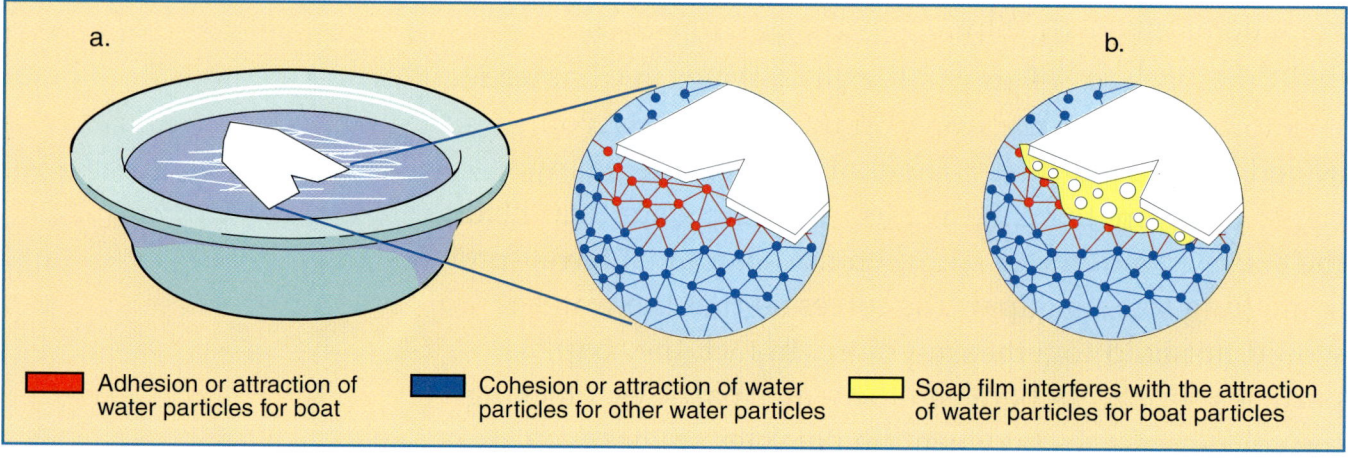

Adhesion or attraction of water particles for boat

Cohesion or attraction of water particles for other water particles

Soap film interferes with the attraction of water particles for boat particles

SIDELIGHT ON TECHNOLOGY
This Wonderful World of Boats

Technology involves people solving problems. One problem that people have worked to solve in a variety of ways is how to get people and things from one place to another—the problem of transportation. Using boats as a means of transportation is an ancient idea. In fact, boats are one of the oldest ways of transporting things and people. The idea of boats emerged about the same time as the idea of sleds. Between 5000 and 3500 B.C. (that is between 7,000 to 5,500 years ago), people began using donkeys and oxen to pull sleds of cargo. It was also at this time that people first built boats in the form of simple rafts. Later, people began to build canoes and dugouts.

People used these early boats as a means of transportation along rivers, streams, and lakes. This greatly increased the contact among people and advanced the trading of goods. It was not until people invented the sailboat in 3200 B.C. that transportation across the ocean was possible. The invention of the sailboat is one of many technologies that helped people overcome obstacles in transportation.

Try to imagine a world without boats. What are some things that might be different in your life if there were no boats?

CHAPTER 13 Your Designing Ways

Is a Boat a Boat?

A boat is a boat is a boat, right? "No," you say? Well, if you disagree that a boat is a boat, you are right in one way. There are different types of boats: sailboats, speedboats, yachts, cruise ships, battleships, canoes, rafts, motorboats, houseboats, steamboats, tugboats, freighters, rowboats, paddle boats, aircraft carriers, garbage scows, junks, oil tankers, sloops, bilanders, barques, schooners, catamarans, gondolas, sampans, sculls, sharpies, kayaks, barges, Yankee clippers, ferries, dinghies, dories, dugouts, skiffs, ketches, punts, feluccas, outriggers, galleons—get the picture? If you agree that a boat is a boat, you are also right. There may be different types of boats, but all boats are similar.

FIGURE 13.3 Even though boats come in a diversity of shapes and sizes, in what ways are they all similar?

How can a paddle boat be like a speedboat? How can a garbage scow be like a cruise ship? When you put images of these pairs in your mind, you probably see two vessels that look and act very differently. Imagine the slow, somewhat clumsy crawl of a paddle boat next to the sleek, slim figure of a speedboat cutting through the water at top speed. Or think of a garbage scow reeking of fumes in a busy harbor loaded down with a month's worth of waste next to the glamorous silhouette of a cruise ship gliding along calm waters. It may seem that these boats are not the least bit the same! Well, not in appearance, anyway. But they share these three criteria:

- First, a functional boat has to float.
- Second, a boat must stay level and not tip over in the water.

- Third, a boat must have some means by which it moves through the water.

If manufacturers did not meet these criteria when designing boats, then we would not have any boats at all! These three criteria make them boats.

Once boat builders meet these common criteria, then other criteria along with some constraints, or limits, make each boat a unique vessel. For example, a boat builder may decide to build a boat that meets the following criteria:

- The boat floats.
- The boat will not tip over in the water when it is used under normal conditions.
- The boat is engine-powered.
- The boat will provide recreation for people.

The constraints that will limit this boat builder's decisions are the following:

- the materials he or she will use,
- the depth of the water in which the boat will move,
- the fact that most engines require a fuel-powered motor, and
- the fact that many people define recreation as relaxation and fun.

The final decisions such a boat builder would make, then, are that the boat will be made of steel, the motor will use diesel fuel, and the boat will be similar to a hotel with recreational facilities. This boat builder, then, ends up with a cruise ship.

In the previous investigation, you were boat designers working with specific constraints and criteria. You were told to make a small-scale boat with one criterion: the boat was to perform as well as, or better than, the teacher's boat.

You were given these constraints:

- The boat was to be propelled by the fuel the teacher used.
- You needed to use the same materials the teacher used.
- You could take only five minutes to design and build your boat.

You did not have to worry about whether your boat would float because you already knew that it would.

You did not have to worry about whether or not your boat would tip over because the boat was flat and carried no cargo. You did, however, experiment with **propulsion**, that is, what made the boat move. Some boats have sails and move using the wind, some have steam engines, some have coal-burning engines, and some have electric engines. Some boats are powered by humans using poles, oars, or paddles. What you probably discovered in your experiments is that the size and shape of the boat affects how the boat moves.

What if someone asked you to design and build a boat that floated, did not tip over, and was propelled somehow? Could you do it? What if you also had to decide what job your boat would perform? Would it carry cargo or passengers? Would it be for recreation or work? Guess what? That is exactly what you will do later in this chapter! First, however, you will need to practice with different methods of propulsion.

Sails, Propellers, and Gas

In this investigation, you will experiment with three major forms of boat propulsion. When you are finished, you should know several ways to move objects across water without pushing them yourself. You will build boats that are more complex than the miniature boats but not as refined as real boats. You will use what you learn in this investigation to design and build the best boat you can in the next investigation.

All Parts

Materials for Each Team of Three:
- one 2.5 × 10 × 20-cm block of wood with a small hole drilled through it
- 1 stick of modeling clay
- access to a tub or sink full of water
- 1 metric ruler

Part A

Materials for Each Team of Three:
- 1 toothpick, craft stick, or new, sharpened pencil
- 3 sheets of white, unlined paper

Part B

Materials for the Entire Class:
- 1 roll of transparent tape

Work cooperatively in your same teams of three. Continue to practice the social skill *Share your thoughts and ideas.* Use the roles of Communicator, Tracker, and Manager. Create a large work area with your desks or at a lab table. You also will need a test area that includes a tub or sink full of water.

CHAPTER 13 Your Designing Ways

Materials for Each Team of Three:
- one $4 \times \frac{1}{4}$-in. rubber band
- 2 paperclips
- 1 plastic bead
- 3 push pins
- 1 paper propeller you choose from the Propellers sheet
- 1 aluminum foil pie plate or baking pan
- 1 pair of scissors
- strips of electrical tape (Get these strips when you come to the step in which you need them.)
- 3 pairs of safety goggles

Part C

Materials for the Entire Class:
- 1 roll of masking tape

Materials for Each Team of Three:
- one 10-in. round balloon
- 2 Alka-Seltzer tablets
- a water source

FIGURE 13.4 As you build your sails for your boats, experiment with different sizes and shapes.

- 1 flexible straw
- one 250-mL squeeze bottle
- 1 pair of scissors

Process and Procedure

Part A—Sails

1. Obtain all of the materials for Part A.
   → This includes materials that are common to Parts A, B, and C.
2. Make a sail for your wood block by piercing 2 holes through a piece of paper with a toothpick, craft stick, or pencil.
   → See Figure 13.5. The toothpick, craft stick, or pencil is called the mast.
3. Attach your sail to the base of your block with a piece of modeling clay.
   → See Figure 13.6.
4. Create a wind that is strong enough to sail your boat across a tub or sink of water.
5. Experiment with different sizes and shapes of paper, different masts, and different placements of the mast on the block to make the most effective sail for your boat.
   → Be sure to identify and control your variables as you experiment with the materials. Remember to be open to others' ideas.
6. Construct a data table to record your attempts: paper sizes and shapes, mast sizes, and mast positions.
   → You should include a space for recording what you observe with each change in sail, mast size, and placement of the mast.

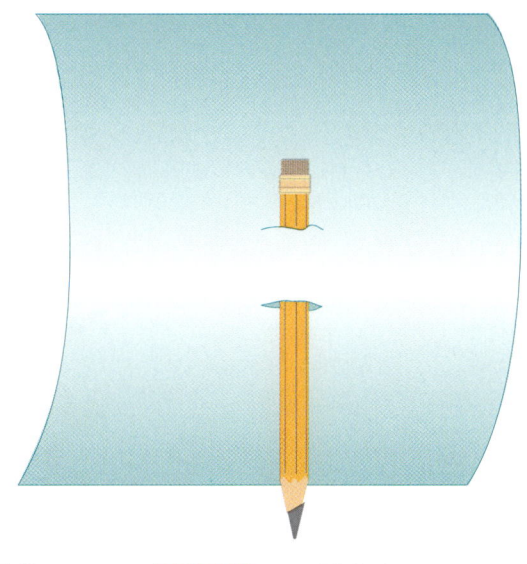

FIGURE 13.5 This is an example of what a mast and sail look like. Make sure the pencil point is down.

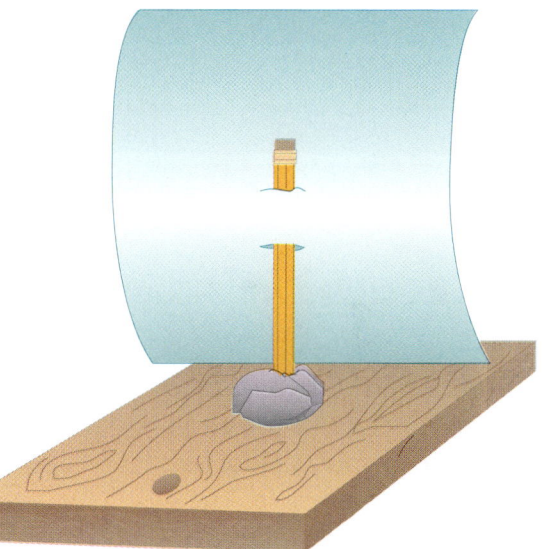

FIGURE 13.6 This is one way to attach your sail and mast to your block by using modeling clay. Make sure the pencil point is pushed down into the lump of clay.

7. Record your data in your data table.
   → Record every sail size, sail shape, mast size, and mast position that you try.
8. Put a star beside the type of sail that worked best.
9. Return all the Part A materials except those that you will use in Parts B and C.

Part B—Propellers

 For this procedure, you will need to wear eye protection, such as safety goggles.

1. Obtain all of the additional materials for Part B.
2. Trace a paper propeller onto an aluminum foil pie plate or baking pan, and cut out the propeller.
   → Choose any propeller pattern.

 Propellers cut from aluminum foil pans might have sharp edges. Be careful not to cut yourself.

3. Set the push pins into the wood block as in Figure 13.7.
   → Be sure the push pins are as far down into the block as they can be. If they are not, they can fly into the air later when you attach the rubber band.

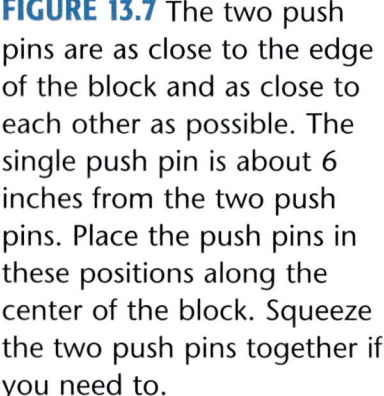

FIGURE 13.7 The two push pins are as close to the edge of the block and as close to each other as possible. The single push pin is about 6 inches from the two push pins. Place the push pins in these positions along the center of the block. Squeeze the two push pins together if you need to.

330 CHAPTER 13 Your Designing Ways

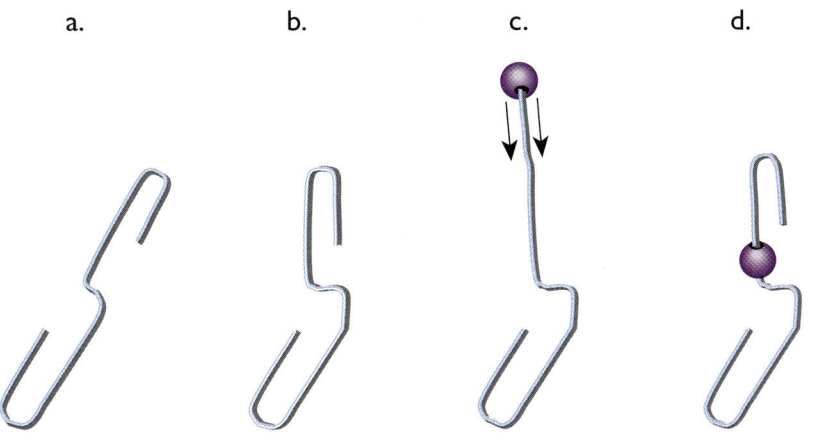

a.   b.   c.   d.

FIGURE 13.8 (a) Unbend the paperclip. Notice that one end is smaller than the other. (b) Bend the paperclip up at the small end. The small end of the clip should form a right angle to the large end of the clip. (c) Unbend the hook at the small end and slip the bead onto this end of the paperclip until it stops at the kink. (d) Bend the smaller end back into a hook.

4. Follow the steps in Figure 13.8a–d to prepare the paperclip and bead assembly.

5. Hook the small end of the paperclip through the rubber band.
   ➡ Squeeze the small end of the paperclip to keep the rubber band from slipping off.

6. Attach the paperclip assembly to the board as in Figure 13.9.
   ➡ Be sure you are wearing eye protection!

7. Twist the rubber band by winding the large end of the paperclip 30 to 35 times.

8. Release your hold on the large end of the clip to see if it spins back like a propeller.

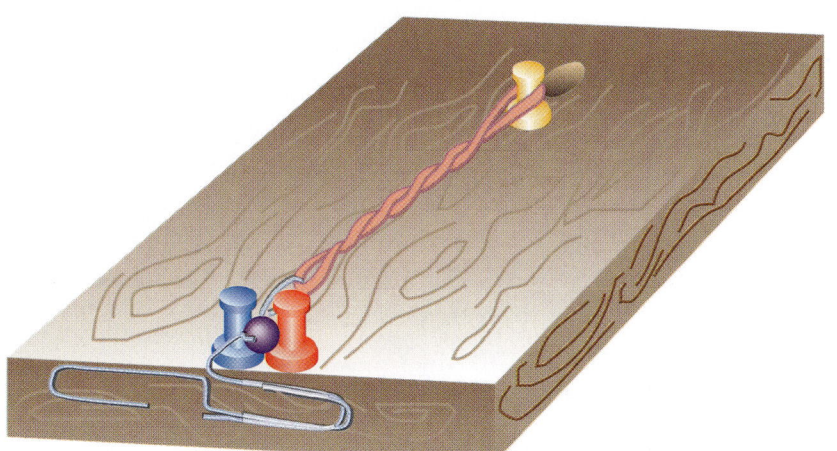

FIGURE 13.9 Attach the small hook of the paperclip onto a rubber band stretched from the single push pin. Squeeze the hook shut. Place the bead on the opposite side of the two push pins.

CHAPTER 13 Your Designing Ways

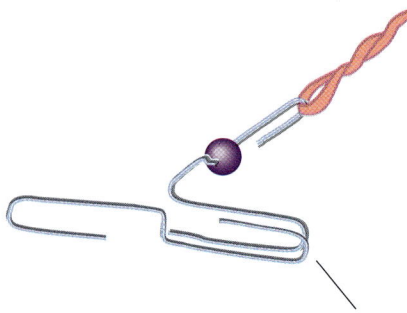

Tape these ends together.

FIGURE 13.10 Tape the small end of the new paperclip to the large end of the clip holding the bead. Try not to use too much tape.

→ If it spins like a propeller, proceed to Step 9. If not, adjust the pin, bead, or rubber band until you make it spin.

9. Tape another unbent paperclip to the small end of the paperclip as shown in Figure 13.10.
   → Use electrical tape.
10. Tape your aluminum propeller to the paperclip.
    → Use electrical tape. Match and bend the aluminum propeller on the paperclip to achieve the best fit, as in Figure 13.11.
11. Gently twist the blades in opposite directions to shape the propeller as shown by the arrows in Figure 13.11.
12. Wind the propeller and release it after you have placed the wood block in a tub or sink full of water.
13. Observe what happens.
    → Notebook entry: Record your observations.
14. Construct a data table in which to record your information.
    → Your data table should include the type of propeller, how many times you wound the

FIGURE 13.11 Experiment with the shape and twist (or pitch) of the propeller. Try different amounts of pitch and different directions of pitch.

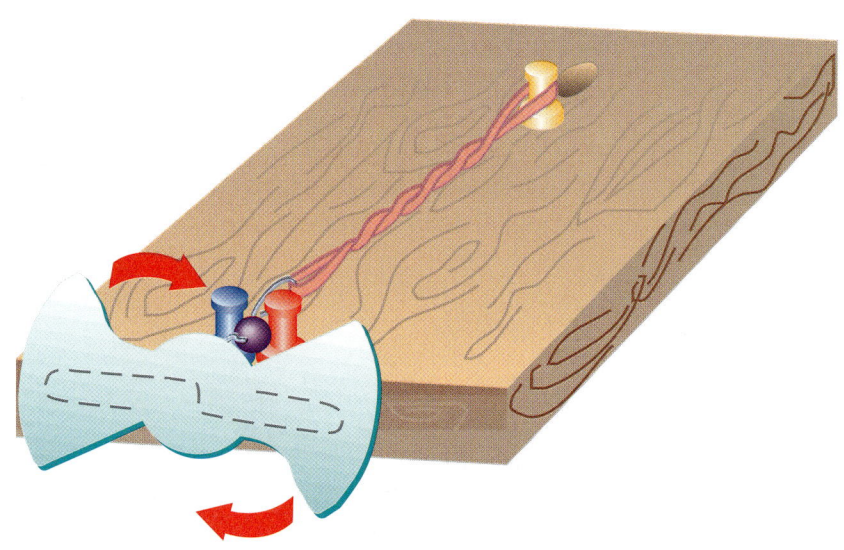

CHAPTER 13 Your Designing Ways

propeller, how many twists of the blade make one full turn, and how you shaped the propeller blades.

15. Experiment with different propellers to see which one provides the most propulsion.
   ➡ Remember to control variables as you try to determine what type of propeller works best.

16. Return the materials for Part B, but keep the materials that you will need for Part C.

Part C—Gas

1. Obtain the additional materials for Part C.

2. Blow up the balloon. Then let the air out.
   ➡ The Manager should do this and check to be sure there are no holes in the balloon. If there are any holes or leaks, obtain a new balloon.

3. Break each of the Alka-Seltzer tablets into 4 pieces.
   ➡ Have the Tracker do this.

4. Stuff all 8 Alka-Seltzer pieces completely into the balloon.
   ➡ The Tracker should do this.

5. Fit the neck of the balloon over the end of the flexible straw that is nearest the joint.
   ➡ Have the Communicator do this.

CHAPTER 13 Your Designing Ways   333

6. Wrap tape around the straw and the neck of the balloon to seal the balloon against the straw.
   → While the Communicator holds the neck of the balloon tightly in place around the straw, the Manager should wrap the straw and neck of the balloon 6 times with tape to make a tight seal between the balloon and the straw. See Figure 13.12.

7. Blow through the opposite end of the straw to inflate the balloon half-full.
   → As the Tracker blows, the other Team Members should help by listening for air leaking out of the seal. If the Tracker cannot blow up the balloon and air is leaking out of the seal, the Manager should squeeze the tape tighter around the balloon and straw or should apply more tape to make a better seal. It is important to make sure that there is no air leaking out of the straw and balloon.

FIGURE 13.12 Create a seal between the balloon and the straw by wrapping masking tape at least six times around the edge of the balloon's neck and the straw.

8. Let the balloon deflate.
9. Place the open end of the straw through the hole in the block, from the top of the block to the underside.
    → Have the Tracker do this. The balloon should be on top of the block.
10. Bend the straw so that the tail of the straw is parallel to the underside of the block.
    → The Tracker should do this. The flexible part of the straw should be in the hole.
11. Wrap a thin rubber band around the block and straw to hold the tail of the straw against the underside of the block.
    → Have the Communicator do this. Be sure the straw is not kinked in any place, especially at the flexible part.
12. Cut off about 8 cm from the tail of the straw.
    → Have the Manager do this. The entire apparatus should look like the picture in Figure 13.13.
13. Fill the 250-mL squeeze bottle with water.
    → Have the Communicator do this.
14. Squeeze water into the tail of the straw until the main part of the balloon is full of water.
    → While the Tracker turns the block down, holding it so the tail of the straw is pointing up, the Communicator should vigorously

Side View

FIGURE 13.13 This is what your gas apparatus should look like now. Notice how the straw bends in the hole of the block.

and quickly squeeze the water into the main part of the balloon. They should do this until the balloon is full of water. If water is added too slowly, the gas will form and dissipate too soon. There will be no gas left to propel the block. See Figure 13.14.

15. Remove the squeeze bottle from the straw and immediately seal the open end of the straw with your index finger.
   - As the Communicator removes the bottle, the Manager should cover the end of the straw with his or her finger as quickly as possible. Try not to let too much gas escape.

16. Lower the block into the water, balloon side up.
   - The Manager should keep his or her finger covering the end of the straw to prevent any air from escaping.

17. Wait until the balloon inflates one-third to one-half full and then uncover the end of the straw.

FIGURE 13.14 Do this step quickly. You might not have time to fill the balloon completely with water if you squeeze the water in too slowly.

→ If the balloon does not inflate with gas, you probably have a leak between the balloon and the straw. You will need to begin Part C over again if this is the case.

18. Observe what happens to the block.
    → Estimate how far the block traveled before stopping and how many seconds it took to stop.
    → Notebook entry: Record these observations.

19. Discuss with your teammates how you could make the block go farther or faster.

20. Experiment with any of the materials from this investigation to design a better gas engine.
    → Be sure to control all variables but one to ensure a fair test each time.

21. Record the observations and results of your tests in a data table.
    → Your data table should include information about the container you used for the engine, how many tablets you used, how much water you used, how far the block traveled, and how fast the block traveled.

22. Draw a star beside your entry for the engine that gives you the best results.

23. Return all of the materials to their appropriate place.

Wrap Up

As a team, decide on answers to these questions and be prepared to share your answers with the rest of the class. Record your answers in your notebook.

1. Which method of propulsion worked best for your team? Use data to explain your response.
2. Explain how easy or difficult it was for your team to share thoughts and ideas during the investigation.

EXPLORE investigation

Anchors Away!

In this investigation, you will have a chance to use what you know about boats from Small-Scale Boats, what you learned in the reading Is a Boat a Boat? and what you discovered in Sails, Propellers, and Gas. You will try to build a boat that meets specific criteria and accounts for specific constraints.

Work cooperatively in your teams of three. You will need a Manager, a Communicator, and a Team Member. In this investigation, concentrate on the skill *Be open to others' ideas*. You again will need a large work space, and you will share a sink or tub with other teams. When you are with members of another team, try to use their names.

338 CHAPTER 13 Your Designing Ways

Materials for Each Team of Three:

- Anything from the materials provided by your teacher. You may not use any other materials.

Process and Procedure

1. Read the following challenge:

   Build a boat that floats, does not tip over in the water, has a means of propulsion other than pushing it, and serves one specific purpose. Your teacher has priced the materials that you will use and will tell you how much you are allowed to spend on your materials.

2. As a team, decide what type of boat you want to build.
   → For example, decide whether you will build a ferry boat, cargo boat, cruise ship, speedboat, or any other type of boat you can think of.

3. Construct a data table with 3 columns. Label the first column "Criteria," the second "Constraints," and the third "Decisions."
   → Fill in this data table with appropriate entries as you design your boat.

4. Design, build, and test your boat in a tub or sink full of water.
   → Check your data table to be certain that the boat functions according to your goals.

5. Redesign and retest your boat until you are satisfied that your boat is the best it can be.

6. Put away the materials.

FIGURE 13.15 All boats, no matter how well designed they are, require maintenance.

Wrap Up

Prepare a presentation as described in Question 1 and complete Questions 2 and 3 with your team. Make sure that each of you can explain your answers and that each of you takes part in the presentation.

1. Present your criteria, constraints, and decisions data table to the rest of the class and show them your boat in action. Ask the class for suggestions that would make your boat even better.

2. Point out to your teammates at least one specific instance in which you noticed how someone was being open to others' ideas.

3. Discuss why concentrating on the unit skill was important for this investigation.

Technological Problem Solving

At this moment, you probably are surrounded with products of modern technology. You are probably at a desk, seated on a chair, with some sort of writing implement handy, and you most likely have your notebook beside you. The classroom itself may be furnished with a computer, a television, an overhead projector, a chalkboard, lab space with sinks, laboratory equipment, cabinets, shelves for books, a trash can, a clock, and perhaps a bulletin board.

These items are just a few of the products in your everyday life that you might take for granted. Until the last investigation, perhaps you had not thought too much about boats, either. Yet, like boats, each of the everyday products around you went through a process of design. Engineers and other designers use a design process every time they design a new object. The things in your classroom required hours of planning, engineering, and testing before the designers felt that they had met their product goals.

Your team developed its own design process during the last investigation Anchors Away! You might not have been aware of it, so this connections activity will help you identify the unique steps of your team's design process. Work on this section with your teammates.

I. Cards

- With your teammates, sort through the deck of Design Process Cards your teacher gives you.
- Discuss the meaning of each card so everyone understands what it says.
- Select the cards that best fit the steps you used when you designed your boat. (Which card seems to state what you did first, second, third, and so on? Which cards did you repeat?)
- Remove cards you did not use. Use blank cards to write in any steps that you do not find among the written cards or to copy a step you repeated.
- Create a new deck by reassembling, in order, the cards that you used and the new cards you wrote. Save any spare cards in a separate pile.

II. Flow Charts

- Arrange the cards in a pattern (like a chart, map, or diagram) to show the order of the steps in the process that your team used in designing your boat.
- Tape the cards into place on a poster board or large sheet of paper.
- Be creative as you make this display.

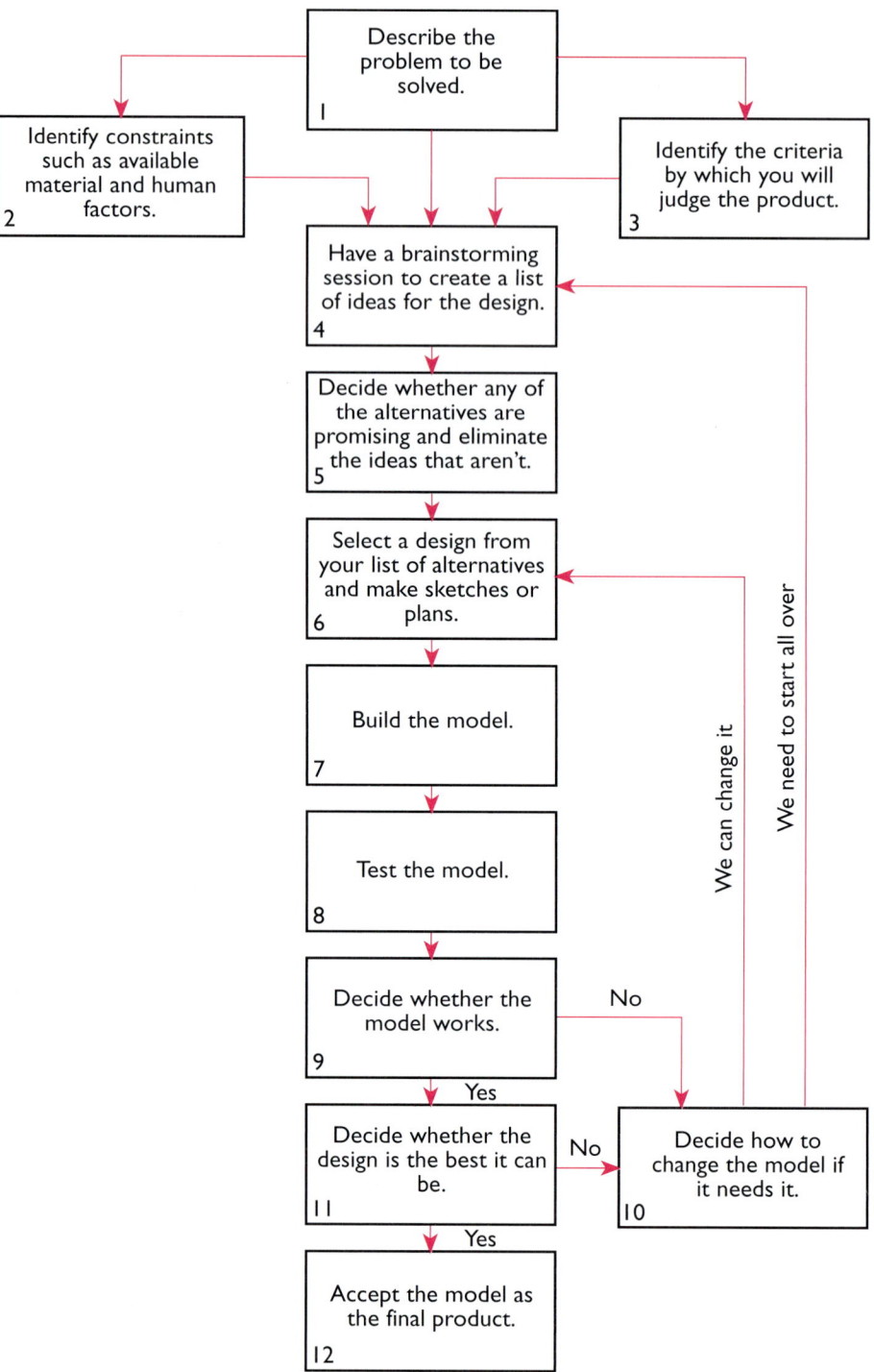

FIGURE 13.16 This is the way the characters organized their cards into a design process. They felt that this best described the process they used when they designed their boats in Anchors Away!

III. Presentations

- Present and explain your finished chart to the rest of the class.
- Listen to other teams explain their design processes. Ask questions if you do not understand something.
- Other teams' design processes might remind you of a step or two that you did but forgot about. After listening to other teams, you may want to change your chart by rearranging the cards or by adding cards from your spare cards pile.

IV. An Example of One Design Process

- Copy your team's design process chart into your notebook.
- Ask your teacher where you can store your poster.
- Review and discuss the characters' design process that follows.

FIGURE 13.17 This was the characters' first step.

FIGURE 13.18 The characters combined two cards to describe what they did next.

FIGURE 13.19 The characters conducted a brainstorming session that helped them fill in their decisions column.

344 CHAPTER **13** Your Designing Ways

FIGURE 13.20 The characters talked through their plans and drew some sketches. This helped save materials and kept them within their budget.

FIGURE 13.21 After their first model was complete, the characters tested it in water.

FIGURE 13.22 The characters' model sank, so Marie suggested ways to change their basic design.

CHAPTER 13 Your Designing Ways 345

FIGURE 13.23 The characters decided that their design was the best and decided to stop changing it.

Making Connections

Answer these questions in your notebook.

1. In the Chapter 12 reading Why Products Fit, the characters played the parts of paper towel designers. What step in their design process (Figure 13.16) did the characters reach in that scene?
2. What step in *your* design process did they reach?
3. Why didn't the characters finish their process?
4. Think back to Unit 1 when you designed the ultimate TV. Decide which of the steps in your flow chart you used when you designed it.

You and your team went through a design process just as Marie, Al, Isaac, and Ros did. As you discovered, and as the characters vividly illustrated, design is not just a haphazard series of events. It depends on people working together and following a process for technological problem solving the way your team and the characters did.

Actually, the chart of the design process that you created, also known as a flow chart, represents how designers go about creating products. Often a design process is as simple as thinking, creating, testing, and modifying. Taking the time to follow a design process means taking less time in the long run to design a product, particularly if you also know how to cooperate and work together as a group.

We have used the word "technological" quite a bit. Remember that **technology** refers to more than just products. It is also the process that one goes through to develop those products. It is the process you just went through as design engineers.

Toys for Tots

Now that you have created your own diagram for technological problem solving (your flow chart of the design process), you are ready to use it to design something else. In this investigation, you will have an opportunity to design a toy.

Materials for Each Team of Three:
- Any materials that you choose from those provided by your teacher. You may not use any additional materials.

Work cooperatively in your teams of three. Use the roles of Communicator, Manager, and Team Member. Practice the social skill *Share your thoughts and ideas.* Gather with your team at a table or at your desks. You will need plenty of work space.

CHAPTER 13 Your Designing Ways

Process and Procedure

Part A—The Task

1. Read the following challenge:

   Use your team's process for technological problem solving to design and construct a toy. The toy should meet the following criteria:
   - interesting to a child who is between 3 and 8 years of age, and
   - safe to use.

   ➡ See Figure 13.24. You can revise your chart any time as you design your toy. Make changes on your chart in your notebook.

2. With your teacher and the rest of the class, discuss the following:

   a. What are two constraints that your team has for designing a toy?

   b. How can you determine the human factors of children 3 to 8 years old that will affect the design of your toy?

3. As a class, develop a chart that all teams will use to record the human factors that will affect the toy design.

FIGURE 13.24 In order to pass the "safe to use" criterion, your toy must meet all the points on this checklist.

A safe toy is one that won't cause injury when a child plays with it. A designer has made a safe toy if the toy meets the following standards:

 It has no small parts that young children could swallow easily, inhale, or choke on. (Do not use balloons or parts of balloons in your toy design.)

 It has no strings that are 12 inches or longer and has no strings that together can form a circle 14 inches around.

 It has no sharp points on edges (e.g., protruding nails, metal edges, or glass).

 It contains no toxic paint or other chemicals.

 It contains no flammable materials.

4. In your teams, construct the chart of human factors in your notebooks.
   → Make the chart neat and leave enough room to record any additional human factors that will affect your design.

5. As a class, develop an operational definition for "interesting to a child."
   → That is, decide how the teams in your class will measure whether a toy is interesting to a child.

Part B—The Process

1. Choose an age for which your team will design a toy.
   → It can be for any age from 3 through 8. Pick an age that you will not have too much difficulty researching.

2. Devise a plan to conduct research on the age you chose.
   → Is each of you also practicing the role of Team Member?

3. Have the Communicator check your plan with the teacher.
   → Inform your teacher of the age you chose and how you will collect information from children of that age.

4. Conduct a brainstorming session to determine a list of things that children in that age group might find interesting.
   → Would your age group prefer a toy that is a challenge like a puzzle or a game or something creative like an art project or building kit?

CHAPTER 13 Your Designing Ways

5. Complete the chart of human factors. Follow these steps to determine the human factors for the age group you chose:
    a. Agree on the characteristics of the children who are going to play with your toy. For example, are the children large or small? Will they be active or quiet when playing with the toy?
    b. Make a list of the human factors for which you want to gather information.
        ➤ Some possible factors to consider are hand and finger size, strength, and color preferences.
    c. Select several children who will provide you with the information you need.

    The children could be brothers, sisters, cousins, or friends of one of your team members. If you have an elementary or preschool nearby, you might gather data from the students there if your teacher has arranged this opportunity for you. Make sure that you obtain permission from the parents, guardians, or teachers of the child or children that you plan to interview or measure.

    d. Only gather information for the human factors that affect your design project.
        ➤ Remember, the children are doing you a favor!
    e. Keep the identity of the people you interview or measure confidential. Tell them that they will remain anonymous.
6. Construct a data table that lists the criteria, constraints, and decisions that will be important in the design of your toy.
7. Fill in the columns in your table.

➡ Do this as you design your toy according to your own design process. Remember to adhere to the criteria, constraints, and human factors that you listed previously. Also remember that you are also limited by your operational definition of "interesting." Finally, you may use only the materials that your teacher provides.

8. Review your toy with your teacher to make sure that it passes all the safety criteria.

9. Return to the group of children who helped you determine the human factors for your chart. Have them test your toy to see if your design has fulfilled all of the criteria.

➡ Before allowing children to play with your toy, be sure it has passed a safety inspection in your team and then a safety inspection by your teacher.

FIGURE 13.25 Be sure your toy has passed a safety inspection before any children play with it. Each time you modify your toy, you will need to inspect it and test it again.

10. If your toy does not fulfill all of the criteria, return to your design process and modify the toy until you are confident that it is better.
    → Each time you modify your toy, you will need to return and test the toy again with the same children as before.
11. After your toy meets all of the criteria, find another team that is at the same step in this investigation and proceed to Part C.

Part C—Does It Meet the Constraints?

1. Trade toys and charts of human factors with another team.
2. Manipulate or play with the other team's toy to determine whether or not the toy accounts for all of the human factors the team listed on its chart.
3. Meet with the team who evaluated your toy and whose toy you just evaluated to discuss your toys.
   → Take turns discussing the good qualities of each other's toy and, in a respectful manner, point out any human factors in the chart that you think the other team did not account for. Help figure out ways to account for these human factors.
4. Present your toy to the rest of the class.

Wrap Up

Discuss these questions with your team, and then write your own answers in your notebook.

1. Explain any changes to the design process you made in your chart.
2. Copy your new chart onto a clean page in your notebook.
3. Use your own rating or grading system to evaluate your progress in using the skill of sharing your thoughts and ideas.

SIDELIGHT ON CAREERS

Ergonomics—The Science of Human Factors

Ergonomics is the branch of science that studies how people interact with the products, equipment, environments, facilities, and procedures they use at work and in their homes. Ergonomics specialists apply information about human factors—capabilities, limits, characteristics, and behaviors—to the design process to make sure that things are safe, comfortable, and easy for people to use.

Specialists in ergonomics work in many different industries, including aerospace, computers, communications, and consumer products. They also work in various businesses, such as general research and development, health care, management consulting, and architectural services. Ergonomics specialists have bachelor's, master's, or doctoral degrees in a variety of subjects, including psychology, engineering, computer science, and industrial design.

Designing even commonplace products like a toothbrush can involve ergonomics. Not long ago, a major toothbrush manufacturer hired an ergonomics company to design an improved toothbrush. Specialists at the ergonomics company discovered that no human factors research had ever been used in toothbrush design, so they began an extensive study. First, they read about dental care and interviewed dentists. Next, they distributed a questionnaire concerning dental care habits to 300 adults and analyzed the information they received. They then began collecting measurements of hands, teeth, and mouths of consumers. To find out how users handle a toothbrush, the design team studied films of people brushing their teeth. They looked at how people held and moved the brush and how much time people spent brushing different parts of their mouths. From these studies, the design team determined the criterion that the toothbrush handle should make the brush easy to maneuver. A constraint they identified was that the handle could not be too wide to fit into the standard bathroom toothbrush holder.

After they examined the results of laboratory studies of plaque removal, the team designed two prototype (model) toothbrushes and manufactured enough for testing. Test subjects (the people trying the toothbrushes) compared the two prototypes to two common toothbrushes that were available already. Both prototypes

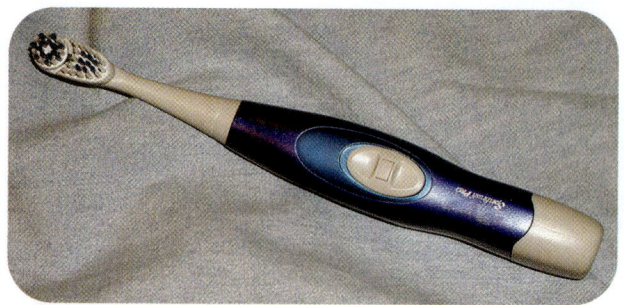

removed plaque better, but the test subjects preferred the bristle head of one prototype and the handle design of the other. So the design team combined the features into a single product, the Reach toothbrush.

EVALUATE  connections

Human Factors as a Design Constraint

Work on this section in your cooperative teams of three.

You have now had a chance to design toys for young children and to evaluate how well other teams designed toys. As a class, you listed things that were important to consider when designing and building toys. No doubt many of the following human factors are on your list:

- Physical characteristics—eye to hand coordination, height, hand size, and muscle strength of children of different ages. These physical constraints limit the ability of a child to manipulate and play with various features of a toy. A constraint is something that sets limits on what you can do.

- Mental characteristics—reading level, math skills, and logical-thinking skills of children of different ages. These mental constraints limit the ability of a child to understand and figure out the operation of a toy or game.
- Behavioral characteristics—for example, small children are likely to put objects in their mouths. These behavioral constraints limit the ability of some children to play safely with a toy.

Physical, mental, and behavioral characteristics are usually the three types of human factors that design engineers try to account for as they design products. Each of these human factors sets a limit on what the designers can do. Your teacher will tell you how you will work to accomplish the following two tasks.

1. Break down your list of human factors into three separate lists: physical, mental, and behavioral. Write the lists in your notebook.

2. Compose a paragraph that summarizes what this chapter was all about and what you learned. Write the paragraph in your notebook and be prepared to share your paragraph with the class.

FIGURE 13.26 Which age child would play on this playground?

SIDELIGHT ON TECHNOLOGY
Lego Success Story

In the United States, nearly two-thirds of all homes with children under age 15 have a Lego set. In Europe, the figure is even higher. How did a company that began more than 70 years ago making wooden toys become such an international success?

The story begins with Ole Kirk Christiansen, a carpenter in the Danish village of Billund, who started a wooden toy company in the 1930s. He started out by making wooden cars, trucks, trains, airplanes, and animals. He continued to add toys to his line, and soon he was making wooden building blocks. He named his toy workshop Lego, from a Danish phrase that means "playing well."

After World War II, when manufacturers began using plastics, Ole Kirk made a plastic baby rattle. In 1949, Lego produced its first plastic bricks. A toy trade magazine in Denmark warned that "plastic will never take the place of good, solid wooden toys." Fortunately Ole Kirk ignored its warning.

Ole's son Godfried Kirk Christiansen, known as GKC, was the one who developed the Lego system. GKC had an inspiration. He established criteria that he thought would mean certain success for toys he designed. One criterion was that the toy should be useable year-round. Another was that the toy should appeal to both boys and girls. GKC thought of one particularly clever criterion—that the more of the toy that people owned, the greater its value.

GKC searched through Lego's hundreds of products trying to find a toy that met his criteria. The toy he decided on was a plastic interlocking brick. The company packaged an assortment of the little plastic bricks, and this new Lego system quickly became the company's best selling product.

Not long after the company introduced the Lego system, it came out with a second line called Duplo. Duplo bricks are eight times larger than Lego bricks. This makes them easy for younger children to handle and too big to be swallowed. And, when children get older, they can use their Duplo bricks with Lego bricks because the two sets fit together.

In 1960, a fire destroyed Lego's wood toy factory, and after that the company concentrated on making only plastic bricks. Today the Lego company sells its bricks in more than 115 countries. The Danish town of Billund is now home to three Lego factories and Legoland Park, a theme park that features exhibits of Lego constructions. A 50-foot-high replica of Mount Rushmore stands in the park. It is made of 1.5 million Lego bricks and 40,000 Duplo bricks. A replica of the port of Copenhagen required three million bricks and two years of labor by eight model makers. Except for Copenhagen, Legoland is Denmark's biggest tourist attraction.

CHAPTER 14

Why Are There So Many Products That Do the Same Thing?

In the last chapter, you learned about design and technological problem solving. You learned that most designers use a process that includes thinking, testing, and modifying.

Why are there so many different products on the market that serve the same purpose? Think about screws, for example. The basic function of screws is to hold things together. But why are there so many different types of screws? Now think of how many different types of telephones or watches there are.

In this chapter, you will explore the relationship between design and diversity. You will study how criteria and constraints affect the diversity of products. Then you will be able to propose an answer to the question, Why are there so many products that do the same thing?

ENGAGE • One Problem, Different Decisions

EXPLORE • Designing with Shapes

EXPLAIN • Similarity and Diversity in Designs

ELABORATE • Ideas That Fly

EVALUATE • Explaining Design Diversity

359

ENGAGE investigation

One Problem, Different Decisions

Think about this question: How many ways can there be to open a can or bottle, hold hair in a ponytail, or attach two pieces of paper together? In this investigation, you will analyze several different products that are designed to do the same thing.

Materials for the Entire Class:
- the products your teacher provides

Process and Procedure

1. Observe the products your teacher displays.
   → Notebook entry: Record your observations. Write descriptions of the products including the names of the items, their general purpose, and how they each accomplish their purpose.

2. With the rest of the class, discuss the questions your teacher asks you.

Your teacher will conduct a short demonstration. You will work as a team with all of your classmates to answer the questions your teacher poses.

FIGURE 14.1 How many ways are there to design cell phones?

CHAPTER 14 Why Are There So Many Products That Do the Same Thing?

Wrap Up

Write an answer to the following in your notebook.

1. List two other products that have the same goal but differ in how they look or how they accomplish the goal.
2. Describe some of the different designs for the products you listed in Question 1.

Designing with Shapes

When architects and computer engineers design buildings or machines, they use basic building blocks, such as bricks or computer chips. The end product depends on the purpose or the function of the building or the computer. It also depends on the availability and cost of materials. In this investigation, you will design various things using different shapes as building blocks and different numbers of those shapes. Then you will compare your designs with the designs from other groups.

Work cooperatively in your teams of three. Use the roles of Manager, Communicator, and Tracker. Concentrate on the unit skill *Disagree with the idea, not the person*. To help you create numerous designs as a team, work at your desks or at a table. Clear a space so that you each have equal access to a large piece of construction paper.

FIGURE 14.2 As you look at these photographs and those that follow, think about what criteria and constraints the designer of each structure had in mind.

Materials for Each Team of Three:

- 1 copy of each of the following:
  Art-1
  Furniture-1
  Robot-1
  Art-2
  Furniture-2
  Robot-2

- shapes, as many copies as you need
- 9 sheets of construction paper, any color
- 1 glue stick or bottle of school glue
- 3 pairs of scissors

Process and Procedure

Part A—7-Piece Designs

1. Obtain the materials.
2. Cut out the shapes from the Art-1 page.
   → This page has 7 shapes.
3. As a team, decide how to arrange your shapes to design a work of art.
   → Your main objective is to arrange the shapes in a design that is pleasing to the eye. You must use all 7 shapes.
4. After you agree on an artistic design, glue the shapes in that design onto a piece of construction paper.
   → Label the design with the Team Members' names, the date, and the word "Art."
5. Cut out the shapes from the Furniture-1 page.
   → This page has 7 shapes.
6. As a team, decide how to arrange the shapes to represent furniture in a family room or living room.
   → You must use all 7 pieces.

7. After you agree on a design for a family or living room full of furniture, glue the shapes in that design onto another piece of construction paper.
   → Label the design with the Team Members' names, the date, and the word "Furniture."
8. Cut out the shapes from the Robot-1 page.
   → This page has 7 pieces.
9. As a team, decide on a way to arrange the pieces to create a robot.
   → You must use all 7 pieces. Remember that if you disagree, disagree with the idea, not the person.
10. After you agree on a design for a robot, glue the shapes in that design onto another piece of construction paper.
    → Label the design with the Team Members' names, the date, and the word "Robot."
11. Compare each design you made with 7 shapes with each design from the other teams.
    → Notice the major similarities and differences among the designs.
12. Answer these questions in your notebook:
    a. When you compared your designs with other teams' designs, which looked most like one another: the art designs, the furniture designs, or the robot designs?
    b. Which designs looked the most different from one another?
    c. In what way did the procedures you used make your designs look similar to, or different from, the other teams' designs?

CHAPTER 14 Why Are There So Many Products That Do the Same Thing?   363

FIGURE 14.3 In designing furniture for a doll house, the big difference is scale.

Part B—15-Piece Designs

1. Using the 15-piece design pages Art-2, Robot-2, and Furniture-2, design a work of art, a family room or living room full of furniture, and a robot.
   - In Part B, there are 15 pieces. As in Part A, you must use all of the pieces (no more and no less) for each design. You cannot repeat any of the 7-piece designs previously created by any of the teams.
2. Glue your designs onto pieces of construction paper as before.
   - Label each design with the Team Members' names, the date, and what the design represents.

3. Compare your designs with those of other teams.
   → Notice the similarities and differences among the designs from team to team.

4. Compare your team's 15-piece art designs with your team's 7-piece art designs.
   → Notice the similarities and the differences.

5. Compare your team's 15-piece and 7-piece furniture designs.

6. Compare your team's 15-piece and 7-piece robot designs.

7. Discuss the following questions as a team.
   → Notebook entry: Record your team's answers.

   a. From team to team, did the designs show more or less diversity
      - in the 15-piece art designs compared with the 7-piece art designs?
      - in the 15-piece furniture designs compared with the 7-piece furniture designs?
      - in the 15-piece robot designs compared with the 7-piece robot designs?
   b. How do you explain these patterns of diversity?

FIGURE 14.4 Compare the designs of these two pieces of furniture.

CHAPTER 14 Why Are There So Many Products That Do the Same Thing?   365

➡ Notebook entry: Record your team's answers.

a. From team to team, did the designs show more or less diversity
- in the 15-piece art designs compared with the 7-piece art designs?
- in the 15-piece furniture designs compared with the 7-piece furniture designs?
- in the 15-piece robot designs compared with the 7-piece robot designs?

b. How do you explain these patterns of diversity?

Part C—Unlimited Designs

1. You again will design a work of art, a room full of furniture, and a robot. This time, however, use the shapes on the page titled Shapes.
   ➡ Use as many shapes for each design as you choose, but do not repeat any of the designs your team previously created in Parts A or B.

2. Glue your designs onto construction paper.
   ➡ Label them with the Team Members' names, the date, and what the design represents.

3. Compare your team's new designs with the new designs of other teams.

4. Compare your team's new designs with your team's 15-piece and 7-piece designs.

5. Discuss the following questions as a team.
   ➡ Notebook entry: Record your team's answers.

   a. Among the teams in the class, did the new designs follow the same pattern of similarity and diversity as the 15-piece or 7-piece versions? Why or why not?

   b. In general, was there more diversity across teams with these new designs than with the 15-piece designs? Why or why not?

FIGURE 14.5 Windows come in many shapes and designs. What reasons might the designers have had for each design?

Wrap Up

Write answers in your notebook to these questions after you discuss them as a team. Prepare to share your answers and observations from all three parts of this investigation with the rest of the class.

1. Think of a reason that this investigation did not require you to make a work of art with seven pieces, then a room full of furniture with 15 pieces, and then a robot with an unlimited number of pieces.

2. In your teams of three, compose a paragraph that summarizes what you discovered in this investigation. You should include the following points in your discussion:

CHAPTER 14 Why Are There So Many Products That Do the Same Thing?

- How do more detailed criteria affect the similarity or diversity of designs that different people produce?
- How do constraints limit the similarity and diversity in those designs?

3. Prepare to share your paragraph with the rest of the class.

EXPLAIN  reading

Similarity and Diversity in Designs

When you look at a cluster of buildings, a parking lot full of cars, a shelf full of books, or other groups of products within a particular group, they might look very similar. There are times, however, when products within a particular group look very different.

FIGURE 14.6 Study the diversity in the design of these doors. What reasons might the designers have had for each design?

CHAPTER 14 Why Are There So Many Products That Do the Same Thing?

Take athletic shoes, for example. If you went into a shop that specializes in basketball shoes, you would see different brands of shoes; but in general, they all would look basically the same and have the same basic features. If you went into a shop that specializes in a variety of athletic shoes, however, you might see a wide range of shoes that resemble each other much less. Low impact aerobic shoes look different from basketball shoes, which don't look much like running shoes, which do not resemble tennis shoes. Yet we still think of all those shoes as part of a group of products that we call athletic shoes.

What is it about designers and the design process that accounts for the degree of diversity in groups of products? How much diversity you observe really hinges on the decisions that designers make during the design process. Because designers make many decisions as they develop a product, design is a creative process.

STOP & THINK

1. What are some of the decisions a designer has to make when designing a product?
2. Complete the following statement by filling in the blanks using the words "more" or "less." The more decisions a designer is able to make, the _____ creative he or she can be with the design of the product, and the _____ unlike others of its kind that product will be.

Think for a moment about the investigation Designing with Shapes. In that investigation, you used shapes to design a number of things. When you could design anything that looked pleasing to the eye, your teams probably developed a wide diversity of designs.

When you used the shapes to design a robot, however, the class designs were probably more similar to each other than the art designs were. Also your furniture designs were probably more similar to other teams' furniture designs. This pattern is apparent because the procedure directed you to use the shapes for a more specific purpose when you designed a robot or furniture than when you designed art. Art also has a purpose, but it has fewer limitations than a robot or furniture.

FIGURE 14.7 Backpacks look similar to each other because their function is specific and the criteria are exact.

In general, the more specific you are in defining the function of the product you design, the more similar that product will be to other products with the same function.

Describing the function of the product is the same as listing the criteria for the product. So, the more exact you are about the criteria of the product you design, the more similar that product will be to other products with the same criteria.

For example, if you ask an automotive engineer to design a vehicle and do not list any specific criteria for the vehicle, you would likely get a transportation machine of almost any type. If you ask the engineer, however, to design a vehicle with these two criteria—it must transport six people and have enough cargo space for six suitcases—you probably will get something close to a station wagon, a minivan, or an SUV.

One short statement makes this point clear: *Criteria limit the diversity of a design.* In the previous example involving car design, when the criteria became more specific, the engineer had fewer choices to make. In other words, the more criteria there are for a product's design, the less creative the designer can be.

Criteria, however, are not the only aspects of design that limit a designer's creativity. In Chapter 12, you learned that both criteria *and* constraints affect the final product decisions a designer makes. Think back again to Designing with Shapes. You probably noticed this general trend in your team's designs: the more pieces you were allowed to use for each design, the more creative you could be.

Recall from Chapter 12 that materials are a design constraint. The use of only seven pieces constrained

CHAPTER 14 Why Are There So Many Products That Do the Same Thing? 371

your designs much more than when you could use as many pieces as you wanted. The more pieces you had, the more creative you could be. The more constraints you have, then, the less creative you can be, and the less diverse your product will be. We could expand our previous phrase to say, *Criteria and constraints limit the diversity of a design.*

Making decisions involves creativity. The more choices you can make for your product, the more creative you will be in the design. When you have fewer choices, your creativity is limited, and your product has a better chance of looking like someone else's that was designed for the same purpose.

3. Go back to the previous Stop and Think and check your answer to Question 2. If you feel that you should change it, do so now. Explain the statement as you now have completed it.

4. Think back to One Problem, Different Decisions. Working individually decide whether the products your teacher showed you could be any more diverse and still function the way they are supposed to. Then decide whether the designs were limited strongly by criteria and constraints or whether the designers were able to make many creative decisions. Justify your decisions. Write your decisions and justifications in your notebook. Your teacher might call upon you to present them to the class.

investigation ELABORATE

Ideas That Fly

If you have ever gone to an air show or visited an airport, you probably noticed that airplanes come in many shapes and sizes. Why do you think there is such a great diversity of airplanes? It has something to do with the idea you just read about: *Criteria and constraints limit the diversity of a design.*

In this investigation, you will work in teams to design airplanes. You will have to adhere to specific criteria and constraints. When you are finished, you will compare your team's airplane design with those of other teams.

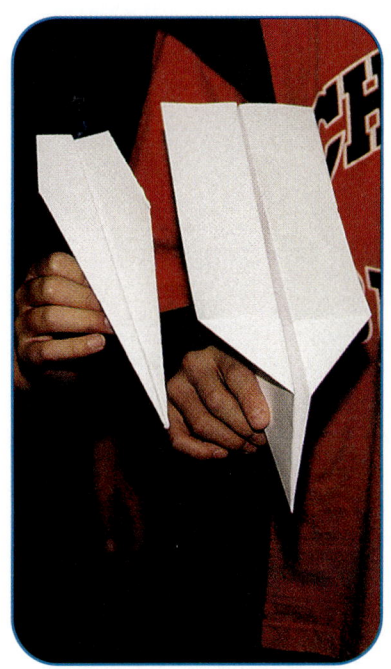

Part A

Materials for the Entire Class:
- 1 roll of masking tape
- 1 roll of transparent tape

Materials for Each Team of Three:
- 10 sheets of $8\frac{1}{2} \times 11$-in. paper, unlined
- 2 paperclips
- 1 meter stick or metric measuring tape
- 3 pairs of safety goggles

Part B

Materials for Each Team of Three:
- any materials from Part A
- 3 sheets of construction paper
- 1 bottle of school glue or 1 glue stick
- 10 rubber bands
- 3 craft sticks

Work cooperatively in your teams of three and use the roles of Manager, Tracker, and Communicator. Because you will be flying paper airplanes and will be in and out of your seats, use the social skill *Stay with your team*.

CHAPTER 14 Why Are There So Many Products That Do the Same Thing?

- 10 sheets of $8\frac{1}{2} \times 11$-in. paper, unlined
- any other materials your teacher provides
- 3 pairs of safety goggles

Process and Procedure

Part A—The Challenge

1. Read each of the following sets of design criteria and constraints.

2. Predict which design will lead to the greatest class diversity and which will lead to the least class diversity.

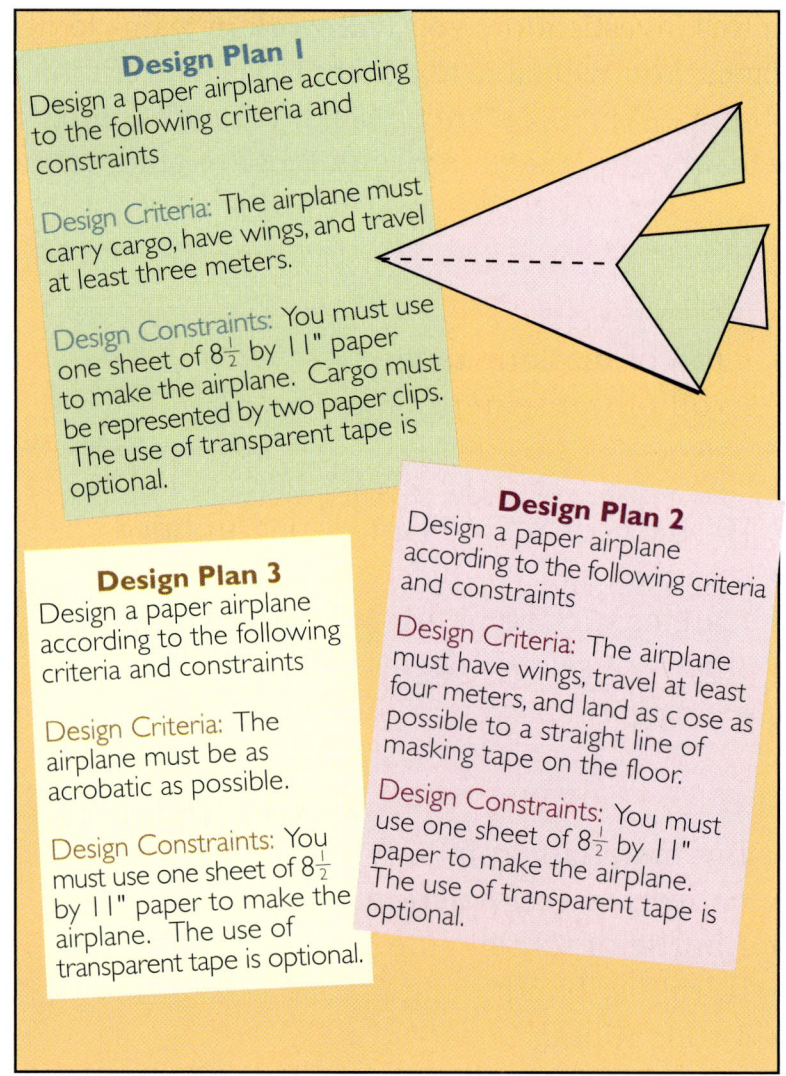

Design Plan 1

Design a paper airplane according to the following criteria and constraints

Design Criteria: The airplane must carry cargo, have wings, and travel at least three meters.

Design Constraints: You must use one sheet of $8\frac{1}{2}$ by 11" paper to make the airplane. Cargo must be represented by two paper clips. The use of transparent tape is optional.

Design Plan 2

Design a paper airplane according to the following criteria and constraints

Design Criteria: The airplane must have wings, travel at least four meters, and land as close as possible to a straight line of masking tape on the floor.

Design Constraints: You must use one sheet of $8\frac{1}{2}$ by 11" paper to make the airplane. The use of transparent tape is optional.

Design Plan 3

Design a paper airplane according to the following criteria and constraints

Design Criteria: The airplane must be as acrobatic as possible.

Design Constraints: You must use one sheet of $8\frac{1}{2}$ by 11" paper to make the airplane. The use of transparent tape is optional.

→ Think back to Designing with Shapes and to the reading Similarity and Diversity in Designs. Justify your predictions and write your predictions and justifications in your notebook.

3. Share your predictions with the rest of the class.
4. Check your predictions by designing each type of paper airplane.
   → Use the following steps to design your plane.
   a. Follow your flow chart of the design process from Chapter 13 as you design and build your plane. Revise your chart if necessary.
   → Notebook entry: Record the decisions you make at each step.

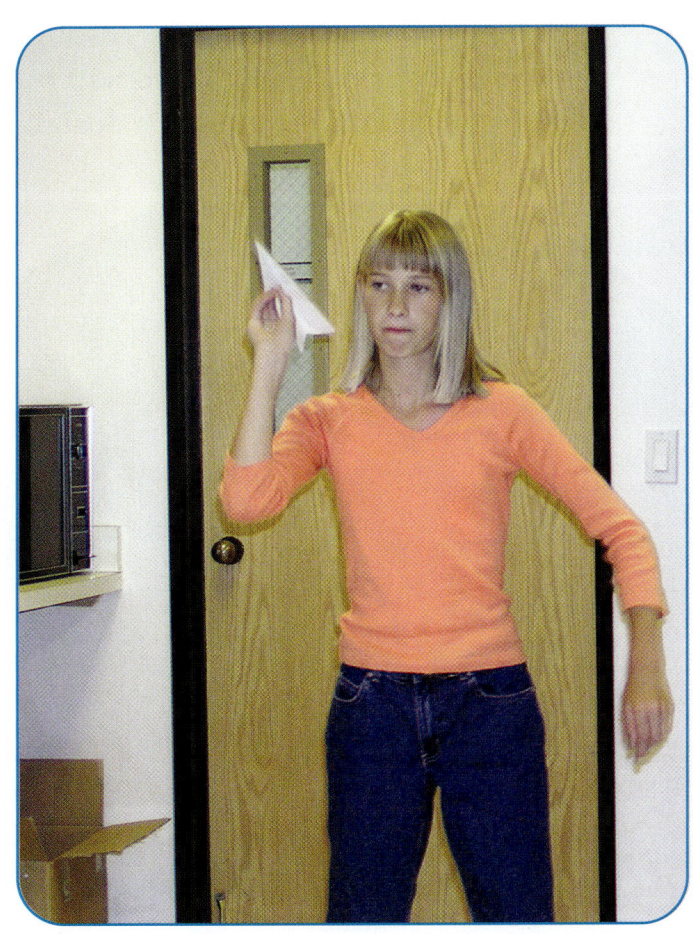

b. When you are satisfied that you have designed and built the best possible airplane for each design, put it in the area your teacher specifies.
→ There will be separate areas for Design 1, Design 2, and Design 3. Be sure you put your names on your plane for the upcoming contest.

c. Each person on the team should be able to explain the process that your team used to design your plane.

 You must wear your safety goggles for Step 5.

5. Participate in each design contest according to your teacher's instructions. You will have a separate contest for each design.
→ Before you send your plane through the air, your teacher will select 1 person from your team to explain your team's design process.

FIGURE 14.8 How much diversity is there in the design of kites?

Wrap Up: Part A

Complete the following tasks and answer the questions as a class.

1. Compare the planes from Design Plan 1 to the planes from Design Plan 2, the planes from Design Plan 2 to the planes from Design Plan 3, and the planes from Design Plan 1 to the planes from Design Plan 3. Which pair is the most different? Which pair is the most similar?
2. Where did you notice a greater diversity—among the planes of a specific design or among the planes from different designs? How do you explain the diversity you observed?
3. Do your observations verify your earlier predictions?

Part B—Free Designs

1. As a class, choose the criteria from Design Plans 1, 2, or 3 to use for designing another plane.
2. This time use only this constraint: The main body of the plane must be paper.
    - ⟹ You can use any type or size of paper and any other materials that you wish. Every team should use the criteria from Design Plan 1, 2, or 3 according to the class decision.
3. Design your plane by following your flow chart for the design process.
    - ⟹ Notebook entry: Record all the decisions you make. Make changes to your flow chart if necessary.
4. Once you are satisfied that your plane meets the criteria that the class chose, put it aside for the flying contest.
5. Take turns presenting and flying your team's plane as your teacher directs.

 Wear your safety goggles and follow your teacher's instructions as you fly your planes.

Wrap Up: Part B

Answer the following questions as a class.

1. How much diversity did you observe among the planes you and your classmates just flew, each of which was designed with the same criteria? Explain your observations.

2. How would you rate your team's ability to stay together as a group: excellent, good, fair, poor?

Explaining Design Diversity

In the investigations Ideas That Fly and Anchors Away! you gained much experience designing products. Now use this experience to convince someone who knows nothing about design that criteria and constraints limit the diversity of a design. Work individually and use any of the methods below to accomplish your task.

- Write a paragraph and read it to the class.
- Produce a radio or television commercial. (You can ask other students to join you.)
- Write a play and present it to the class. (You can ask other students to join you.)
- Create a pictorial diorama.
- Use a computer and graphic arts software to create a display or slide show.
- Create an advertising brochure complete with photographs or sketches.

FIGURE 14.9 Think about the four items pictured here in sequence. Can you trace the development of ideas?

CHAPTER 14 Why Are There So Many Products That Do the Same Thing?

CHAPTER 15

Masters of Design

Consider yourselves designers! In this chapter, you will review what you have learned about technology and design. You also will have a chance to use your knowledge to evaluate something at your school and to redesign it for a different purpose. This will be your chance to change your school! You also will redesign your school environment to make some aspect of it more accessible to a person with a physical challenge. Finally, you will have an opportunity to design a classroom newspaper. The articles that you write will help you and your teacher evaluate some of what you have learned so far this year about science and technology.

EXPLORE
EXPLAIN
- Let's Talk Technology—Again

ELABORATE
- Evaluating Your Environment
- Enabling the Physically Challenged

EVALUATE
- A Science and Technology Gazette

EXPLORE EXPLAIN

reading

Let's Talk Technology—Again

Technology is a complicated sounding word, but by now you should be familiar with the topic. After all, you have designed boats, toys, works of art, furniture, robots, and airplanes. Now consider exactly what you have learned about technology.

First, you learned that product designers must set goals for their product. Their goal is to make their product the best at accomplishing a particular job. Such goals are called criteria. After designers set their criteria, they must determine which factors limit their ability to create the perfect product. Limiting factors are called constraints and usually have to do with types of materials, cost, time, budget, and human factors. Once designers have identified their criteria and constraints, they make a series of decisions that lead to the development of a product.

FIGURE 15.1 What criteria and constraints do you think the designer of this skurfer had to consider?

1. What criteria did you and your classmates use to evaluate and rank paper towels?
2. What criteria did you and your classmates use to evaluate and rank breakfast cereals?
3. List the constraints you think the designers of paper towels and breakfast cereals worked with when they designed their products.

In Chapter 13, you used a process to design boats and toys. Your process allowed you to decide, build, test, and modify your product in much the same way designers do when they make a product.

 4. Review your flow chart of the design process. How did this chart help you design toys and paper airplanes?

In Chapter 14, you explored how criteria and constraints affect the creative process of design. You can account for the similarities and differences you see in many products by the number of criteria and constraints the designer used, and the choices the designer made. The more choices designers can make, the more diverse products tend to be.

 5. Consider the criteria and constraints you used when you designed art, robots, and furniture in Designing with Shapes. How did these criteria and constraints affect your designs?

Now, you will learn how comfortable you are with technology by completing the following two activities. These activities will test the limits of your design skills, so be sure to draw on all of your previous experiences and knowledge.

connections ELABORATE

Evaluating Your Environment

Everything around you—desks, chairs, lockers, drinking fountains, gym equipment, the library, and the lunchroom—was designed by someone or by a group of people. How well did the designers do? Did they account for the human factors of middle school students? In this

CHAPTER 15 Masters of Design

FIGURE 15.2 How well did the designers of this drinking fountain account for human factors?

connections activity, work in your teams of three to evaluate the design of something in your school environment. Follow these steps in conducting your evaluation. Then report your findings to the class.

1. In your team, conduct a brainstorming session to choose a play, work, rest, or eating environment in your school to evaluate. Possibilities include gym equipment, a study area in your library or classroom, a portion of the lunchroom, a commons area in your school, a locker room, or a waiting area in the office or health room.

2. Evaluate the environment that you have chosen. Your evaluation should answer the following questions and include other things your team thinks important.

   a. What criteria did the designers have in mind? (Although you cannot be certain about these criteria, you can make some good guesses.)

   b. What constraints were the designers working with? (Again, although you cannot know all of the constraints, such as the exact cost of materials or labor, you can make some guesses about them. You should be able to figure out many of the constraints by pretending you had to design the environment.)

c. How well did the designers accommodate various human factors?

d. How might you improve the designs?

e. If you were in charge of the school, would you buy similar equipment or look for something new? Why?

3. Prepare a report for the class. Include your answers to the questions in Step 2. Each Team Member should write a portion of the report and take part in the presentation to the class.

Enabling the Physically Challenged

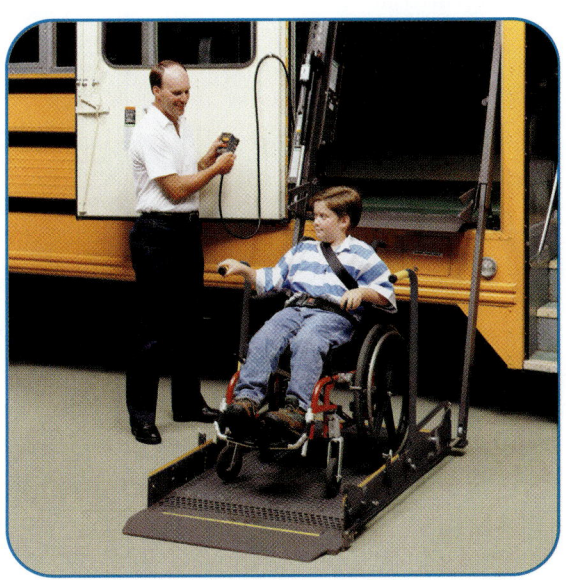

You probably have seen products that help physically challenged people use environments that would otherwise be inaccessible to them. Close your eyes for a few moments as your teacher guides you through two situations in which you imagine that you are physically challenged and trying to cope in an environment that does not account for your needs. Now stop and imagine . . .

In this investigation, you will redesign the environment you chose in Evaluating Your Environment to make it more accessible to physically challenged students.

Working Cooperatively

Remember that all teammates should fulfill the duties of a Team Member. In addition, use the roles of Manager, Communicator, and Tracker. *Be open to others' ideas* and if you disagree, *Disagree with the idea, not the person.* You will first meet at your desks. Then your team will work in the environment you will be redesigning.

Materials for Each Team of Three:
- large sheet of poster board or large drawing paper
- 6 markers, assorted colors

Process and Procedure

1. As a team, conduct a brainstorming session to decide on the physical challenge that you want to accommodate in the environment you chose in Evaluating Your Environment.
   → This could be one that someone at your school has or one that a friend or relative has. It might be one that you have.
2. Conduct your research of the challenge you choose.
   → If you or any Team Member is unsure about how to conduct research, read How To #6, How to Conduct a Research Project.
3. Return to the environment you evaluated in Evaluating Your Environment.

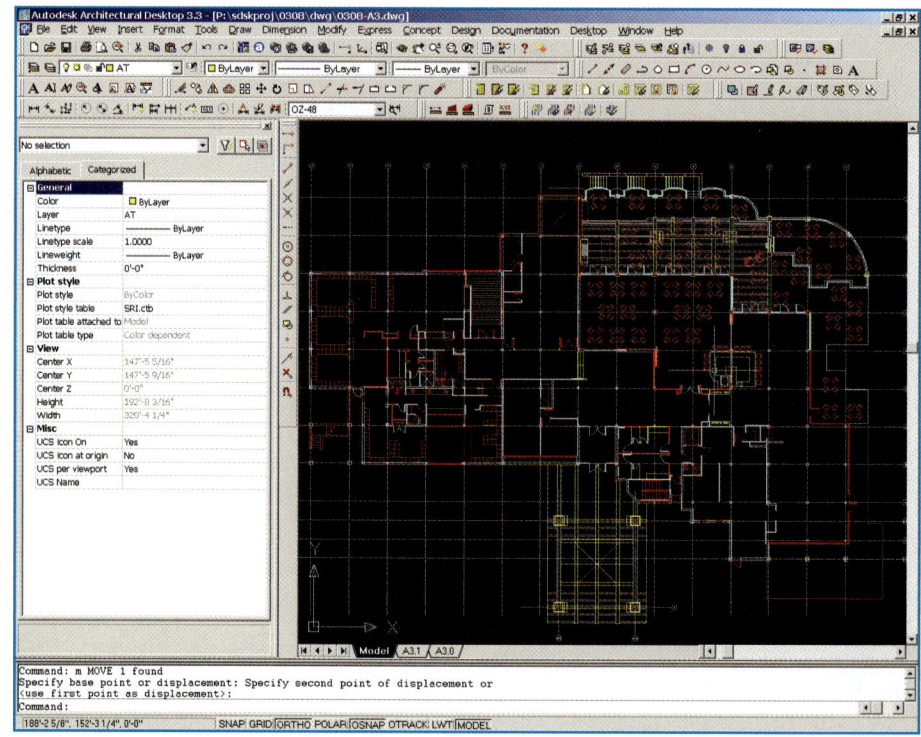

FIGURE 15.3 Architects and engineers often use computer programs and blueprints when they are designing or redesigning an environment.

CHAPTER 15 Masters of Design

4. Make a list of the human factors that you now need to consider to accommodate a student with the physical challenge you researched.
   → Notebook entry: Record your list.

5. Redesign the environment to accommodate the challenge that you chose.
   → Use the same criteria and constraints that you identified when you evaluated the environment before. Also consider the human factors you identified in Step 4. The Tracker should make sure that your team follows the flow chart for your design process.

6. Sketch your final design on a piece of drawing paper or poster board.
   → Notebook entry: Use your notebook for any of your rough drafts. Take your time with your final sketch.

Wrap Up

Complete the following as a team.

1. Present your design to the class. In your presentation, identify the changes you made in the environment and explain why you made those changes. As before, all Team Members should participate in the presentation. After you finish, ask the class for suggestions about how you can modify your design to better accommodate the physical challenge that you chose.

2. Discuss how well you think your team has worked together during this unit and how well your Team Members used the unit skill *Disagree with the idea, not the person*. Compose a page-long summary about how you know whether you have been successful at cooperative learning so far this year.

EVALUATE connections

A Science and Technology Gazette

Imagine that you are a writer for your school newspaper. Recently the editor has asked you to design and develop a special edition called *The Science and Technology Gazette*. The purpose of this special edition is to help you, your teacher, and the school community evaluate what you have learned so far about science and technology.

The task has two important parts:
- the design of the newspaper and
- the writing of the articles.

FIGURE 15.4 In the 1950s, newspaper publishers used linotype typesetting machines such as this one. The linotype was invented in the late 1800s.

The Corbis-Bettmann Archive

As you consider the design of the newspaper, create a table for criteria, constraints, and decisions. As you create your table, think about the components and design of newspapers with which you are familiar. Record this table in your notebook. Have your teacher review your table before you begin writing articles.

Your newspaper must include the following articles:

1. A lead article titled "Science Is a Way of Knowing about the Natural World."
2. Another major article titled "Technology Is a Way of Adapting to the World."
3. An article that presents the biography of a scientist or technologist who is studying or has studied something about diversity or limits.
4. An article that describes a career in science or technology.
5. A science fiction story that begins, "It is the year 2100. It has been many decades since broadcast systems and television technology took an enormous leap forward. Let me tell you what it is like now." You should include some of your own ideas about the *ethics* of broadcast systems and television technology. Ethics means telling right from wrong.
6. An advertisement that describes the design process for a product.
7. An editorial on cooperative learning and your own learning style.

In addition, you may include other articles about things that you have learned. You also may include art, photographs, or a comic strip.

After you have completed your gazette, ask two classmates to review it for you. Then revise it and prepare a final version to share with your teacher and your classmates.

SIDELIGHT ON TECHNOLOGY

Putting Human Factors to Use for the Physically Challenged

You probably have seen signs designating parking spaces for the handicapped. The signs, a profile of a human figure in a wheelchair, reserve close-in parking for physically challenged drivers or passengers. It is easy to assume that a car parked in a handicapped person's parking place is there because one of its passengers is physically challenged. But many times it is the driver who is challenged and sometimes in a wheelchair. Have you ever wondered how a person in a wheelchair can drive a car? Thanks to modern technology, people in wheelchairs do drive.

One important criterion in the design of vehicles for people who use wheelchairs is "ease of access." The designers have to assume that the driver will be alone when entering the car. The driver must get from the wheelchair into the car and store the wheelchair without help. The manufacturers decided to

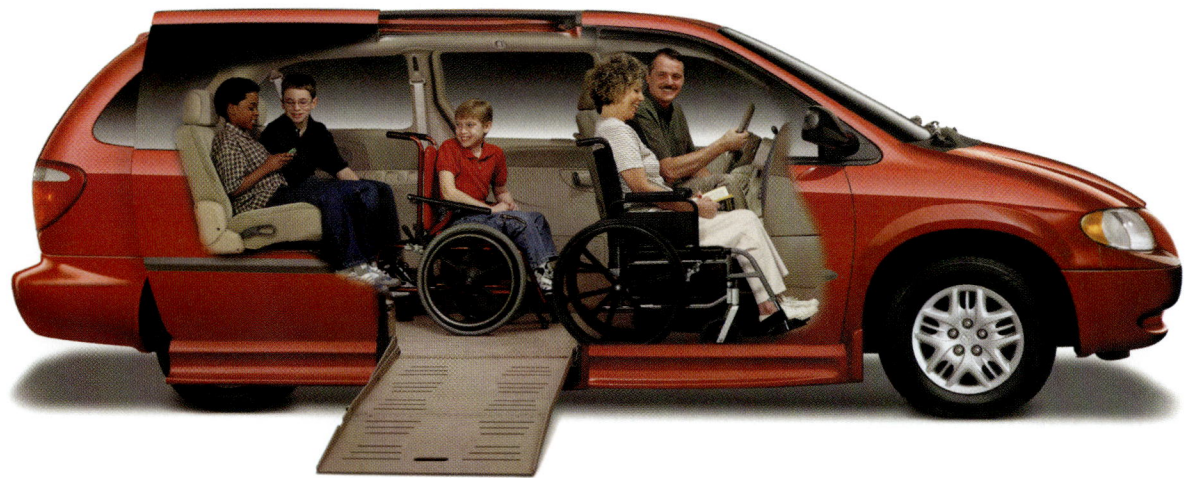

Ease of access and safety are important design criteria whether the physically challenged person is the driver or the passenger.

make these cars with two doors rather than four so that the doors can be extra long. They also make the doors at least 36 inches high (the height of most wheelchairs). This gives the driver enough room to get in and then pull in the wheelchair.

One important constraint in designing a vehicle for the physically challenged is the extra space required for the hand controls needed to drive the car. Sometimes the driver cannot use his or her legs. Based on that constraint, manufacturers decided to make these vehicles full-sized rather than compact-sized.

Other decisions the manufacturers made based on criteria and constraints include the following: an automatic transmission and power steering for easier handling, power brakes so a person can apply them easily by hand, multi-adjusting power seats so the person can move the seats by pressing a button, cruise controls so the person does not need to press the accelerator arm all the time, and power door locks and windows so the person does not have to lean over to operate them.

The manufacturers base all these decisions on the associated human factors. The American Automobile Association publishes a book for physically challenged drivers. This book describes how drivers can find out where to purchase special equipment and learn about existing design modifications. Trying to accommodate the diverse needs of people is one important use of technology.

Sidelight on Careers

Science Nonfiction

So you wrote a science fiction story. Many authors before you have done the same. Some authors specialize in science fiction. Some authors who wrote science fiction never realized that the things they imagined might come true. Consider the following examples:

- In the 1950s, Ray Bradbury's book *Fahrenheit 451* told of large video screens, high-speed automobiles, interactive TV sets, and the government burning books it did not approve of.
- In the 1800s, Jules Verne wrote books titled *Twenty Thousand Leagues under the Sea* and *Journey to the Moon* in which he told of adventures in underwater ships and giant artillery shells that could fly to the moon.
- In the 1940s, George Orwell's book *1984* told of a society in which the government secretly placed listening devices in

buildings. Using the devices, the government could listen in on what people were doing or saying.

Do any of these science fiction stories ring true? If you read these books and science fiction works written by other authors, you might find more examples of science fiction turned nonfiction. What did you write about in your science fiction story? Were you careful? You never know. You might have written science nonfiction!

UNIT 4
LIMITS OF ENERGY IN SYSTEMS

Energy is something that you experience every day. You have it. You use other systems that have it.

If you think back on your science classes, you have probably already had some experience exploring energy. What is energy? Where does it come from? In this unit, you will begin to answer such questions. What you learn might change how you think about and use energy.

In this unit, you will continue to examine diversity and limits, but this time with a focus on energy. In Chapter 16, you will learn how to recognize clues that tell you energy is involved in a system. In Chapter 17, you will explore ways that technological systems can help you harness one energy source—the Sun. In Chapter 18, you will expand your understanding by exploring other energy technologies. Finally, in Chapter 19, you will explore your personal choices for wise energy use. All of these chapters will focus on the question, How can we make the best use of energy?

Chapter

 **16** Exploring Energy in Systems

 17 Solar Sources: Energy from the Sun

 **18** Energy Benefits and Costs

 19 The Power to Choose

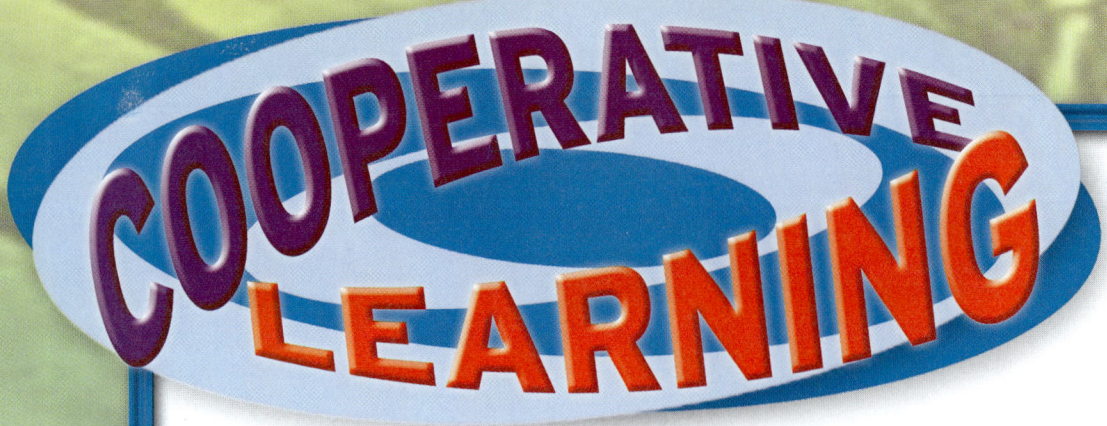

The characters are discussing the Unit 4 social skill *Choose an explanation that includes the ideas of all teammates.* After three units of cooperative learning, you too might realize how difficult it is to come to agreement in your team. This social skill is not really about agreement, however. It is about communication. Take a moment to discuss how this skill depends on communication. Then construct a T-chart for this unit skill.

Al raised another issue about the activity skills. You now have a good idea of what cooperative learning is all about. With many of the cooperative activities, you have had the opportunity to evaluate yourselves on the use of specific social skills. In this unit, you will use what you know about how you work to decide on the skills you should practice. Take a moment to try to list all of the activity skills you have practiced in Units 1 and 2. Then add some skills that you might have thought of on your own, or that you can think of now, that you consider important.

Now come to a class consensus about what skill you will concentrate on for the first chapter of this unit. Decide how you will evaluate yourselves on your use of this skill.

As you proceed through this unit, you will have the chance to choose a skill just for your team to practice. This will allow you to personalize cooperative learning for your team.

CHAPTER 16
Exploring Energy in Systems

In previous units, you explored how things are different, why they are different, and the technology that allows things to have similar functions, but look different. In this chapter, you will investigate the role energy plays in systems that change. You will use skills that you have developed in previous units to construct an operational definition of energy. This definition will help you address the question, What is energy?

ENGAGE
EXPLORE
- What Do You Already Know about Energy?

EXPLAIN
- Thinking More about Energy

ELABORATE
- Heat In, Heat Out
- What If Energy from the Sun Were Blocked?

EVALUATE
- Tracing the Flow of Energy

ENGAGE EXPLORE

investigation

What Do You Already Know about Energy?

You experience energy every day, but have you ever observed it directly? Energy is an abstract idea that has many concrete manifestations. This investigation gives you an opportunity to explore energy using a variety of different systems. First, we ask, What is a system? Do you know? How would you explain it to your teammates? A **system** is a set of things so connected or related that they form a whole. So what are some examples? Lots of things are systems. For instance, a cell phone is a system, and so is a television, and a computer. Even the CD-ROM drive is a system; it's a system within a system. The systems you are about to explore will allow you to experience energy long enough to observe some of the changes it can cause.

WORKING COOPERATIVELY

Work cooperatively in your teams of three using the roles of Manager and Communicator. In addition to the unit skill, practice the skill that your class agreed upon. Work at a table or with your desks pushed together so that you can hear and see all Team Members.

FIGURE 16.1 Look at this collage. What type of energy does each system contain? What do you already know about each system?

400   CHAPTER 16 Exploring Energy in Systems

FIGURE 16.2 These pipes are used to transport hot water from a geothermal energy source. Does this system contain energy?

Materials for the Entire Class:
- 10 systems that contain energy

Materials for Each Team of Three:
- 3 pairs of safety goggles

Process and Procedure

Part A—The Social Skill
1. As a team, discuss the social skill you will practice.
2. Decide on the specific strategies you will use to practice the skill.
   → Notebook entry: Record these strategies.
3. Create a T-chart for this skill, if appropriate.

Part B—Describing Energy
1. Observe the 4 systems that your teacher will demonstrate to you.
   → As you observe these systems, try to gather as much evidence as you can to infer that energy is present.
   → Notebook entry: Record the evidence and inferences you collect.

FIGURE 16.3 What type of energy helps to move these boats?

FIGURE 16.4 What evidence indicates that energy is present in this system?

2. Work with your teammates to complete the following description of energy.

   If a system has energy, it _____.
   → Notebook entry: Record your description in your notebook and be prepared to explain it to the rest of the class.

3. Put on your safety goggles, tie back your hair if it is long, and clear all papers off your table or desks.

4. Examine the system assigned to your group.
   → When you operate these systems, follow the safety guidelines your teacher established.

    Remember to wear your safety goggles at all times.

5. Answer the following questions about your system.
   a. What are the parts of the system?
   b. What evidence indicates that energy is present in your system?
   c. What is the input to the system?
   d. What is the output from the system?
   e. What evidence indicates that the system changed the input into the output?
      → It is acceptable to guess at the answers to these questions if you are unsure.
      → Notebook entry: Record your answers.

6. Make a class presentation of the evidence you used to determine whether energy was involved in your system.
   → In your presentation, include your answers to the questions in Step 5.

7. As you watch other teams present their systems, make notes about any additional observations that might help you revise your original description of energy.

Wrap Up

Discuss these questions with your teammates. Write your own answers in your notebook and be prepared to contribute them to a class discussion.

1. Use your method of evaluation to rate your use of the social skill.

2. Think about your first description of energy. After observing several different types of systems, how would you modify your description?

3. Make a list of six systems that contain energy. For each of these systems, describe the evidence of energy that you would expect to observe. (These systems should be different from any of the ones your class used during this investigation.)

4. Does the evidence that you and your classmates gathered suggest that there are different types of energy? Explain your answer.

EXPLAIN reading

Thinking More about Energy

Earlier we asked, What is a system? Think about some examples of systems. Now we want you to think about energy in a system.

To complete the statement, "If a system has energy, it _____." you had to decide what evidence indicates that energy is present in a system. Perhaps someone in your class completed the statement with "makes noise." Other students might have said "moves." Both of those are good ways to complete the statement. Yet, does either one define what energy is? Not exactly.

You can see a flash of light, you can feel that something is hot, and you can hear a loud bang. All of those are evidence that energy is present in a system. Yet you cannot touch or hold the energy itself. Instead, you observe a change occurring and infer from that change that energy is present. For example, you can hold a lightbulb, but you cannot hold the light that it gives off.

FIGURE 16.5 What inferences can you make about the energy in this system?

1. What types of evidence might indicate that energy is present in a system?
2. How many of your five senses are involved in the description of energy that you proposed in the previous investigation?
3. Is it possible to recognize the presence of energy without using any of your five senses? Explain your answer.

Topic: Energy
Go to www.scilinks.org
Code: physical404

404   **CHAPTER 16** Exploring Energy in Systems

Chances are that in the previous investigation you used many of your senses to gather evidence about energy in systems. Your description might have included such observations about energy as "has heat," or "grows," or even "moves." When you completed the statement "If a system has energy, it _____" with these observations, you were actually creating an operational definition: If a system has **energy**, it has the ability to do something or it *is* doing something. You can use this operational definition to determine whether a system contains energy.

FIGURE 16.6 A ball moving through the air is an example of kinetic energy. Kinetic energy is the energy of movement.

4. According to the operational definition, "If a system has energy, it has the ability to do something or it *is* doing something," what evidence might indicate that energy exists in the following systems?
   a. 404A softball pitcher throws a softball through the air.
   b. Water flows over the top of a dam.
   c. Wood logs burn in a campfire.

Systems can contain many different forms of energy. Consider the example of wood logs burning in a campfire. Using your senses, you might detect heat and light—two forms of energy—coming from the burning logs. Another form of energy you might observe is the pop and crackle of the burning logs. That form of energy, sound energy, is released by the burning logs. You can detect all of these forms of energy—heat, light, and sound—using your senses.

5. What are other examples of systems that involve light energy, heat energy, or sound energy?

Topic: Forms of Energy
Go to www.scilinks.org
Code: physical405

CHAPTER 16 Exploring Energy in Systems

You can detect other forms of energy using your senses. Consider the example of the softball pitcher throwing a softball. You can see the ball changing its position as it moves through the air. This is a type of mechanical energy called kinetic energy. **Kinetic energy** is the energy of movement. Energy that an object has because of its position is called **potential energy**. Water behind a dam has potential energy. When it flows over the dam, its potential energy is converted to kinetic energy.

People have learned to harness the energy of moving water to grind grains such as wheat and corn. For example, people used to build flour mills next to dams (see Figure 16.7). These flour mills were powered by a large waterwheel. The waterwheel turned a series of notched wheels called gears. When one gear turned, the notches transferred the mechanical energy from that gear to the next gear. The series of gears was connected to a large grinding stone. As the gears turned, the stone ground the grain into flour. In this system, the turning of the waterwheel, the grinding stones, and the connecting gears all involve mechanical energy.

But where does this energy come from? What is its source? What makes the gears move? In the example above, the source of the energy used to grind the wheat and corn is the moving water.

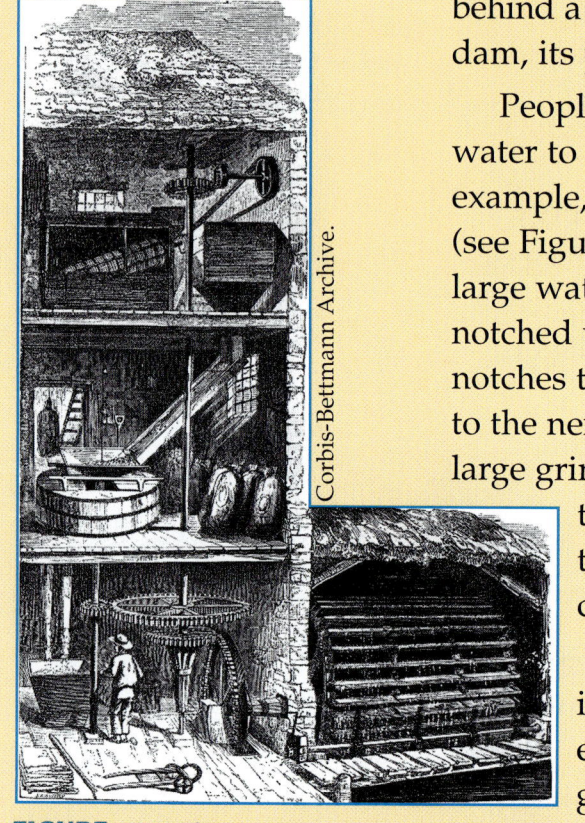

FIGURE 16.7 This is an example of an old-fashioned flour mill. What forms of energy are in this system?

6. What other forms of energy might be present in the waterwheel?
7. What are other examples of systems that involve mechanical energy?
8. What is the energy source in these examples?
   a. Wood logs burning in a campfire.
   b. A scooter going down the street.
   c. A runner completing a marathon.

CHAPTER 16 Exploring Energy in Systems

Can a system change energy from one form to another? To answer that question, think about how you build a fire. First, you build a stack of wooden logs. Inside the stack, you put several scraps of paper, small twigs, branches, and other kindling materials. These are called **input**, meaning the things that go into a system to make it work. Next, before you light the fire, step back and consider the question, Is there any evidence of light, heat, sound, or mechanical energy in this system?

You probably cannot detect much evidence of energy in your stack of materials. As far as you can tell, nothing is changing. Now you strike a match and light the kindling. Suddenly, the fire leaps into action. Flames spread rapidly to the other pieces of kindling. A piece of pitch in one of the logs pops with a loud snap. Several sparks shoot brightly into the air. Your face grows warm and you step backward. Your eyes begin to smart from the smoke that billows in your direction. Now you are satisfied; you have plenty of evidence that the system contains energy. You can sit back and enjoy the pleasant effects of heat, light, sound, and movement. What you are seeing, those things coming out of the fire, is called **output**. Every system has input and output.

Unfortunately, you did not have a large supply of logs to build your fire. Gradually the flames die down. Eventually, the fire goes out. All that remains is a pile of smoldering ashes. Then you remember the original question, Can a system change energy from one form to another?

9. Where do you think the different forms of energy in the fire system came from?

STOP & THINK

10. What are all of the inputs to the fire? What are all of the outputs from the fire? Construct a diagram of the inputs and outputs of the fire.
11. What happened to the different forms of energy when the fire went out?

Think about the diagram you constructed. In the case of the fire, the energy source is the logs and the energy input is chemical energy stored within the logs themselves. **Chemical energy** is the energy that holds the particles of an object together. When you break the particles apart, you release the chemical energy. (This does not mean that if you break the object apart into smaller pieces you release chemical energy. You have to break the particles themselves apart.) When the fire has released all the chemical energy in the logs, the fire goes out.

The fire is an example of a system in which the input is different from the output. Chemical energy in the logs becomes heat, light, sound, and movement. The system actually changes the energy from one form to another (see Figure 16.8). In the process, the total amount of energy

FIGURE 16.8 This fire system contains an energy input and an energy output.

remains the same. In other words, all of the chemical energy transformed by the system becomes heat, light, and movement. Then these energy outputs spread out over a large area. For example, after the fire goes out, your face gradually becomes cool and the air is no longer as warm as it was. Yet this heat energy is not lost. It simply becomes spread out over a larger and larger area until eventually you can no longer detect it.

12. Is chemical energy released when you tear a piece of paper in half or chop a piece of wood into many small pieces of kindling?
13. What are some examples of other systems that contain chemical energy?
14. What evidence might indicate that chemical energy is present in a system?

As you have seen in the examples of the waterwheel and the campfire, people can use systems to harness different forms of energy for their own use. One of the most common forms of energy that people use in the 21st century is electrical energy.

You probably recall from Chapter 10 that Democritus thought that all materials are made of very small particles, which he called "atoms." He was correct, but he didn't know that atoms are made up of even smaller particles called protons, neutrons, and electrons. Protons and neutrons clump together in the center of an atom and electrons move around them. Sometimes, electrons are exchanged between atoms. **Electrical energy** is produced when electrons flow from one atom to another.

Electrical energy flows through the electrical wires every time you plug in and turn on a television or a

CHAPTER 16 Exploring Energy in Systems

FIGURE 16.9 A chain of falling dominoes is a good model for the movement of electrons that creates electrical energy.

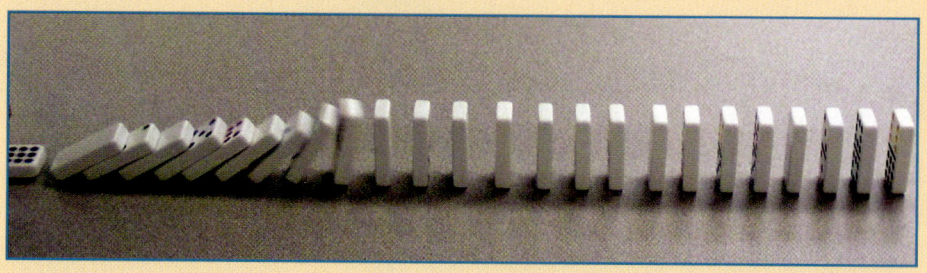

computer. Electrons don't flow through a wire like water through a hose, however. They don't enter the wire at one end and leave at the other. Instead, they move only a short distance, pushing against the electrons in the neighboring atom. They pass their movement along the wire, much like one falling domino causes a whole chain of dominoes to fall.

Electrical energy is a common and useful form of energy because it is easy to change into other forms of energy. For example, many people in the United States have stoves that change electrical energy to heat energy that they use to cook food. Most cities in the world have streetlights that change electrical energy to light energy. An electric fan changes electrical energy to the energy of movement. A radio changes electrical energy to sound energy.

Later in this unit, you will examine several systems that transform electrical energy. You also will trace the flow of electrical energy backwards through various systems to your local power plant. As you study these systems and other forms of energy, keep in mind that energy can change from one form into another.

FIGURE 16.10 Electrical energy can change into other forms of energy. The electrical energy used to fuel streetlights changes to light energy.

15. What are some examples of other systems that contain electrical energy?
16. What types of evidence might indicate that electrical energy is present in a system?

410 CHAPTER 16 Exploring Energy in Systems

investigation ELABORATE

Heat In, Heat Out

Many systems require the balance of heat input and heat output. The Earth, your body, and cars are a few examples. In this investigation, you will build a system that keeps heat input and output in balance. Balancing the heat input and output of a system is not an easy thing to do, especially when the heat input comes from a source as large as the Sun. This investigation gives you a chance to see whether you can outwit the Sun.

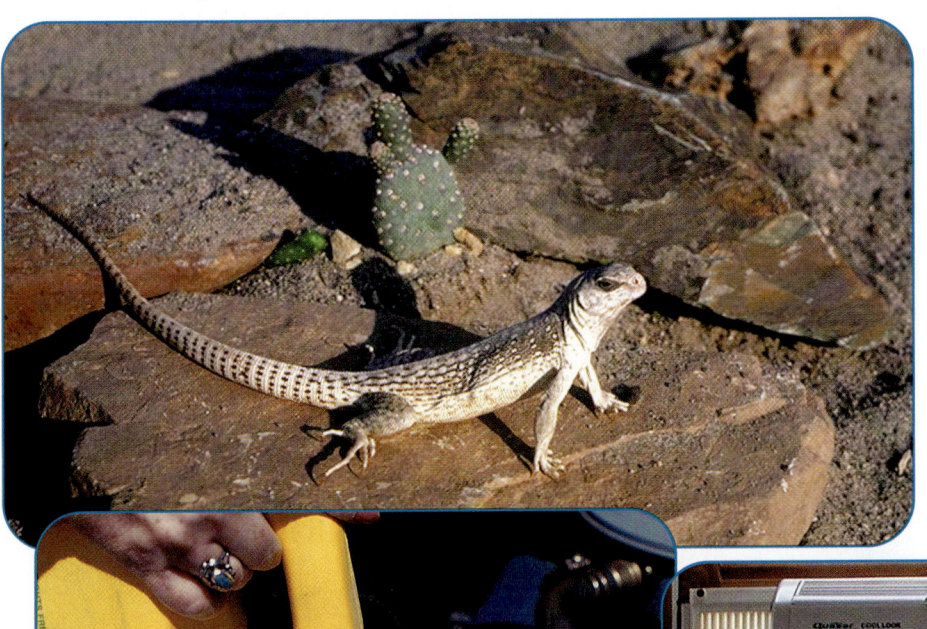

FIGURE 16.11 These systems require that heat input and heat output be in balance. What might happen if the heat input or output were out of balance?

CHAPTER 16 Exploring Energy in Systems

Working Cooperatively

Work cooperatively in your team of two. You will need a Communicator and a Manager. As you complete the investigation, *respect the working environment of others*. Work side by side at a table or with your desks pushed together.

Materials for Each Team of Two:
- 1 shoe box, with or without a lid
- 1 thermometer
- 1 pair of scissors

Materials for the Entire Class:
- 100 sheets of construction paper, assorted colors
- 1 roll of plastic wrap
- 1 roll of aluminum foil
- several rolls of masking tape
- additional materials needed by individual teams

Process and Procedure

1. Consider the following challenge:
   - ➡ Construct a shoe-box system that increases in temperature as little as possible when you leave it in direct sunlight for 20 minutes.
   - ➡ Make sure that everyone on your team understands this challenge. You will monitor the temperature of your system using a thermometer that you place through a hole in the side of the box, as shown in Figure 16.12. The members of your team may be part of your shoe-box system. You must allow the heat input from the sun to reach your shoe-box system. This means that you cannot shield the box from the sun or cover the top with anything.

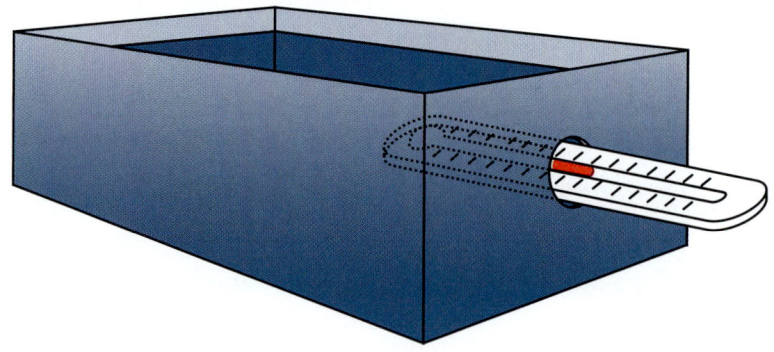

FIGURE 16.12 Place a thermometer inside your shoe box as shown here. The thermometer will indicate the temperature inside your box throughout the course of the test.

CHAPTER 16 Exploring Energy in Systems

2. As a class, discuss how you will test your systems.
   - Be sure to consider the variables you will need to control in order to make this test the same for all teams. (For example, should all teams place their systems in direct sunlight, or can some teams place their systems in partial shade?) Also, consider the variable you want to compare.
   - Notebook entry: Record the variables your class chose to control and the variable that you will compare.

3. With your teammates, conduct a brainstorming session to create a list of possible designs.
   - Remember that to keep the temperature from increasing; you must make sure that the heat input is equal to the heat output.

4. Choose 1 or 2 ideas from your list that you would like to try.
   - Notebook entry: Record your plan.

5. Construct your system.
   - Both Team Members should take part in the construction. The Manager should get materials as you need them.

6. Test your system with the rest of your classmates.
   - If you have never used a thermometer before, refer to How to Read a Thermometer, How To #7.

7. Modify your system if necessary and test it again.
   - Notebook entry: Record the changes you make to your system.

8. With your teammate, think of another part that you could add to your system and test it again.

Wrap Up

Work on this section as a team. Write the answers to the questions in your notebook after you have discussed them with your partner.

1. How can you decide which team was the most successful at balancing heat input and heat output?
2. What did you add to your system in Step 8, and how did the addition change the results of your test?
3. What other questions do you now have about systems? How can you find out answers to your questions?

  connections

What If Energy from the Sun Were Blocked?

By now, you have explored energy in a number of different types of systems. For the entire system of the Earth, the ultimate source of energy is the Sun. What if its energy were blocked? In this connections activity, work individually to think about how a natural disaster might affect different communities on Earth.

Read the following scenario:

Imagine that Earth is entering a phase of instability that no one predicted. Throughout the world, hundreds of volcanoes are erupting with great force. Earth's atmosphere is thick with volcanic debris and dust. As much as 75 percent of the sunlight now is blocked from reaching Earth's surface. This period of eruptions is

FIGURE 16.13 When a volcano erupts, it throws ashes high into the atmosphere. The ashes can form a dense layer in the sky and partially block the sun.

expected to continue indefinitely, and it is likely that virtually all sunlight will soon be blocked from reaching Earth's surface.

Think about the following questions and record your own answers in your notebook.

1. In what ways do living systems on Earth depend on energy from the Sun?
2. What might be the effect if only 50 percent of the sunlight were blocked from reaching Earth? What might be the effect on a tree, on a shark, and on a squirrel?
3. Imagine that the trend continued and eventually all sunlight was blocked from reaching Earth's surface. What might be the effect on a tree, on a shark, and on a squirrel?

EVALUATE connections

Tracing the Flow of Energy

Now it is your turn to trace the flow of energy through a system. Examine the four systems shown in Figures 16.14 through 16.16. For each system, trace the flow of energy into and out of the system and answer these questions. Write your answers in your notebook and be prepared to share them with the class.

FIGURE 16.14 This young plant has just emerged from the soil and is beginning to grow. Can you trace the energy through this system?

FIGURE 16.15 Soon this pizza will be ready to eat. What role does energy play in this system?

416 CHAPTER 16 Exploring Energy in Systems

1. What are the parts of the system?
2. What evidence supports the idea that there is energy in the system?
3. What inferences can you make about the forms of energy in the system?
4. What evidence supports the idea that the system is changing energy from one form to another?

FIGURE 16.16a–b (a) How is energy involved in this melting icicle? (b) Can you find evidence that energy in this system is changing from one form to another?

Solar Sources: Energy from the Sun

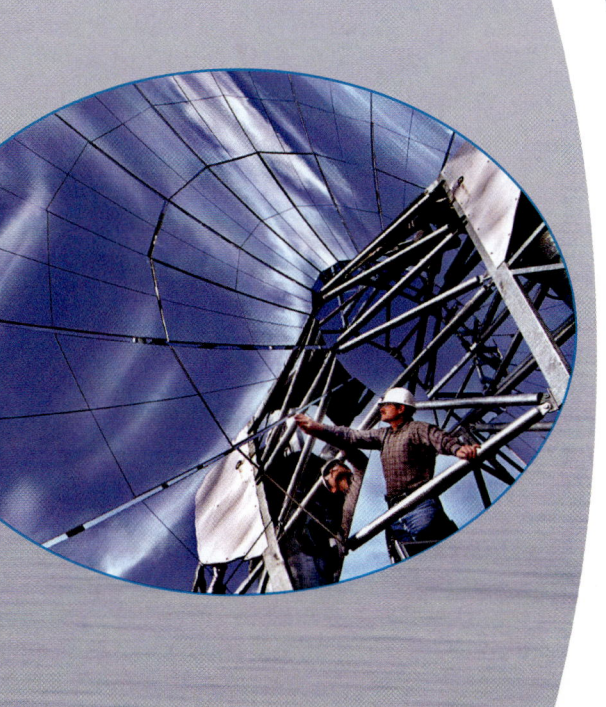

Have you ever heard of the World Solar Challenge? It is a race between cars that are powered only by sunlight. Designers of different solar cars from around the world meet to compete in a race across the Australian outback. Competitors come from 13 different countries and include corporations, high school and college students, and family teams. The first World Solar Challenge was held in 1987. Individuals who have designed vehicles such as these, or solar-powered airplanes or facilities, have learned a lot about ways to collect the Sun's energy and make use of it.

In this chapter, you will experiment with ways to collect the Sun's energy and use it to solve problems. Along the way, you will develop a better understanding of how technological systems solve energy problems.

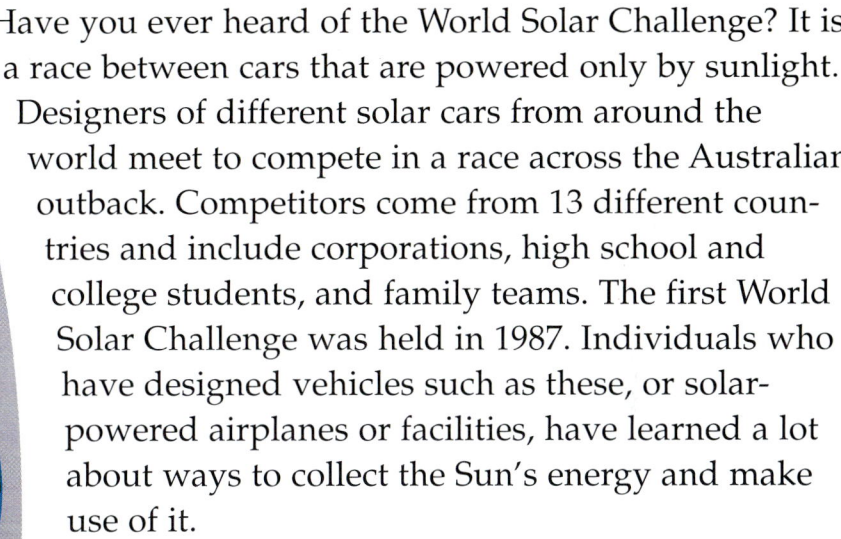

ENGAGE — Sun Seekers

EXPLORE — What Is Skin Cancer?
— Collecting the Sun

EXPLAIN — Systems for Solving Problems

ELABORATE — Hotter and Hotter

EVALUATE — Water-Heating Systems

419

ENGAGE investigation

Sun Seekers

What a great day! It was your first day of summer vacation and you spent the entire day outside. You soaked up some sunshine. You played a great game of Frisbee. But now, on your way home, you begin to feel the effects of the long day in the sun. Your eyes begin to ache. Your skin feels tight, dry, and painful. You may feel dizzy and tired. How can the Sun's energy cause changes in your body's systems? In this investigation, you will explore possible answers to this question.

Working Cooperatively

Work cooperatively in your teams of three. You will need a Tracker, a Communicator, and a Manager. You will work together at your desk, and then you will work outside. Before you begin, you might consider how the social skill you chose applies to outdoor experiences.

Materials for Each Team of Three:
- 6 pieces of sun-sensitive paper
- at least 3 samples of sunscreen, each with a different SPF rating
- 3 medicine cups or other small containers to hold the sunscreen samples

FIGURE 17.1 What actions do you take to prevent sunburn?

CHAPTER 17 Solar Sources: Energy from the Sun

- 1 clock or watch
- 3 scraps of paper, smaller in size than the sun-sensitive paper
- transparent tape
- a water supply that you can take outside

Topic: Solar Cells
Go to www.scilinks.org
Code: physical421

Process and Procedure

1. Read the Background Information.
   → Take turns reading this information aloud to your teammates. This information is located after the procedure for this investigation.

2. Make the sun detector shown in Figure 17.2 by using the following steps.
   → You will need to do these steps as quickly as possible to avoid exposing the sun-sensitive paper to any bright or direct sunlight. Each Team Member should make his or her own sun detector.

   a. Tape 1 piece of sun-sensitive paper to your notebook.

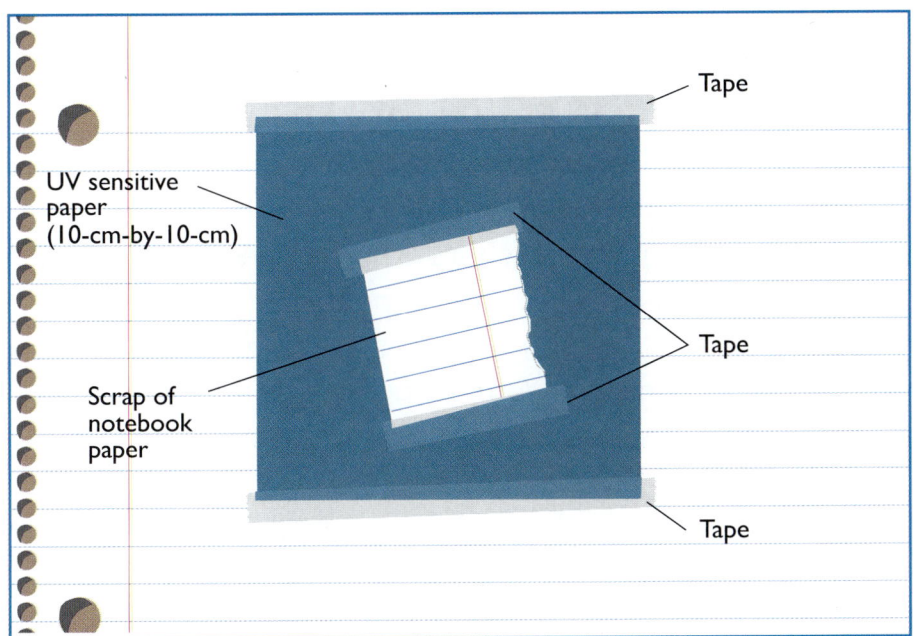

FIGURE 17.2 Construct the sun detector shown here.

CHAPTER 17 Solar Sources: Energy from the Sun

  b. Tear a small scrap of paper from your notebook.
   → The scrap must be smaller in size than the sun-sensitive paper.
  c. Place the scrap of notebook paper on top of the sun-sensitive paper.
   → The scrap of paper should not move or shift about.
  d. Close your notebook.
3. Take your sun detector outside and place it in direct sunlight when your teacher tells you to do so.
 → Notebook entry: Record the time and note the weather conditions. You could use words and phrases like those you hear on the news, such as "partly cloudy" or "mostly sunny."
4. Observe the sun detector for 5 minutes.
 → You can begin Steps 6 and 7 while you are waiting and observing.
5. Develop the sun-sensitive paper by removing it from your notebook and gently rinsing it in water.
 → Before you rinse the paper, be sure that you remove the scrap of notebook paper.
 → Notebook entry: Record your observations.
6. Plan an experiment that will help you find out how 3 sunscreens with different SPF ratings can protect a person's skin from the Sun.
 → Be sure that your experiment is a fair test of the sunscreens. If necessary, refer to How to Conduct a Fair Test, How To #8. Remember to consider what variables you will need to control. For this test, you will not need to include the scrap paper in your sun detector. You can apply the sunscreens directly to the sun-sensitive paper.
 → Notebook entry: Record your plans.

7. Construct a data table for your experiment.
   → If you have questions, refer to How To #2, How to Construct a Data Table.

8. Perform your experiment.
   → Notebook entry: Record the results in your data table.

9. Share your results with the rest of the class.
   → Make sure that each member of your team understands your results well enough to report them to the class.

Background Information

Sunburns are more than just painful inconveniences. They permanently damage your skin and eyes. Every time you get a sunburn, you increase your risk of developing skin cancer. This is because your skin has a "memory."

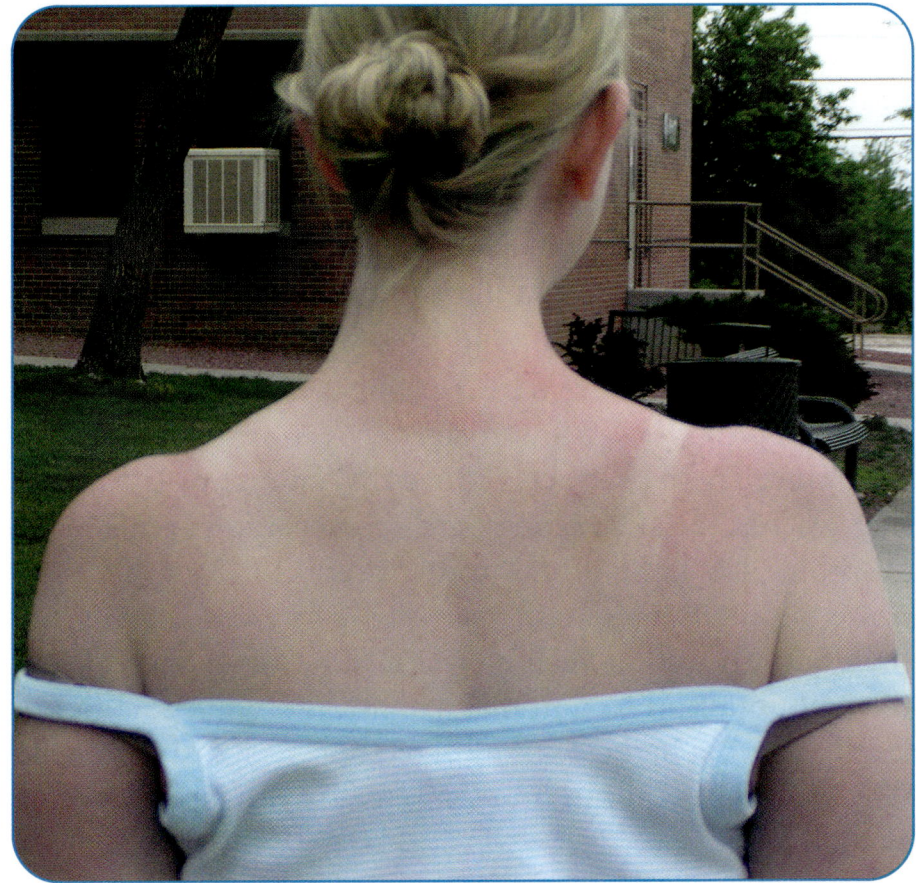

FIGURE 17.3 This photo shows you what a mild to moderate sunburn can look like.

Topic: Radiation from the Sun
Go to www.scilinks.org
Code: physical424

It remembers every exposure to the Sun's light. Some medical experts suggest that under some conditions, even 20 minutes of exposure to sunlight is enough to cause changes in your skin that could lead to skin cancer. Although people have different levels of sensitivity to sunlight, people of all skin types can experience sunburns.

Solar radiation is a process by which energy from the Sun can be transported. There are different types of radiation, including ultraviolet, infrared, and visible light. We can observe visible light with our eyes in the colors of a rainbow, but we need instruments to detect the energy from other parts of the spectrum. Let's consider some of the dangers and benefits of solar radiation.

Sunlight can also cause potentially dangerous changes to other parts of your body. High levels of exposure to sunlight are associated with certain types of eye diseases. In addition, some studies have found that overexposure to the sun can weaken the body's ability to fight some viruses, such as the *Herpes simplex* virus, a virus that produces cold sores.

Sunlight is composed of many different types of light. The part of sunlight that is most damaging to your body is known as **ultraviolet radiation** (UV radiation). You cannot see UV light with your unaided eye.

To protect themselves from changes caused by UV light, people can use sunscreens to cover exposed areas of their skin. These products reduce the amount of UV light that reaches the skin. Some products accomplish this by absorbing UV light; other products reflect it. Sunscreens come in different strengths, which are rated according to their Sun Protection Factor, or SPF. SPF ratings indicate the amount of protection that a sunscreen will provide. Until 1986, the highest strength of sunscreen available was SPF 15. Today, you can find sunscreens with an SPF rating as high as 50.

Wrap Up

Imagine that you are a dermatologist, a doctor who specializes in skin. A new patient has come to you with a severe sunburn. After treating the sunburn, you would like to advise your patient on skin care. Write a summary that describes what you would say. Be sure to include answers to these questions in your summary. Be prepared to read your summary to the class. Also be prepared to report on the social skill you have been using. Why did you choose this skill? Have you improved in your use of this skill?

- Which SPF ratings offer you longer protection from the Sun's UV radiation?
- What evidence do you have to support this choice of SPF rating?
- How might weather conditions affect the amount of time you can be outside without risking a sunburn?
- How might the season of the year make a difference in the amount of time you can be outside without risking adverse effects from UV radiation?
- How might your skin type make a difference in the amount of time you can be outside without risking a sunburn?
- What other factors might make a difference in the amount of time you can be outside without risking adverse effects from UV radiation?
- How can you protect yourself from other changes caused by the Sun's energy?

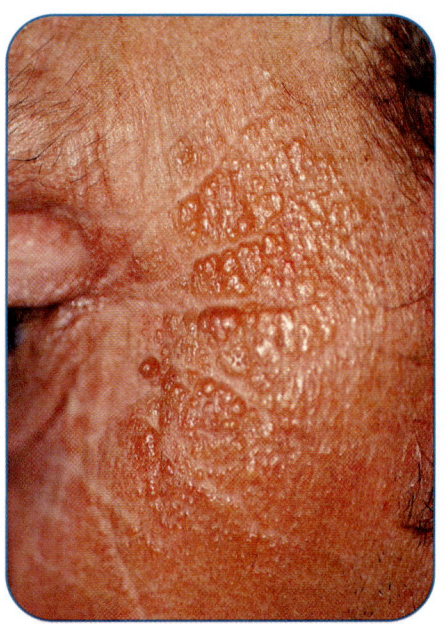

FIGURE 17.4 Ouch! Look at this photo of an extreme sunburn. If you were a dermatologist, what skincare advice would you give this person?

CHAPTER 17 Solar Sources: Energy from the Sun

Explore connections

What Is Skin Cancer?

The spectrum of the Sun is a continuous distribution of energy: from ultraviolet through visible light, into the infrared and beyond. The packets of energy, called **photons**, that are in the ultraviolet range are the most energetic. When they are absorbed by our skin, they can cause damage to the cells of our skin, or sunburn. Fortunately, all of the far ultraviolet photons are absorbed by the oxygen in the atmosphere before they reach us. Also, most of the near ultraviolet photons are absorbed by ozone, a layer of gas (O_3) in the atmosphere. That is rather amazing because ozone molecules represent only about .00004 percent of the atmosphere.

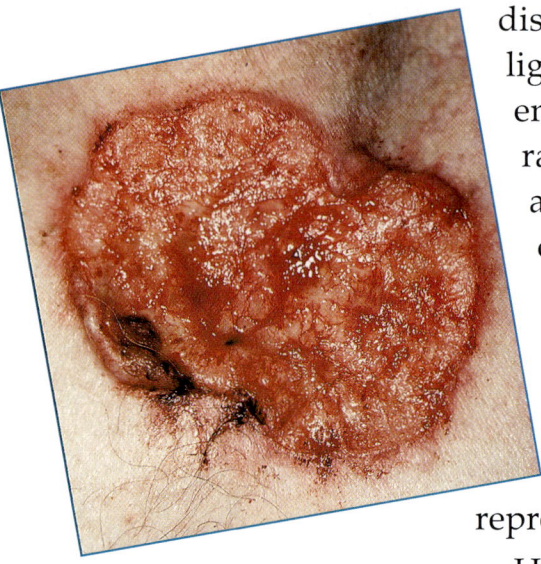

FIGURE 17.5 This is what a carcinoma can look like.

How does exposure to sunlight affect the balance of your body's systems? Exposure to the Sun's UV radiation can cause sunburn to your skin and eyes. It also might lead to skin cancer. As a class, read and discuss the following information on skin cancer. Then make a list of things that you can do to maintain your body's healthy balance and decrease your risk of getting skin cancer.

Skin cancer can be a life-threatening disease, but it is one that is preventable. There are three types of skin cancer. Two of the types are called carcinomas. Carcinomas are the most common and are usually curable when found and treated early (see Figure 17.5).

Carcinomas usually affect just the cells in the upper skin layers. Doctors can remove these types of skin cancers by excising them (cutting them out). Dermatologists can do this in their offices using local anesthesia.

CHAPTER 17 Solar Sources: Energy from the Sun

A third type of skin cancer is more dangerous. It is called melanoma. It is the least common, but most dangerous, type of skin cancer (see Figure 17.6). It affects cells that are in the deeper layers of skin. These cells are important because they produce the skin's pigments, particles of color that lead to skin colorations such as freckles. Melanomas account for about 7,800 deaths in the United States each year. When doctors diagnose this more serious form of skin cancer, they must use more aggressive procedures to treat the patient. Treatment usually involves hospitalization and surgery as well as chemotherapy or radiation therapy.

Your risk of developing a malignant melanoma is at least doubled if you have had one or more severe sunburns when you were a child or teenager. Yet, according to one study of teenagers in suburban Virginia, 33 percent of teenagers never wear sunscreen and only 9 percent always use sunscreen. So despite our knowledge of the risks, a great number of teenagers may be in danger of developing skin cancer later in life. Another nationwide survey found that nearly 72 percent of teenagers got sunburned at least once during the summer, even though about one-third of them said they were wearing sunscreen. How do you think people could get sunburned if they put on sunscreen before going out into the sun?

The American Cancer Society (ACS) suggests several ways to reduce your chances of getting skin cancer. Cover up with long-sleeved shirts and long pants that the sun cannot penetrate. Wear a wide-brimmed hat to protect your face and a bandanna to cover your neck. Use sunscreens. The ACS recommends sunscreens with an SPF rating of 15 or higher. Apply the sunscreen at least one hour before going into the sun. Apply more sunscreen after you swim or perspire. Do not use sunlamps, tanning beds, or tanning pills. Understand how different factors affect the amount of UV radiation reaching your skin and take appropriate steps to protect yourself.

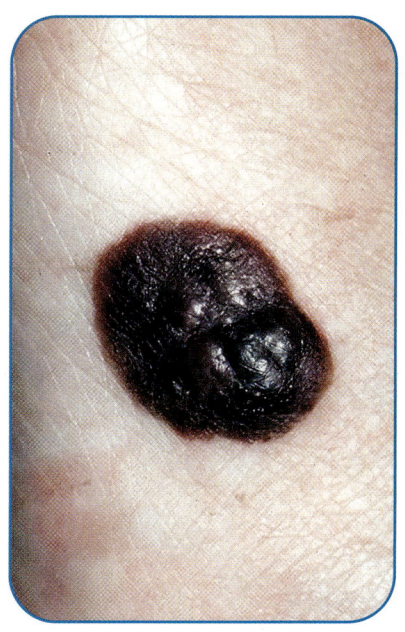

FIGURE 17.6 This is what a melanoma can look like.

FIGURE 17.7 How is this person protected from overexposure to the sun?

You also can wear sunglasses to protect your eyes. When you choose a pair, make sure that they offer UV protection. Most sunglasses sold today do. Older sunglasses, however, may not and some actually increase the chances that your eyes will burn. That is because the glasses shade your eyes from visible light and cause your iris to let in more light, including dangerous UV radiation. If you have questions about your sunglasses, you can have them checked at a local eyeglass shop to determine how much UV radiation they actually block.

One other way to protect yourself is to know your own limits. Do you know how much time you can expose your skin to the sun without getting burned? Depending on your skin type, you might burn after only a few minutes of full sun. On the other hand, you might be able to stay outside for over an hour without getting burned. You can use the information in Figure 17.8 to help you determine your exposure limits.

Differences in skin type come from differences in the amount of melanin (MEL uh nin) particles in the skin. Melanin is a dark brown pigment that absorbs both visible

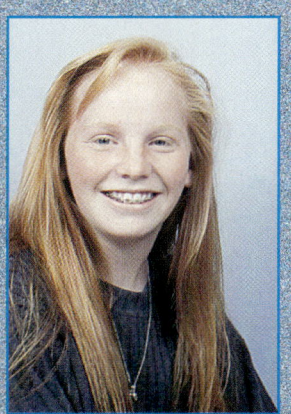

Type I
Extremely sun-sensitive skin. Eyes are usually blue, hair is often red, and unexposed skin is white. This skin type tans little and burns easily. Sunburns are often painful and cause the skin to peel.

Type II
Very sun-sensitive skin. Unexposed skin is white. Eyes are usually blue, hazel, or brown. Hair is usually red, blond, or brown. This skin type burns easily. Sunburns can be painful and cause the skin to peel.

Type III
Sun-sensitive skin. This skin type often develops a light brown tan after exposure to sun. When sunburns do occur, they can be moderately severe.

Type IV
Minimally sun-sensitive skin. Unexposed skin is white or light brown. Eyes are usually dark colored. Hair is usually dark brown. This skin type tans easily and burns very little.

Type V
Sun-insensitive skin. Unexposed skin is brown. This type rarely burns and tans easily and substantially.

Type VI
Sun-insensitive skin. Unexposed skin is black or dark brown. This skin type burns only with severe exposure to the sun.

Source: American Academy of Dermatology

Average time for unprotected skin to redden
(in minutes, measured at noon during mid-summer):

40-44° latitude (Northern California, Kansas City, New York City, Indianapolis)		20-25° latitude (Brownsville, Miami, Hong Kong)	
TYPE	MIN.	TYPE	MIN.
Type I	20	Type I	12
Type II	25-30	Type II	15-18
Type III	35-40	Type III	20-24
Type IV	40-50	Type IV	30-35
Type V	50-60	Type V	35-40
Type VI	70-75	Type VI	50-55

FIGURE 17.8 Dermatologists use the scale shown here to help people determine how long they can be in the sun without risking a sunburn. According to this scale, how long can you safely expose your skin to the sun?

light and UV radiation. If you have dark skin, you have enough melanin particles to absorb a large amount of UV radiation. These particles act as a natural protection and you will be less likely to get a sunburn, except under extreme conditions. If you have light skin, however, you have fewer particles of melanin to protect you from UV radiation. You sunburn easily. Yet your skin also can produce extra melanin particles that provide extra protection and make your skin look tanned.

The American Cancer Society recommends that you consult a nurse or physician if you notice any changes in your skin, such as an irregularly shaped mole or a smooth or scaly thickening. All forms of skin cancer are treatable if you detect them early.

What can we do about ultraviolet radiation in our everyday lives? Remember this rather simple scientific fact: the amount of ultraviolet absorbed by ozone depends on the number of molecules of ozone in the path of the sunlight. If sunlight passes through the atmosphere on a slanted path, it passes through more molecules of ozone before it reaches you than it does if it shines straight down. In the early morning and late afternoon, the path of the sunlight is slanted. So if you stay out of direct sunlight in the middle of the day, you reduce the most intense exposure to ultraviolet. If you are in the sunlight in the middle of the day, wear sunscreen, a hat, and protective clothing. You may be able to prevent skin cancer if you routinely wear a sunscreen with at least an SPF of 15. Regular use of a sunscreen with SPF 15 for the first 18 years of life might decrease your risk of getting skin cancer by 78 percent. If you have additional questions about skin cancer or any other cancer-related topic, contact the American Cancer Society at their toll free number (800) ACS-2345 or visit their Web site at *www.cancer.org*.

SIDELIGHT ON NATURE

The Problem of Stratospheric Ozone

Ozone is made by the action of ultraviolet (UV) radiation from the Sun. When ordinary oxygen molecules (O_2) absorb UV, the molecule breaks apart into separate oxygen atoms (O). If one of these separate oxygen atoms (O) attaches to an ordinary oxygen atom (O_2), the result is ozone (O_3). When ozone absorbs UV, the ozone also breaks apart. Generally, in the atmosphere there is a dynamic balance with a fairly small number of ozone molecules. Other reactive molecules that contain nitrogen, chlorine, or hydrogen contribute to this balance of ozone in the atmosphere as well. Most of the ozone is found in a layer of the atmosphere about 25 kilometers (15.5 miles) above Earth's surface. This region is called the **stratosphere**.

We need this ozone layer to protect us from too much UV radiation. A few years ago, atmospheric scientists became aware that activities of humans increase the number of reactive molecules in the stratosphere, especially molecules that contain nitrogen or chlorine. Supersonic transports could be adding too much nitrogen to the stratosphere for a balance to be maintained, for example. This does not seem to be the case at present, but scientists will keep this in mind as a possibility. Unfortunately, we already have a problem with the chlorine compounds that humans have been adding to the atmosphere.

Industrial chemists made a class of molecules that we call CFCs (chlorofluorocarbons). These compounds were useful in refrigeration, air conditioning, aerosol cans, and many other systems. At first, these compounds did not seem to have any negative effects on the atmosphere. But they began to diffuse slowly into the stratosphere. Scientists think that once the CFCs are in the stratosphere, strong UV frees chlorine atoms from them. These chlorine atoms, in turn, destroy the ozone. These theories about the balance of stratospheric ozone earned Paul Crutzen, F. Sherwood Rowland, and Mario Molina the Nobel Prize in 1995.

The startling discovery of the polar ozone "hole" is strong evidence to support this theory. That discovery was made by a team of British scientists in Antarctica and was later confirmed by NASA satellites. More recent research indicates that, for a few weeks each spring, chlorine compounds are particularly effective in destroying the ozone in the cold, polar stratosphere in both the Antarctic and the Arctic.

Fortunately, the polar ozone hole recovers quickly as the season progresses. Satellite measurements, however, indicate that average midlatitude ozone has decreased about 3 percent in a 10-year period. Industrial chemists have developed substitutes for CFCs, and international agreements limit their general use. The CFCs that already are present in the lower atmosphere, however, will continue to circulate and diffuse into the stratosphere for some time. The ozone layer has been damaged, but we hope we have stopped making the problem worse. It also is important to remember that other natural variations continue to be just about as great as the variation that we observe in the stratosphere. The seasonal variation of ozone in the midlatitudes averages about 20 percent.

The stratospheric ozone layer is our principal protection from solar UV. The ozone layer blocks about two-thirds of the Sun's UV radiation before it ever reaches Earth's surface. If the ozone layer becomes thinner, then more UV will reach Earth's surface and your skin. For every 1 percent decrease in the number of ozone molecules, physicians expect to see 20,000 more cases of skin cancer.

People should use sunscreen and wear hats and long sleeves to protect their skin from too much UV radiation. It is also important to take advantage of the early morning and late afternoon periods when the Sun's rays reach Earth through a slanted path. During these periods, the Sun's rays come in contact with more ozone molecules, which in turn absorb more of the UV before it reaches you. Avoiding the midday, direct sun rays (between 10 A.M. and 4 P.M. standard time) will eliminate much of the threat of UV. For the same reason, the amount of UV radiation in winter and at high latitudes also is less.

Changes in the ozone layer will affect many animals and plants. For example, high levels of UV radiation can cause genetic changes in microscopic marine plants know as phytoplankton and in marine animals such as fish larvae. Such changes could seriously disrupt the food web in Earth's oceans because phytoplankton are a food source for many marine animals. If large numbers of phytoplankton die or become extinct because of genetic changes caused by increased amounts of UV radiation, other marine animals also might die or become extinct.

The puzzle of ozone in the atmosphere has many pieces. Scientists will be studying each piece of the puzzle as well as the big picture for many years. In the process, we will continue to learn about the dynamic balance of the Earth and how humans play an important role in it.

investigation EXPLORE

Collecting the Sun

Al, Isaac, Marie, and Ros face an interesting problem (see Figure 17.9). Now it is your team's chance to help Al by creating a system to heat water for his cocoa. Using only the Sun as a source of energy, your team will compete with other teams to see who can raise the temperature of 250 milliliters of water the most within 20 minutes.

WORKING COOPERATIVELY

Work cooperatively in your teams of three. You will need a Communicator, a Tracker, and a Manager. Concentrate on the unit skill as you work. Work at your desks pushed together or at a large table. Make sure everyone has equal access to the experiment.

FIGURE 17.9 How would you create a system to heat water for Al's cocoa? In this investigation, you will have a chance to design a system that solves Al's problem using the Sun's energy.

CHAPTER 17 Solar Sources: Energy from the Sun

Materials for the Entire Class:
- 1 clock with a second hand or 10 stopwatches
- at least 10 Styrofoam cups
- at least 20 aluminum pie pans
- 1 roll of aluminum foil
- 1 roll each of several colors of plastic wrap
- at least 10 shoe boxes
- at least 10 cans of various sizes
- at least 10 paper cups and bowls of various sizes
- at least 10 sections of newspaper
- 1 roll of waxed paper
- at least 10 cardboard boxes of various sizes
- 1 package of construction paper or poster board of various colors
- 1 roll of masking tape
- at least 10 pairs of scissors
- at least ten 2-L plastic bottles
- 3 to 4 bottles of food coloring, assorted colors in dropper bottles
- 10 to 20 scraps of boards of various lengths
- a water supply, room temperature

Materials for Each Team of Three:
- 1 Celsius thermometer
- one 250-mL beaker

Process and Procedure

1. Examine the materials for constructing a water-heating system.
2. As a class, discuss how you could conduct a fair test to compare the teams' water-heating systems.
   ⟶ You might want to refer to How To #8, How to Conduct a Fair Test, as you work on this step.

3. With your team, conduct a brainstorming session to think of ways you could use the materials to create a system that would meet the following challenge:
   → Use sunlight to heat 250 mL of water in a beaker to the highest temperature possible within 20 minutes.
   → You must create the system using the materials your teacher supplies. Remember to use the process you learned earlier this year to design your water-heating system.
   → Notebook entry: Record your ideas.

STOP: Remember to use your social skills!

4. Decide on 1 system that your team will construct.
   → Notebook entry: Draw or describe what your team's water-heating system will look like.

5. Construct the water-heating system.

 If you use the scraps of board, be careful not to get splinters.

6. Construct a data table to record the results of the class tests.
   → This data table should include columns to record time, the beginning temperature of the water, and its temperature at the end of 20 minutes.

7. Conduct the tests to compare your water-heating systems.

8. Compare your team's water-heating system to the systems created by other teams.
   → You might do this step while you are conducting the test and waiting for results.
   → Notebook entry: Record any similarities and differences between your team's system and other teams' systems.

CHAPTER 17 Solar Sources: Energy from the Sun

9. After exactly 20 minutes, measure the temperature of the water in your team's water-heating system.
   ⟶ Notebook entry: Calculate the change in water temperature.
10. Compare your team's results with the results of the other teams.

Wrap Up

Working individually, write answers to these questions in your notebook. When all of your teammates have answered the questions, discuss your answers. You may want to revise your answers after your discussion.

1. What component or components of your team's water-heating system changed during the test?
2. What energy input did your system use to make this change?
3. Which team had the best results with its water-heating system?
4. What parts of the system seemed to be important for getting good results?
5. Was the class test fair? Explain why you think it was or was not.
6. Write a one-sentence summary of how well your team used the Unit 4 social skills during this investigation.

Systems for Solving Problems

Look around you at all the systems you use every day. To get up for school today, you probably used some type of alarm system to wake you. You also probably used some transportation system to get to school, such as a bus and driver or a car pool. Your school probably has a bell system to indicate the beginning and end of each period. Many schools have cafeterias and food service systems so you can get your lunch each day. People create many different systems to solve many different problems.

Like Al, Marie, and Ros, your team just constructed a system to solve a problem (see Figure 17.10). The system you built used energy from the Sun to heat water. Although your team's system might not have performed as well as other teams' systems, the parts of your system still worked together well enough to do something. In Chapter 16, you learned that a system is a collection of objects that work together to do something. In this unit, you can see that "doing something" can include solving a problem.

For the rest of this unit, you will be working with systems that solve problems related to energy. These systems help solve problems associated with the interaction of humans and the environment. Such systems are part of what we call **technology**.

FIGURE 17.10 How did Al, Marie, and Ros's system solve the problem?

CHAPTER 17 Solar Sources: Energy from the Sun

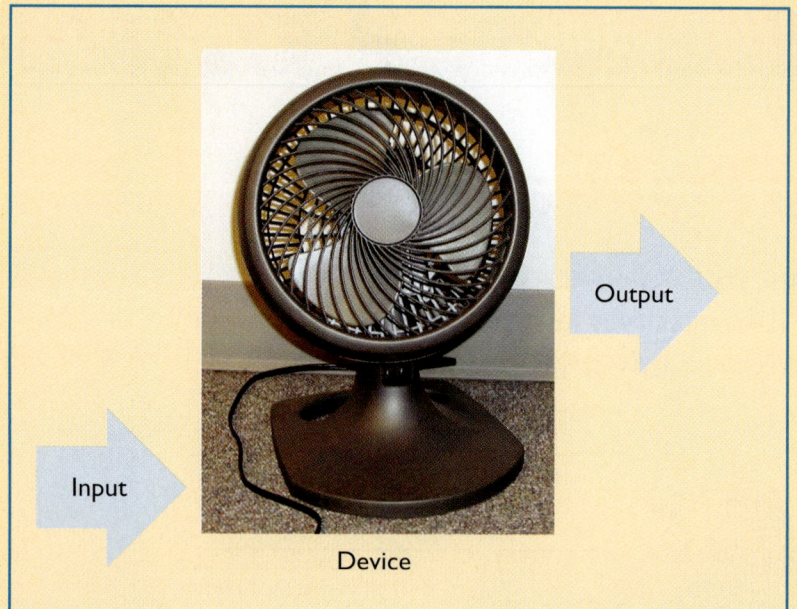

FIGURE 17.11 Can you identify the components of this technological system?

Examine the diagram in Figure 17.11. It illustrates the basic parts of a system for solving a problem. A technological system usually consists of a device that includes all of the components that make up the system, the input to the system, and the output from the system. Not all technological systems are as simple as the one shown in Figure 17.11. As you proceed through this unit, you can practice identifying the devices, the input, and the output of technological systems. This will help you understand these more complex systems and see how people solve their energy problems.

1. Consider the technological system shown in Figure 17.11. What are the parts of this system?
2. What problem is this system designed to solve?
3. What is the input to the system?
4. What is the output to the system?
5. What parts were included in the device your team constructed to heat water in the investigation Collecting the Sun?
6. What was the energy input into and output from your team's system?
7. Did this system solve the problem as well as you expected it would? Why or why not?

investigation ELABORATE

Hotter and Hotter

You now may have several ideas of ways to improve your water-heating system. If you could change your system, how much do you predict that your team's results would improve? In this investigation, you will have an opportunity to test such a prediction. In Part A, each team will design a test of one variable that might affect how well the parts of a system interact to heat water. In Part B, you will combine the information that all of the teams gather about variables to improve your team's water-heating system.

Materials for the Entire Class:

- 1 clock with a second hand or 10 stopwatches
- at least 10 rulers
- at least 10 protractors
- at least 10 Styrofoam cups
- at least 20 aluminum pie pans
- 1 roll of aluminum foil
- 1 roll each of several colors plastic wrap
- at least 10 shoe boxes
- at least 10 cans of various sizes
- at least 10 paper cups and bowls of various sizes
- at least 10 sections of newspaper
- 1 roll of waxed paper
- at least 10 cardboard boxes of various sizes
- 1 package of construction paper or poster board of various colors
- 1 roll of masking tape
- at least 10 pairs of scissors
- at least ten 2-L plastic bottles

WORKING COOPERATIVELY

Work cooperatively in your teams of three. Use the roles of Manager, Communicator, and Tracker. Practice the unit skill, but also choose a skill that every team will practice. This means that you must decide as a class what skill you will practice and how you will evaluate each team's use of the skill.

CHAPTER 17 Solar Sources: Energy from the Sun

- 3 to 4 bottles of food coloring, assorted colors in dropper bottles
- 10 to 20 scraps of boards, various lengths
- a water supply, room temperature

Materials for Each Team of Three:
- 1 Celsius thermometer
- one 250-mL beaker
- your team's water-heating system

Process and Procedure

Part A—The Variables

1. With your teammates, conduct a brainstorming session to create a list of variables that might affect your team's water-heating system.
   - This system is the one you created in the investigation Collecting the Sun. You might recall from Unit 1 that a variable is something that can vary or change in a situation. For example, you might vary the color of the inside of your water-heating system or you might vary the kind of container that holds the water. Both of these are variables.
   - Notebook entry: Record the list of variables.

2. Decide on one variable that your team would like to investigate.
   - Notebook entry: Record the variable.

3. Predict how you think this variable will affect the water-heating system.
   - Notebook entry: Record your prediction.

4. Plan an experiment that will help you test your prediction.
   - In your experiment, you should control all variables except the one you are testing.

FIGURE 17.12 Engineers have designed solar heating systems that collect solar energy and heat homes after the Sun sets.

→ Notebook entry: Record the experiment that your team proposed.

5. Construct a data table.
6. Carry out your experiment.
   → Notebook entry: Record the results in your data table.
7. Report the results of your experiment to the rest of the class.
8. Compare the results of your experiment with those that other teams conducted.
9. Decide which variables you can change to improve your team's water-heating system.
   → Notebook entry: Record your ideas.

Part B—Improving the System

1. With your teammates, plan how to change your water-heating system to improve its output.
   → Notebook entry: Draw or describe what your team's improved water-heating system will look like.
2. Construct your team's water-heating system.
   → You can modify the system you built in Collecting the Sun or you can start over with new materials.

CHAPTER 17 Solar Sources: Energy from the Sun   441

3. Construct a data table.
   → This data table should be similar to the one you created in Collecting the Sun because you will be gathering the same types of data.
4. Test your team's water-heating system during the class testing time.
   → Notebook entry: Record the time and the beginning temperature of the water.
5. Compare your team's water-heating system with the systems of other teams.
   → Notebook entry: Record any similarities and differences between your team's system and the other teams' systems. You might do this step during the test.
6. After exactly 20 minutes, measure the final temperature of the water in your team's water-heating system.
   → Notebook entry: Record your results.
7. Compare your team's results with the results of other teams.
   → Notebook entry: Record your observations.

Wrap Up

Imagine that you and your teammates are design engineers who will present your water-heating system at an engineering conference. Prepare a presentation of your system that includes a description of your system, the method(s) you used to improve the system, and the evidence that shows how much you improved the system. You might make a poster or display, write a short essay, or make an audiotape or videotape for this presentation. Also include a description of how well you used the social skills *Respect the working environment of others* and *Choose an explanation that includes the ideas of all teammates*.

Water-Heating Systems

Do you know people who don't worry about where hot water comes from? They just turn on the tap and out comes plenty of hot water for washing dishes, doing laundry, and taking luxurious hot showers. Hot water does cost money, however. The cost of heating water depends on the type of water-heating system you use. Examine the two water-heating systems shown in Figures 17.14 and 17.15. Use your experiences in this chapter to analyze these systems. Then answer these questions and prepare to share your answers with the rest of the class.

FIGURE 17.13 What ways can you think of to improve your water-heating system?

1. What are the parts of the systems?
2. What problem or problems are the systems supposed to solve?
3. Do you think that both systems solve the problem? Explain your answer.
4. What is the energy input into the systems?
5. What is the output from the systems?
6. What evidence would indicate that the systems transform energy?
7. Compare these systems with your team's water-heating system. How are the systems similar or different?

CHAPTER 17 Solar Sources: Energy from the Sun

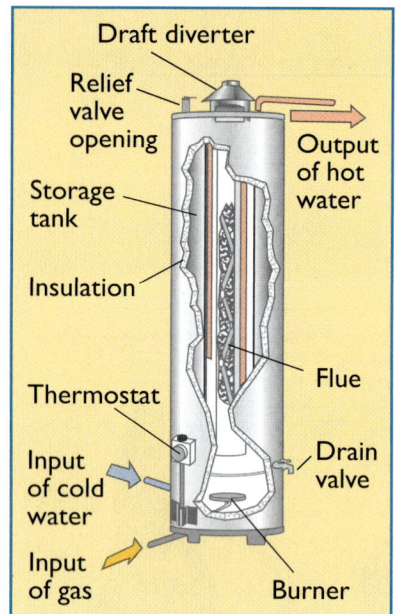

FIGURE 17.14 This is one example of a water heater used in many homes.

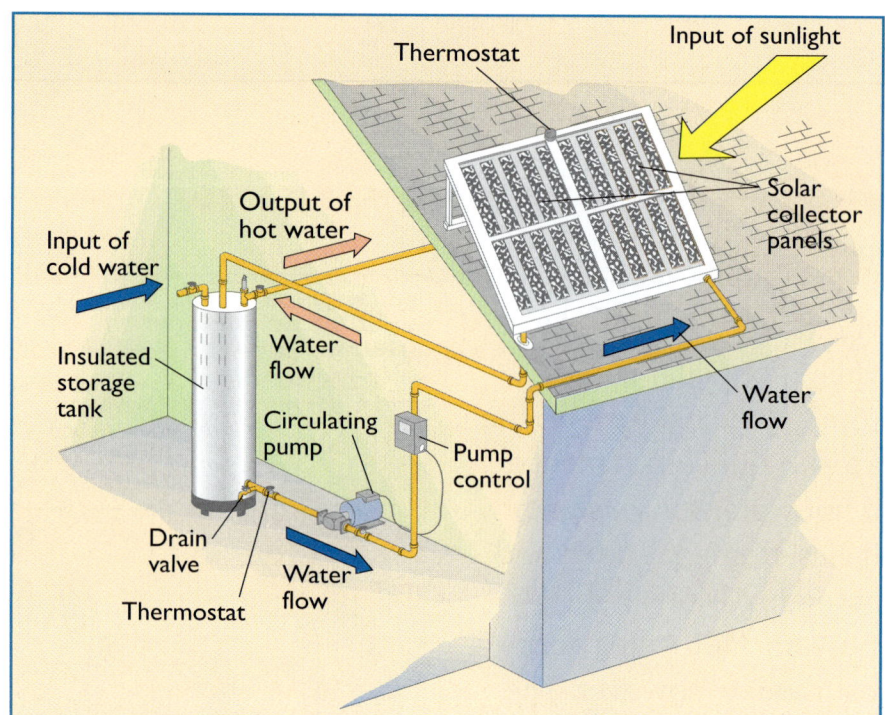

FIGURE 17.15 This is one example of a solar water heater used in some homes.

SIDELIGHT ON TECHNOLOGY
Energy from Garbage

People usually install alternative systems for heating water to solve a problem or address a special concern, such as spending less money to heat their water. Sometimes, the concern is more unique. Consider the story of John Fiarkoski of Omaha, Nebraska, who wanted cheaper hot water but also wanted to dispose of lawn clippings from his lawn care business. The following article describes the system he designed to solve both problems. As you read, think

444   CHAPTER 17 Solar Sources: Energy from the Sun

about whether your family could use a similar system to heat hot water in your home.

Compost + Sun = Hot H_2O

John Fiarkoski reduced the cost of heating water for his house in Omaha, Nebraska, by about 75 percent when he made compost and the Sun part of his hot-water system. Even on the coldest winter days, when the air temperature dips to minus 25 degrees Fahrenheit, a tank buried in Fiarkoski's compost pile supplies water preheated to a tepid 95 degrees.

The heart of the system is a 6 × 8-foot wooden bin that contains the compost pile. The well-insulated bin, which Fiarkoski calls "the digester," is located just outside the house. A solar collector is mounted in front of it, facing south. The system also includes three 30-gallon tanks; two are buried four feet below ground, and the other is at the center of the digester.

As water travels through the system, it is heated to ground temperature inside the first underground tank. Then it flows to the second underground tank, which is wrapped with copper tubing connected to the solar collector. An antifreeze solution, heated by the Sun, circulates through the tubing and transfers heat to the water. In the third tank, the water draws heat from the compost pile before it is finally piped to a gas-fired water heater inside the house.

Fiarkoski, who is 67 and retired from the Air Force, operates an organic lawn care service. He began experimenting with his innovative hot-water system in 1985. He started with just a solar collector but decided he needed a more constant source of heat during the winter. "The idea of using compost as a heat source is as simple as knowing that when freshly mowed grass clippings are heaped in a pile, a large amount of heat is generated," he observes.

Every fall, Fiarkoski loads the digester with layers of leaves, grass clippings, and shredded vegetables from his yard and his neighbors'. He adds manure, plenty of water, and a little lime to slow the composting process and prevent odor. "I live at a busy city street intersection where pedestrian traffic is quite heavy. In four years of operation, I have not received one complaint of an odor problem," he says.

Fiarkoski keeps the pile working through the winter by watering it regularly and keeping it aerated. About midwinter, the pile begins to cool down and he refuels it with a load of fresh manure.

To get the most out of a solar/compost water-heating system "may require adjusting some of your habits," Fiarkoski warns. For example, it is most efficient to use hot water during the middle of the day, when the solar collector is working. "The warm water might not be available at all hours of the day," Fiarkoski says, "but at midday on a sunny day, there should be an ample warm-water supply regardless of the outside temperature."

From: *Harrowsmith Country Life,* July/Aug 1991. Reprinted with the permission of *Harrowsmith Country Life Magazine.*

CHAPTER 18
Energy Benefits and Costs

Do you know where the energy that runs your town or city comes from? In Chapters 16 and 17, you examined how energy in systems transforms. You practiced identifying energy inputs and outputs. Could you trace this energy back to its source? Perhaps you know that most electricity is generated at power plants. You even might know that power plants use energy from coal, oil, rivers, or nuclear reactions to produce electricity. Most people, however, don't know *how* the power plant generates electricity.

In this chapter, you will experiment with a few systems that generate electricity. You will investigate the benefits and costs of using different energy inputs to generate electricity. You will find out about some of the problems related to the different electricity-generating systems and learn why there are no easy answers to such problems.

ENGAGE
- Generating Electricity

EXPLORE
- Doing Things in Reverse
- Paper Clip Lifters

EXPLAIN
- Benefits and Costs

ELABORATE
- Wind Farms
- A Variety of Systems

EVALUATE
- Which System Do We Use?

ENGAGE investigation

Generating Electricity

Look around your classroom. How many of the technological systems that you see depend on energy input in the form of electricity? Without electricity, many modern systems would not operate. Fortunately, in our society, electricity seems abundant and easy to use. You simply plug your stereo into a wall socket and electricity flows into the device.

In this investigation, you will explore ways to generate your own electricity. To detect this electricity, you will need a device called a **galvanometer**. Galvanometers are named after Luigi Galvoni, an Italian scientist who investigated electricity during the 18th century. Galvanometers can measure very small amounts of electricity. Although you will not generate enough electricity to light a lightbulb, you will generate enough to detect with the galvanometer.

WORKING COOPERATIVELY

Work cooperatively in your teams of three. Use the roles of Manager, Tracker, and Communicator. Create a triangular work space with your desks or at a table so that every Team Member has equal access to the project. Continue to practice the social skill that you chose to practice in the last investigation of Chapter 17, Hotter and Hotter.

Topic: Electricity
Go to www.scilinks.org
Code: physical448

CHAPTER 18 Energy Benefits and Costs

Materials for Each Team of Three:

- 12 m of 22-gauge plastic coated or insulated wire
- transparent tape
- 1 small magnetic compass
- one 3 × 5-in. index card
- 1 pair of scissors
- 1 bar magnet
- 1 small paper tube
- one 1.5-V or 6-V battery

Process and Procedure

1. Construct a galvanometer according to the following directions.

    a. Make a 3 × 3-inch square out of the index card by cutting 2 inches off one end.

    b. Create 4 flaps by cutting small squares out of each corner of the index card. Leave enough card in the middle to hold the compass.

    → See Figure 18.1a.

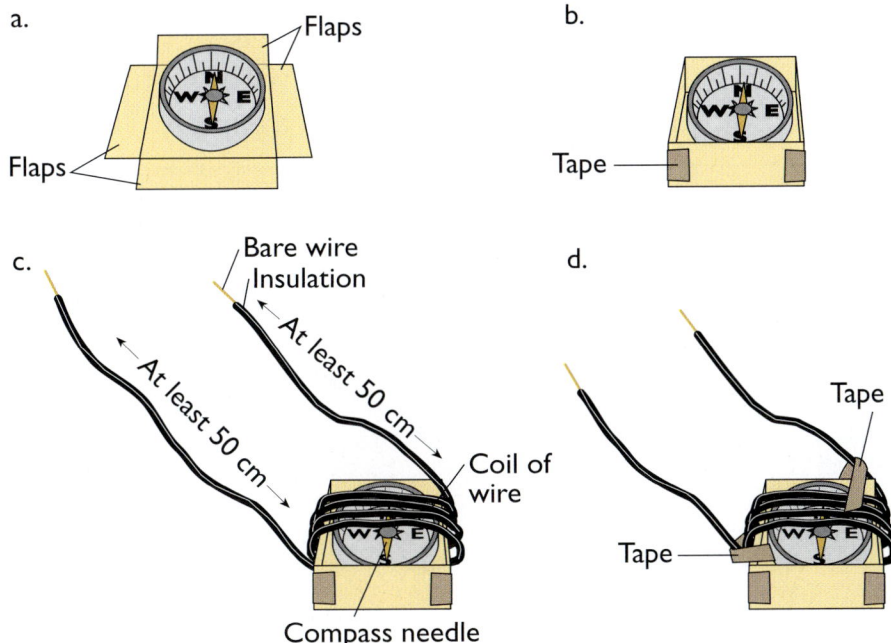

FIGURE 18.1a–d Construct a galvanometer by following steps (a) through (d) shown here. A galvanometer can detect small amounts of electricity.

CHAPTER 18 Energy Benefits and Costs

c. Fold the flaps up and tape them together to form a box.
→ See Figure 18.1b.
d. Place the compass in the box.
e. Wind at least 30 turns of wire around the compass and box.
→ Make sure the wire is in a narrow band so that you can still see the compass needle. Also be sure to leave 2 free ends of wire that are at least 50 cm long.
→ See Figure 18.1c.
f. Secure the wire with transparent tape.
→ See Figure 18.1d.
2. Test your galvanometer by holding the compass steady and touching the free ends of the wire to the 2 terminals of your battery.
→ See Figure 18.2.
→ The Tracker should hold the compass and the Communicator should touch the ends of the wire to the terminals of the battery. Steady the compass on a table or desk so that the needle is aligned with the turns of the wire and is able to move freely.

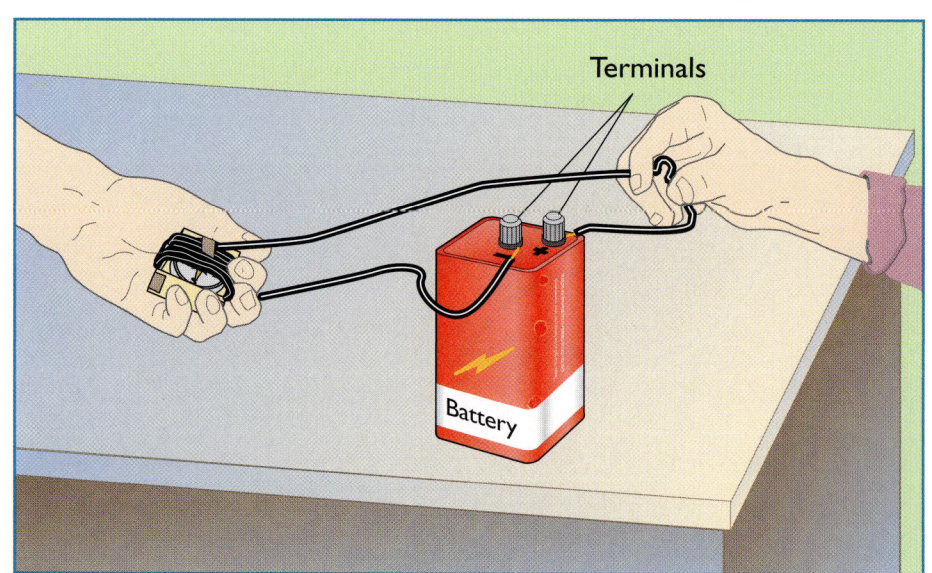

FIGURE 18.2 Test your galvanometer by touching the two ends of the wire to the two terminals of a battery while holding the compass level. What happens to the compass needle?

 To avoid receiving a shock, hold the insulated part of the wire and do not touch the battery terminals. This is a good safety practice. In this experiment, the battery voltage (the amount of potential energy per electron) is only 6 volts, so the danger is low.

3. Observe what happens to the galvanometer as you connect and disconnect the wires from the battery terminals.
   → Notebook entry: Record your observations.
4. Observe what happens to the galvanometer when you connect the 2 ends of wire to the opposite terminals of the battery.
   → Notebook entry: Record your observations.
5. Construct the system shown in Figure 18.3 according to the following directions.
   a. Make a coil of wire that contains at least 40 loops by wrapping the wire around a paper tube.
      → Keep these loops as close together as possible but do not overlap them.
      → See Figure 18.3a. Leave at least 30 cm of wire free on each side.
   b. Cut away any extra paper tube.
   c. Tape the free ends of the coil to the tube to keep the coil from unwinding.
      → See Figure 18.3b.
   d. Connect the free ends of the coil of wire to the galvanometer. Hold the galvanometer as far away from the coil of wire as the connecting wires will allow.
      → See Figure 18.3c.
6. Observe what happens when you hold a strong bar magnet inside the coil of wire. Then observe what happens when you move the bar magnet back and forth inside the coil of wire.
   → See Figure 18.4.

CHAPTER 18 Energy Benefits and Costs

FIGURE 18.3a–c Make the electricity-generating system shown here by following steps (a) through (c).

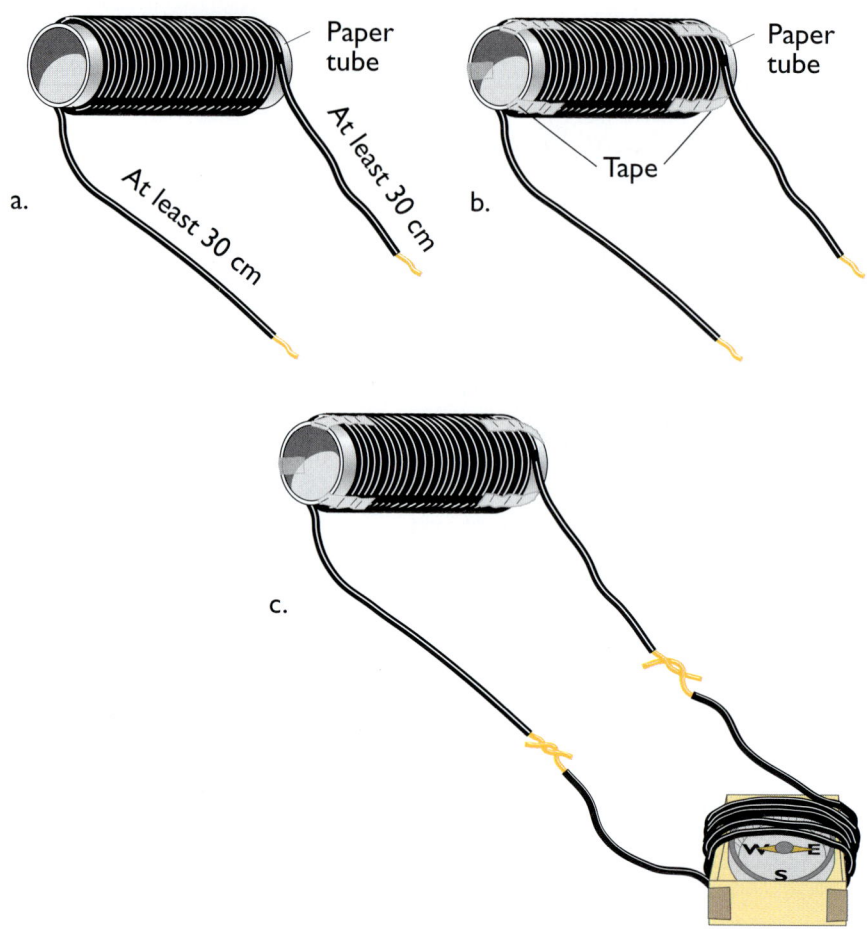

→ Experiment with how far you move the magnet inside the coil of wire. Also observe what happens when you hold the magnet still and move the coil of wire back and forth around it.

→ Notebook entry: Record your observations.

Wrap Up

Discuss these questions with your teammates. Record your answers in your notebook. Be sure that each Team Member can explain your answers in a class discussion.

1. What evidence suggests that the system you built in this investigation generated electricity?
2. What was the output of your system?
3. What was the energy input to your system?

CHAPTER 18 Energy Benefits and Costs

Diagram of bar magnet:

a. Held inside coil

b. Moving (in and out)

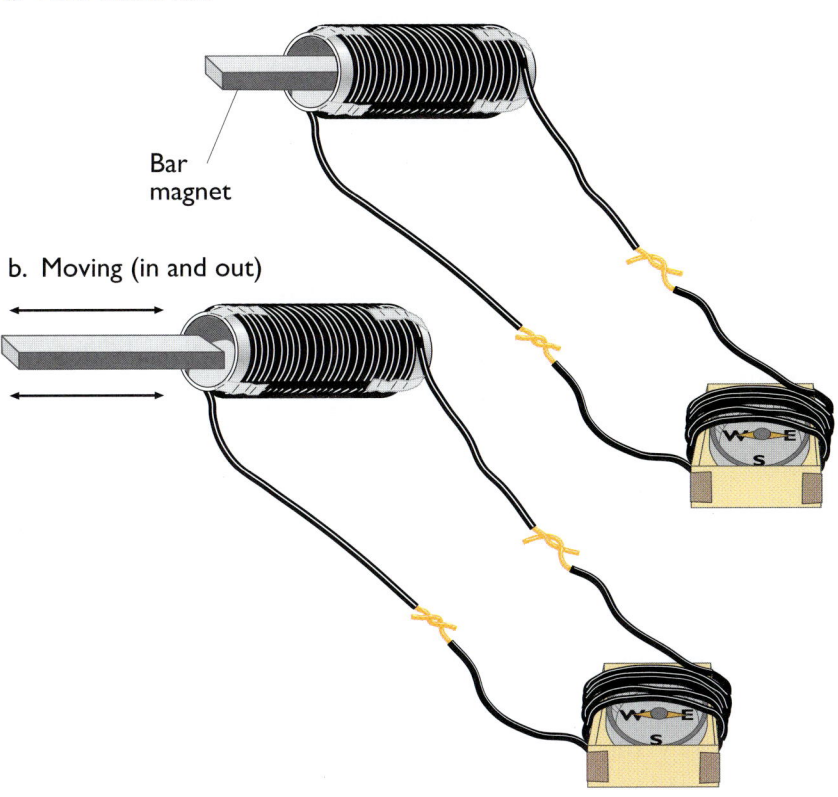

FIGURE 18.4 (a) Hold a bar magnet inside the coil of wire and observe what happens to the galvanometer. (b) Then observe the galvanometer as you move the magnet in and out of the coil of wire.

4. Compare your system with the system shown in Figure 18.5. This system is known as a turbine. How is your system similar to or different from a turbine?

5. On a scale of 1 to 10 (with 10 as the highest), rate your team on your improvement in the use of the social skill.

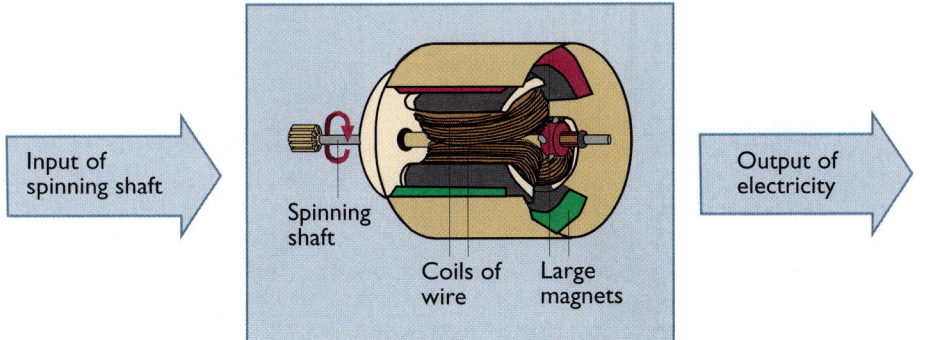

FIGURE 18.5 This system is known as a turbine. It consists of many coils of wire wrapped around a spinning shaft. This shaft is inside very large magnets. When the shaft spins, the system generates electricity. This system is similar to the one that many power companies use to generate electricity

EXPLORE connections

Doing Things in Reverse

Topic: Electrical Current
Go to www.scilinks.org
Code: physical454

As you saw in the previous investigation, you can generate electricity by moving magnets past coils of wire. You have constructed a very small electricity-generating system. Power plants, on the other hand, construct very large systems that generate enough electricity to supply the needs of an entire community. We call both examples of such electricity-generating systems **generators**.

Did you know that you can reverse the input and output to a generator? Galvanometers are one example of such a system. In the galvanometer, you surrounded a compass needle with a coil of wire. (A compass needle is nothing more than a small magnet.) To operate the galvanometer, you allowed electricity from a battery to flow through the coil of wire. As a result of the electricity, the compass needle moved. In this system, the input was electricity and the output was movement.

FIGURE 18.6 The main function of a generator is to supply electricity. What are the benefits of using generators? What are the costs?

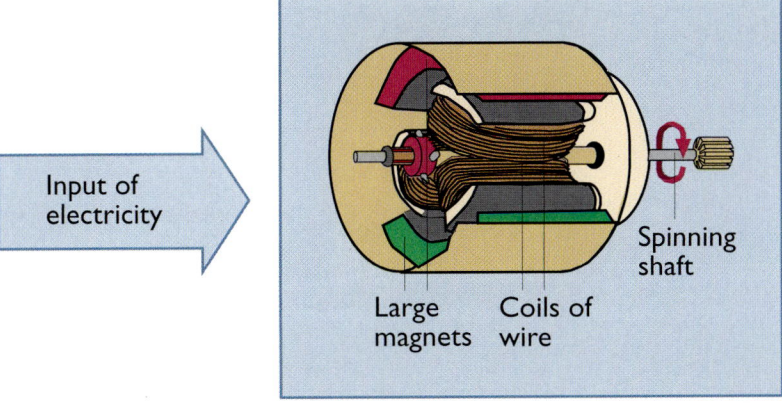

FIGURE 18.7 If you change the arrangement of the components in a turbine, you also change the output. By sending electricity through the coils of wire, you can cause the shaft to spin. This system is a motor.

You can build a variation of this system using the same types of components contained in a turbine but in an arrangement that keeps the magnet still and allows the coil of wire to move (see Figure 18.7). If you send an electrical current through the coils of wire in this system, the coils of wire will spin because the magnets are stationary. If these coils of wire are attached to a shaft, then as the coils spin, the shaft also spins. You can harness the energy of the spinning shaft to perform various types of outputs or work. We call such a system a **motor**.

To summarize, in a generator, you spin a shaft that contains coils of wire inside magnets to generate electricity. If you run the system in reverse, that is, put electricity into the system and allow the shaft to turn, then you have a motor. Motors and generators are essentially the same system with the inputs and outputs reversed (see Figure 18.8).

Observe the electric motor and the generator your teacher will demonstrate. With your class, conduct a brainstorming session in which you list as many systems as you can think of that contain motors and as many systems as you can think of that contain generators. Then select one system from either list and draw a simple diagram that shows input and output of your particular motor or generator system.

FIGURE 18.8 The energy input of a generator is a spinning movement that causes magnets to move past a coil of wire. The output of a generator is electricity. The energy input of a motor is electricity. The output of a motor is a spinning movement.

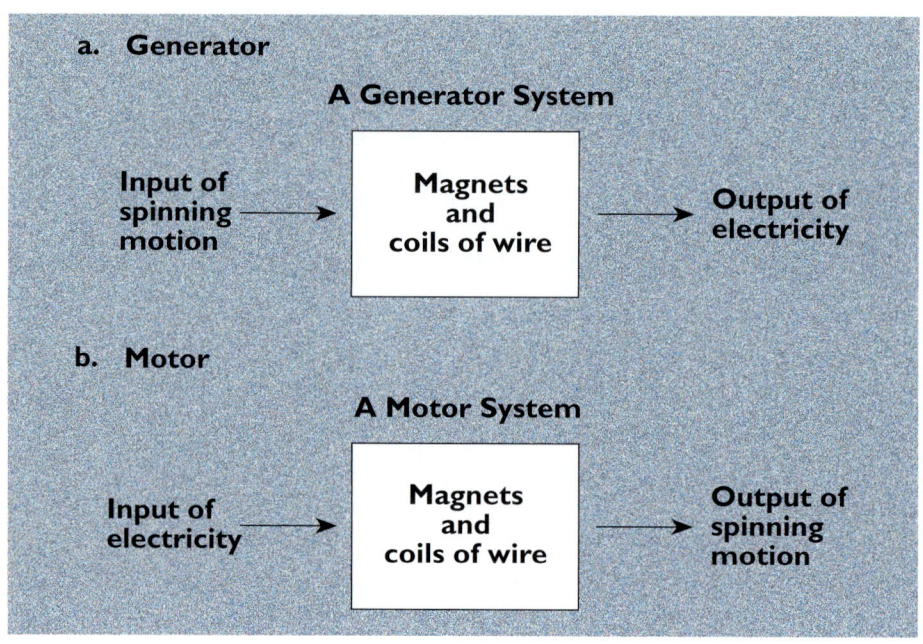

Paper Clip Lifters

In this investigation, you and your teammates will invent a technological system similar to a generator. Instead of electricity, however, the output of your team's system must be the kinetic energy of a spinning motion that can lift a container of paper clips. The better your team's device spins, the more paper clips it will lift. Your team's goal is to invent a system that lifts more paper clips than any other system in the class.

Materials for the Entire Class:

- at least 3 blow dryers
- at least 20 Styrofoam, paper, or plastic cups or bowls
- at least 10 aluminum pie pans
- at least 1 roll of aluminum foil
- at least 1 roll of plastic wrap
- at least 5 shoe boxes or other cardboard boxes
- at least 1 box of plastic straws
- at least 10 cans of various sizes
- several stacks of newspaper
- 1 roll of waxed paper
- at least 20 sheets of construction paper
- at least 2 sheets of poster board
- 2 rolls of masking tape
- 1 box of rubber bands
- at least 5 plastic bottles
- at least 5 scraps of boards
- at least 10 wire hangers
- at least 1 pair of wire cutters
- at least 20 paper tubes
- at least 10 pencils
- at least 2 spools of thread
- at least 2 skeins of yarn
- 1 ball of string
- a water supply
- additional construction materials that your teacher provides

Materials for Each Team of Three:

- 1 box of paper clips
- 1 pair of scissors
- 3 pairs of safety goggles
- 1 small paper cup

WORKING COOPERATIVELY

Work cooperatively in your teams of three. Use the roles of Manager and Communicator. Review the unit skills before you begin this investigation and concentrate on using those skills for this investigation.

CHAPTER 18 Energy Benefits and Costs

Process and Procedure

1. Participate in the demonstration and discussion that your teacher will conduct.
   → This demonstration will show you one system in which the output is a spinning motion that generates electricity.

2. Read the following criteria and constraints for developing a system for lifting paper clips.

   Criteria:
   - Your system must use a spinning motion to lift a container of paper clips.
   - The system must lift the paper clips at least 20 cm off a table or the floor.

   Constraints:
   - The container must be a small paper cup.
   - The energy input to the system cannot be the movement of a human hand.

   You have already learned about criteria and constraints, but to refresh your memory, you might recall that **criteria** are the goals that you want your product to accomplish. These criteria make your product the best it can be. **Constraints** are things that limit you from reaching your goals. These usually involve factors such as money, available materials, and time.

3. As a team, conduct a brainstorming session to list possible ways to build a system.
   → The Manager will survey the materials and report to the other Team Members. If your team is having difficulty, the Communicator should visit with other team Communicators to share ideas. Make sure that you include the type of energy input the system will have.

STOP: Remember to practice your unit skills.

4. Decide on a system.

   → Notebook entry: Record your decision and a list of the materials you will need. You might draw your system design in your notebook.

 Water will conduct electricity. During this investigation, you might be working with electrical appliances and water. Keep the appliances away from water and be sure that your hands are dry when you handle the appliances.

5. Build your system.
6. Use the guidelines shown in Figure 18.9 to check your system for safety.

FIGURE 18.9 Use the guidelines shown here to check your system for safety.

Safety Guidelines

Answer the following questions. If you answer **"YES"** to any of these questions, you must revise your system.

1. Does the system contain any sharp edges or points that could hurt someone?

2. Does the system contain any loose objects or pieces that could fly off and hurt someone while the system is operating?

3. Does the system contain water that spills out or otherwise causes slippery surfaces?

4. Does the system contain any hot surfaces or produce large amounts of heat that could hurt someone?

5. Does the system contain any exposed electrical wires or any wires or electrical appliances that might come into contact with water and create an electrical hazard?

Remember, all Team Members must wear eye protection and stand a safe distance away from the system.

7. Test your system by seeing how many paper clips it can lift 20 cm off a surface.
   → Observe the output and note any problems with your system.
8. Discuss ways that your team might improve its system.
   → Remember to generate alternative ideas.
9. If necessary, revise your system.
   → Remember to check your system for safety before you test it again.
10. Present your team's technological system to the class.
    → Describe how the system works and its energy input. Demonstrate how many paper clips your system can lift at least 20 cm off a table or the floor. Also describe how and where your team used the unit skills.
11. Compare all of your classmates' systems and answer these questions.
    - Which energy input generally lifted the most paper clips?
    - Which energy input seemed to be the easiest to use?
    - Which team's system seemed to be the most efficient?
    - Which team's system would you judge to be the best?

    → Notebook entry: Record your answers.

Wrap Up

As a team, decide on answers to these questions. In your notebook, record your team's answers.

1. What problem was your team's technological system supposed to solve?

2. What energy transformations took place in your system?

3. What were the energy input and output from your team's system?

4. Explain whether your team's system met your design criteria as well as you expected it would.

5. If you had another chance, how would you modify your team's system?

6. Consider the electricity-generating systems you examined in previous activities. Describe how you might change your system so that it could generate electricity.

Benefits and Costs

In the investigation Paper Clip Lifters, you compared several systems for lifting paper clips. Which system did you judge the best? You might have defined the best system as the one that lifted the most paper clips. There are other ways, however, to define best. Even systems that were not the best for winning the contest might be best for other reasons.

FIGURE 18.10 How would you decide which of these transportation systems is best? Under what kinds of circumstances would the system in photo (a) be best? How about (b) or (c)?

STOP & THINK

1. Consider the different systems your class constructed. Imagine that you needed a system that could spin a generator. Which system that your class constructed would you consider to be the best under each of the following circumstances? Explain your answers.
   a. You live in an area that has strong winds.
   b. You live near a waterfall.
   c. You need a system that spins a generator constantly without interruption. (This would provide you with a constant flow of electricity, for example, to operate a computer.)
   d. You need a system that is as simple and inexpensive to build as possible.

As you probably observed, each different system that your class constructed has its own set of advantages and disadvantages. In fact, every technological solution involves both benefits and costs. The **benefits** are the advantages of a particular solution; the **costs** are the disadvantages. The best solution for a problem is the one that gives you the most acceptable costs along with the most desirable benefits.

Often what is acceptable and desirable for one person is not acceptable and desirable for someone else. For example, every student must solve the problem of how to get to school. For some students, the best solution is to take the bus. The benefits of this solution include getting students to school inexpensively. The costs of this solution

FIGURE 18.11 Do you ride the bus to school? What are the benefits and costs of riding the bus instead of having a parent take you to school?

include having to get up early to catch the bus and having to wait at the bus stop. For some students, these costs are too high for the benefits. They would rather have their parents drive them to school so they can sleep late and they don't have to stand outside in bad weather.

Your solution to the problem of getting to school also involves a decision about using energy. If you solve the problem by riding in a bus or car, you depend on energy input from gasoline or diesel fuel. If you solve the problem by walking or riding a bicycle, you depend on energy input from your muscles. The benefits and costs related to each form of energy input differ. Using gasoline-powered vehicles might get you to school more quickly than would walking. Gasoline is expensive to buy, however, and burning gasoline causes air pollution. Using your muscles does not cost anything, gives you exercise, and does not pollute the air. Walking, however, can be tiring, slow, and unpleasant in some types of weather.

Energy-related problems have many potential solutions. Each solution comes with a different set of benefits and costs. It is not always easy to see which solution has the most acceptable costs and the most

FIGURE 18.12 Many people are concerned about the amount of energy that is wasted by unnecessary driving. Air pollution caused by car fumes also is a concern. Many people car pool or ride the bus to help alleviate problems with traffic congestion and pollution.

desirable benefits. To help you evaluate each solution, you could develop a method of rating each benefit and each cost associated with different solutions. After you have rated them, you could then total the scores for the benefits and the costs. Comparing these totals for different solutions can help you determine which solution has the most desirable benefits and the most acceptable costs.

2. Figure 18.13 shows how Marie rated some of the possible benefits and costs for two solutions to the problem of getting to school. According to this rating system, 10 means that Marie feels very strongly about a particular cost or benefit, and 1 indicates that she cares very little about that particular cost or benefit. Based on the totals for each set of benefits and costs, which solution do you think Marie will choose?

FIGURE 18.13 Marie has analyzed the costs and benefits for these two ways of getting to school. According to this data table, which solution do you think Marie will choose?

Problem: How to get to school

Solution 1: Ride my bike

Benefits (score)	Costs (score)
1. I can go when I want. (8)	1. I might get cold and wet in bad weather. (6)
2. It doesn't cause air pollution. (2)	2. My bike might get stolen. (5)
3. It doesn't cost me anything. (10)	3. My helmet would mess up my hair. (3)
Total 20	Total 14

Solution 2: Have my sister drive me to school

Benefits (score)	Costs (score)
1. I'd get there faster. (3)	1. I'd have to help pay for gasoline. (10)
2. I'd stay warm and dry. (6)	2. I'd have to wait for my sister to get ready every morning. (6)
3. I could get up later in the morning. (6)	3. It would cause air pollution. (2)
Total 15	Total 18

STOP & THINK

3. Now make your own chart that shows the costs and benefits of the different ways of getting to school that are available to you.
4. According to your cost and benefit chart, decide on the best way for you to get to school.

CHAPTER 18 Energy Benefits and Costs

ELABORATE connections

Wind Farms

In 1981, wind energy produced enough electricity in the United States to power two homes. By 1989, it produced enough to power all the homes in a city as large as Washington, D.C. or San Francisco. By 1996, the capacity of installed wind energy systems in the United States was 1794 megawatts. In 1981, this capacity was only 10 megawatts. One large system that contributed to this increase in wind energy is located in Altamont Pass in California. This system is a collection of 7,300 windmills known as a wind farm. Each windmill contains a turbine on top of a tower. The turbine is attached to a set of blades. When the wind blows, the blades spin and cause the turbine to spin. Since 1981, the turbines have produced more than 6 billion kilowatt hours of electricity, which is enough electricity to power about 800,000 California homes for a year.

The residents of Altamont Pass have mixed opinions of the wind farms in their backyards. A reporter interviewed several of the residents and recorded what they thought about utility companies producing electricity by using energy input from the wind. Make a list of the benefits and the costs of this solution to the problem of how to generate electricity. Then rate how strongly you feel about each benefit and cost. Finally, use your ratings to decide whether you think wind farms are a good solution for your community's electricity needs. To help you evaluate your decision, answer these questions. Prepare to share your ratings and answers with the class.

1. What do you consider to be the greatest benefit of wind farms?
2. What do you consider to be the greatest cost of wind farms?
3. Why did you rate the benefits and costs the way you did?
4. If the local power company planned to install a wind farm in your neighborhood, would you approve of their plan? Explain your point of view.
5. Why are wind farms an appropriate electricity-generating system to use in Altamont Pass, California?
6. Would wind farms be an appropriate electricity-generating system to use in a large city? Explain your answer.
7. Would wind farms be an appropriate electricity-generating system to install in a heavily wooded area, such as in a forest? Explain your answer.

Topic: Generators
Go to www.scilinks.org
Code: physical467

ELABORATE investigation

A Variety of Systems

Al, Marie, and Ros have created a unique system to operate their radio (see Figure 18.14). Yet, Rosalind is right; there are many systems that will work to spin a generator for producing electricity. Each system uses a different input and different devices to produce the output of electricity. Of course, some systems deliver energy to you without ever producing electricity. Each system comes with a different set of benefits and costs.

In this investigation, you and your teammates will become the class experts about one system for generating electricity. As the class experts, your team will make a

FIGURE 18.14 What are the costs and benefits of this electricity-generating system?

presentation that explains the input, the device, and the output of the system. Your expert team also will explain three benefits and three costs of using this system to generate electricity.

As you research various electricity-generating systems, keep in mind the basic design of technological systems shown in Figure 18.15. Most of the systems you and your classmates will investigate use a turbine as a device and have electricity as the output. In this way, the systems are a lot more alike than they are different. As you do this investigation and listen to your classmates' presentations, look for similarities in the systems.

Topic: Energy Transformations
Go to www.scilinks.org
Code: physical469

Materials for Each Team of Three:
- 1 lemon
- 2 pennies
- 2 dimes
- 1 small knife
- one 5 × 5-cm piece of fine grit sandpaper
- one 250-mL beaker
- 200 mL of ammonium chloride solution
- one 25-cm long piece of copper wire
- one 25-cm long piece of iron wire
- 1 Bunsen burner
- 1 galvanometer
- 3 pairs of safety goggles
- 1 sheet of poster board (Part B)
- 1 set of markers
- 3 copies of the Rating Sheet

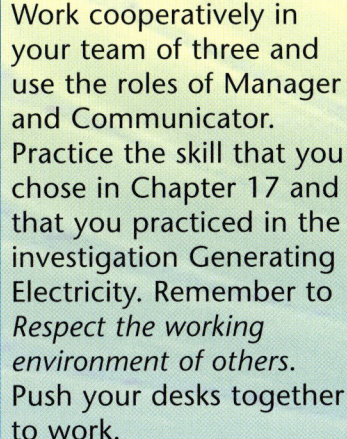

Work cooperatively in your team of three and use the roles of Manager and Communicator. Practice the skill that you chose in Chapter 17 and that you practiced in the investigation Generating Electricity. Remember to *Respect the working environment of others.* Push your desks together to work.

Input → Device → Output

FIGURE 18.15 Technological systems contain an input, a device, and an output.

CHAPTER 18 Energy Benefits and Costs

Process and Procedure

Part A—Three Electricity-Generating Systems

1. Construct and test the lemon battery shown in Figure 18.16 by using the following steps.

    Wear eye protection at all times.

   a. Get 1 penny, 1 dime, 1 lemon, a piece of sandpaper, and your galvanometer.
   b. Use the sandpaper to clean both sides of the penny and the dime. The system works better if the coins are clean. It is not necessary, however, to polish the entire coin.
   c. Roll the lemon back and forth on a table or in your hands, pressing down gently with your palms.
      → This will break the cells inside the lemon and produce lemon juice inside the lemon.
   d. Cut 2 slits in the side of the lemon that are about 1 cm ($\frac{1}{2}$ inch) apart.
      → Keep the slits small, just large enough for one of the coins to fit inside. You only need one set of slits in each lemon.
   e. Insert the 2 coins into the slits in the lemon. Leave about half of each coin sticking out above the lemon peel.
   f. Touch the bare ends of the wire from the galvanometer to the 2 coins.
   g. Alternate touching each wire to the coins and then remove the wires.
      → Notebook entry: Record your observations.

FIGURE 18.16 Construct and test the lemon battery shown here.

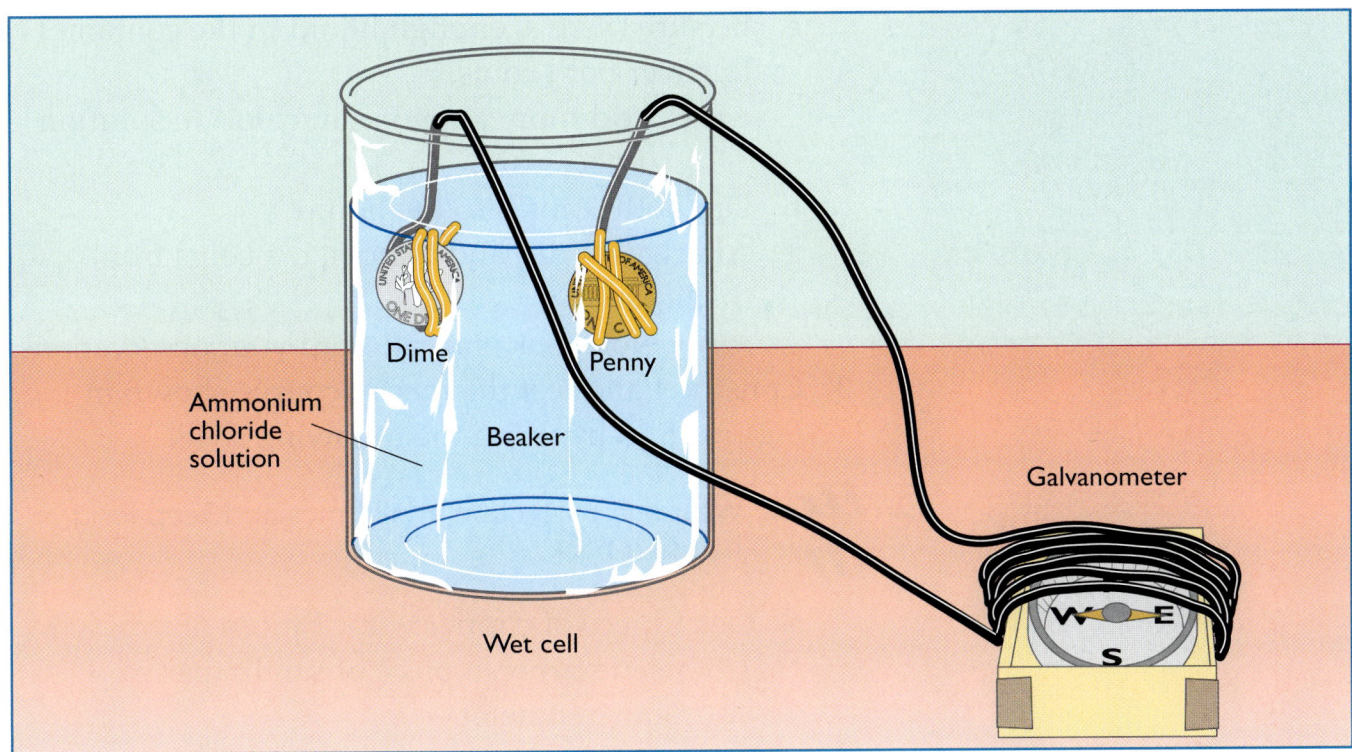

2. Construct and test the wet cell shown in Figure 18.17 by using the following steps.

FIGURE 18.17 Construct and test the wet cell shown here.

 Wear eye protection at all times and keep long hair tied back.

a. Get 1 penny, 1 dime, 1 beaker of ammonium chloride solution, a piece of sandpaper, and your galvanometer.
b. Use the sandpaper to clean both sides of the penny and the dime.
   → The system works better if the coins are clean. It is not necessary, however, to polish the entire coin.
c. Wrap the bare end of one wire from the galvanometer around the penny. If the bare end of the wire is not long enough, strip more insulation from the wire.
d. Wrap the bare end of the other wire from the galvanometer around the dime.

CHAPTER 18 Energy Benefits and Costs

e. Be sure there is enough liquid in the container to cover both coins.
→ Add more ammonium chloride solution if necessary.
f. Dip both coins into the liquid.
g. Alternately dip and remove the coins from the liquid.
→ Notebook entry: Record your observations.

3. Construct and test the thermocouple shown in Figure 18.18 by using the following steps.

 Wear eye protection at all times and keep long hair tied back.

a. Get 1 piece of copper wire, 1 piece of iron wire, 1 Bunsen burner, a piece of sandpaper, and your galvanometer.
→ Review the procedure for How to Operate a Bunsen Burner, How To #9, if you need help.
b. Use the sandpaper to clean the ends of both wires. Clean about 5 cm on each wire. The system works better if the wires are clean.

FIGURE 18.18 Construct and test the thermocouple shown here.

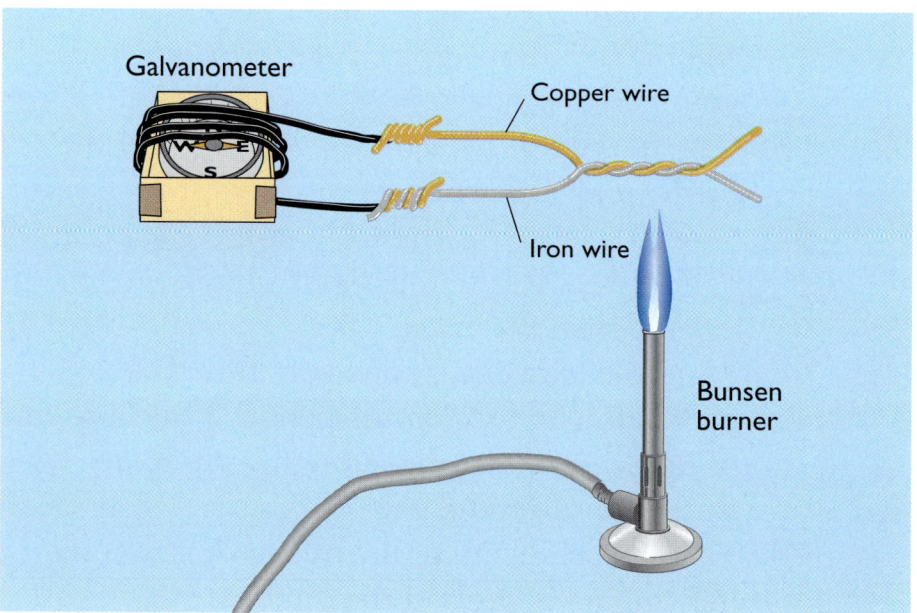

c. Twist one end of the copper wire and one end of the iron wire together.
d. Twist the other bare end of the copper wire to one of the wires from the galvanometer.
e. Twist the other bare end of the iron wire to the other wire from the galvanometer.
f. Hold the copper-iron twist in the flame of the Bunsen burner.
g. Alternately heat and cool the twisted wires.
  ➡ Notebook entry: Record your observations.

4. For each system you constructed, describe the parts of the system, the energy input, and the energy output.
  ➡ Notebook entry: Record your descriptions.

5. Describe what is similar and what is different about each of these electrical-generating systems.
  ➡ Notebook entry: Compare the parts of the system, the energy input, and the energy output.

Part B—The Presentations

1. Your teacher will assign your team a presentation to prepare on 1 or more of the following systems for generating electricity:

FIGURE 18.19 Where might you find information about each of the systems shown on the following pages?

CHAPTER 18 Energy Benefits and Costs

- coal
- petroleum
- natural gas
- synthetic fuels
- nuclear fusion
- nuclear fission
- Solar Thermal Power Systems (STPS)
- photovoltaic conversion
- wet geothermal (steam and geysers)
- dry geothermal (underground hot rocks)
- hydroelectric
- tides
- waves
- wind
- Ocean Thermal Energy Conversion (OTEC)
- biofuels

2. Research and learn about your assigned system.
   → If you have questions about how to conduct your research, you might consult How To #6, How to Conduct a Research Project. In addition, there is a Special Section on Energy Systems at the end of this chapter that will

help you and your teammates begin your research. When you are ready for more information, you can find additional resources in your media center, local library, or on the World Wide Web. Your local utility company also might be able to provide some useful information.

3. Prepare a poster that shows where the energy input for your system comes from, the parts of the device, and the output of the system.

4. On the Rating Sheet, list 3 benefits and 3 costs of using this system for generating electricity.
   → Each Team Member should fill out his or her Rating Sheet.

5. On a scale of 1 to 10, rate how your team values each of the benefits and costs.

6. Prepare a 5-minute presentation that explains to the class how this system works and what the benefits and costs are.
   → Each member of your team must take part in the presentation.

7. Present your assigned system to the class.
   → In your presentation, be sure to describe your team's use of the social skill. As you listen to the other

CHAPTER 18 Energy Benefits and Costs 475

presentations, think about how all of the systems are similar and how they are different.

→ Notebook entry: Record the similarities and differences of the systems.

8. Compile a class chart that summarizes the costs and benefits of each system using the following steps.

   a. Using your Rating Sheets, help your teacher fill in the benefits section of the class chart.
      → If your team listed a benefit of your system and rated that benefit 5 or higher, tell that benefit to the teacher, and he or she should add it to the benefit list.
   b. When the class is satisfied that the benefit list is complete, have your Communicator go to the chalkboard and fill in your team's rating for the benefits you listed for your system.
   c. Repeat Steps 8a and 8b to compile information about the costs of each system.

Wrap Up

After you have made your presentation and constructed your class chart, write a newspaper article that answers these questions. Share your article with your teammates.

1. Identify and describe the one or two systems that have consistently high ratings for benefits and low ratings for costs.
2. Identify and describe the one or two systems that have consistently high ratings for costs and low ratings for benefits.
3. What do the different ratings tell you about the problems a community might have in developing and implementing some of these energy systems?

connections EVALUATE

Which System Do We Use?

In this chapter, you have learned about different ways to generate electricity. You have discussed many of the benefits and costs of using each system. Based on these discussions, you probably have opinions about which systems you would like to see used in your community. Maybe your community already uses your preferred system to generate electricity.

Find out what type of electricity-generating system your community uses. Discuss these questions with a partner and be prepared to share your answers with the class.

1. Describe how the system that your local power plant uses is appropriate for your community.
2. What, if any, are the costs of using this system in your community?

Now imagine that the system your community uses had to be shut down. Which of the other systems for generating electricity do you think would be an appropriate replacement? Prepare a proposal for the one system that you feel is best suited to your community. Make sure that you include a description of how the benefits and costs of this system apply to your community.

Special Section on Different Energy Systems
Biofuels

Biofuels are energy resources—such as trees, crops, weeds, animal manures, and seaweed—that are derived from living organisms. People can even make some types of biofuels from human wastes. In general, biofuel systems transform the chemical energy in plant or animal matter into a liquid, solid, or gaseous fuel. People can then use these fuels for heating and cooking. By using techniques similar to those used in coal-burning power plants, companies also can burn biofuels to produce electricity. Such systems can generate enough electricity to meet the needs of an entire city.

People can cultivate biofuel crops such as corn. Then, using a process called fermentation, they convert the corn into a refined fuel, such as ethane gas. This gas burns more cleanly and efficiently than would the unrefined corn. The process of growing corn and refining it into a gas, however, requires energy input. It also requires land to grow the crop. In some areas, people cannot afford to use land to grow biofuel crops instead of food crops. However, people can grow some biofuels on land unsuitable for food production, such as hillsides and land with poor soil. Growing biofuel crops on such land might help stabilize the soil and prevent erosion. Alternatively, people can grow crops like sea kelp in offshore plantations. Like corn, this kelp can be converted into a fuel such as methane gas.

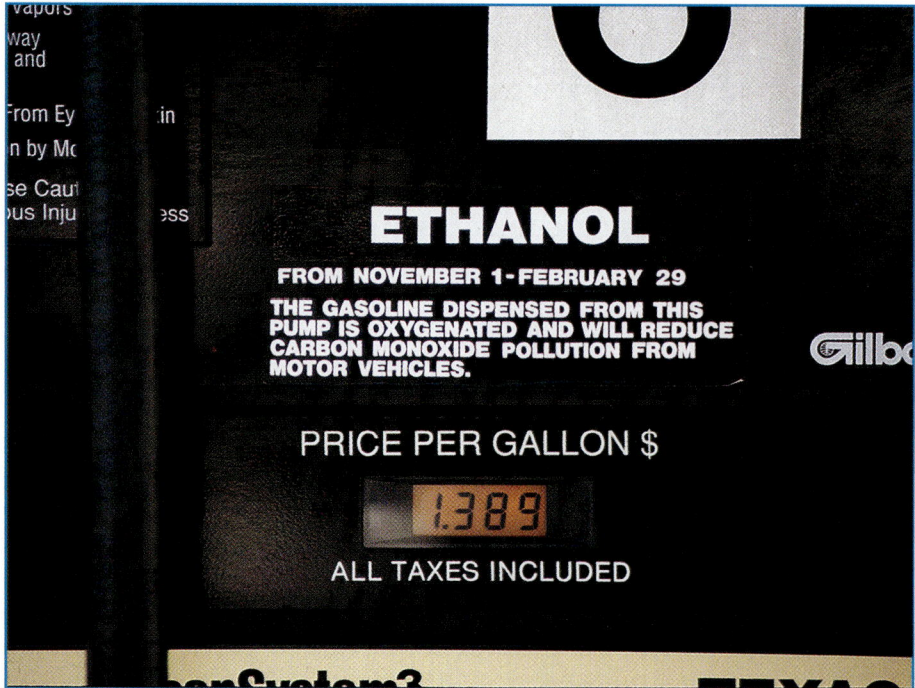

FIGURE 18.20 Ethanol, a substance derived from corn, is added to gasoline. It helps gasoline burn more efficiently.

One advantage of biofuels is that they are a renewable energy resource. Unlike the nonrenewable fossil fuels, people probably will never run out of a supply of materials to make biofuels. For example, after harvesting trees for wood, people can plant more trees. Some people suggest that in the future, biofuels might take the place of fossil fuels such as coal and gasoline. (For a description of fossil fuels, see Coal.)

Biofuel systems also can help solve waste-disposal problems. For example, some systems use bacteria or chemical processes to convert household sewage or garbage into "biogas." In agricultural areas, farmers also can produce biofuels from animal manure and the inedible or unharvested parts of plant crops (for example, rice husks, corn stalks, coconut shells, and peanut hulls). In this case, however, farmers must use large amounts of energy to collect, dry, and transport the materials to a biofuel processing plant. Some scientists suggest that it makes more sense to use the animal and crop wastes to fertilize the soil or to use the wastes as a mulch to spread over areas that are susceptible to erosion.

Sometimes converting animal and plant matter into biofuels is not as effective as using the materials directly as an energy resource. For example, wood is a valuable energy resource for people in many countries. In the United States, we use wood for 3 percent of our energy consumption. In comparison, people in Sweden use wood for 8 percent of their energy needs and people in Brazil use wood for 27 percent of their energy needs. Unfortunately, if managed improperly, the harvesting of wood fuels can result in air pollution and deforestation. (Deforestation refers to the clearing of forests, either through the actions of people or because of natural processes such as climate change.)

Coal

Coal is one of the energy resources we call fossil fuels. Fossil fuels come in three forms: solids such as coal or oil shale, liquids such as petroleum, and gases such as natural gas. Fossil fuels earn their name because they are the remains of ancient organisms that have changed over many millions of years.

To understand how coal formed, scientists study processes that occur in shallow freshwater and marine environments such as swamps. These environments contain a variety of organisms. When the organisms die, their bodies sometimes are deposited in the water. Often, bacteria break down their remains through decomposition. Yet sometimes there are too many dead organisms and not enough oxygen to support the bacteria. As a result, partially decomposed remains accumulate.

FIGURE 18.21 Although burning coal can produce electricity, coal-burning power plants are not very efficient, and they can produce heavy air pollution.

Scientists hypothesize that coal began to form as a result of similar processes from the partly decomposed bodies of ancient organisms that accumulated in shallow bodies of water. Over the years, layers of sediments such as mud, sand, and gravel also built up along with the remains. As these layers accumulated over many thousands of years, the weight of the layers on top compressed the lower layers, causing them to heat up. Over the course of millions of years, the pressure and heat caused chemical changes that turned the remains into solid coal.

Coal is the most abundant fossil fuel in the world. It is also the most abundant fossil fuel in the United States. The United States Geological Survey has identified 1.7 trillion tons of coal deposits in the United States. Some scientists, however, estimate that there may be as many as 4 trillion tons.

Mining companies use strip mining or contour strip mining to extract coal from deposits that are close to the surface. In these processes, the miners make a trench by removing a strip of rock that covers the coal deposit. They place this rock in a pile to one side of the trench. After removing the coal, the miners fill the trench with rock debris from the next trench. Once the mining company finishes removing the coal deposit, they often attempt to restore the land. This is because in the United States, federal law requires mining companies to restore the land so that it is as productive as it was before mining.

Unfortunately, such restoration projects are slow and costly. In addition, chemicals used in the mining process can leach into the ground and surface water as a continuing source of pollution. Sometimes companies abandon the land without completing the restoration project. Sometimes the efforts fail to return the land to a productive state. Nonetheless, there are farms, parks, wilderness areas, and recreational areas on lands that were once surface mines.

FIGURE 18.22 This is a photo of an open pit coal mine. What do you think are the benefits and costs of coal mining? What alternatives to fossil fuels would you use?

Mining companies also mine coal from deposits deep underground. In this instance, miners must dig deep shafts and blast long tunnels and rooms to reach the deposits. That kind of work can be dangerous, both from the risk of collapsing tunnels and other mine disasters and from diseases such as "black lung" that scientists link to breathing air laden with coal dust.

The coal extracted from surface and underground mines is a valuable energy resource. In the United States, power plants burn roughly 70 percent of the extracted coal to generate electricity. In such a power plant, large amounts of heat are produced that usually must be converted into mechanical energy and then into electrical energy. Steam is frequently used in this process. The coal is first crushed into a powder and then burned to produce heat to heat water and change it into steam. The high pressure steam is directed so that it turns the blades of a turbine. This process also has been used with a system of valves to operate an engine (for example, the famous railroad steam engine). Many of these systems are able to extract the heat of vaporization as well as the heat energy of the steam. There is a limit to the efficiency of this energy transfer system, however. About 60 to 65 percent of the original heat energy cannot be transformed into mechanical energy. Instead, it is released into the surroundings. In the United States, about 35 percent of the energy produced comes from coal-burning power plants.

Unfortunately, in the process of transforming chemical energy into heat energy, coal-burning power plants contribute to air pollution. They produce large quantities of carbon dioxide and airborne particles such as ash and soot. We know that carbon dioxide is increasing in the atmosphere and that it absorbs infrared radiation, which in turn promotes the greenhouse effect and contributes to global warming. Other gases such as sulfur dioxide and nitrogen dioxide are released. In the sunlight, these gases are converted to acids, which then contribute to the acid rain problem. (For a description of global warming, see Petroleum.) Fortunately, several technologies are available to reduce the quantity of some of the pollutants. These technologies include electrostatic precipitators and scrubbers. In addition, a process exists that cleans coal before it is burned to remove some of the sulfur and ash-forming minerals.

Dry Geothermal

Geothermal means "earth-heat." Deep inside the Earth, layers of rock are hot and molten (liquefied). In some places, such as the western United States, this hot, molten rock is close to the surface. Here, volcanoes occasionally bring molten rock to the surface. In other places, the molten rock heats large underground reservoirs of water. These reservoirs can form hot springs and geysers at the surface. (Geysers are fountains of steam and hot water.)

Dry geothermal systems transform the heat energy of hot rocks into electricity. (Wet geothermal systems, on the other hand, rely on heat from the hot water in underground reservoirs. For more information, see Wet Geothermal.) In some dry geothermal systems, engineers drill two holes into an area that contains molten or hot rocks close to the

FIGURE 18.23 This is an area in Hawaii where hot, molten rock reacts with moisture in the air and produces steam.

surface. These holes are located several hundred meters apart from one another. Pumps push cold water into one of the holes. Then the heat from the rocks changes the cold water into steam. The steam becomes pressurized and travels to the surface through the second hole. As it reaches the surface, the steam spins a turbine that generates electricity.

Some advantages of dry geothermal systems are that they do not create significant amounts of air or water pollution. They also use a free energy resource, hot rocks. Hot, dry rock is potentially the largest and most widely distributed geothermal resource in the United States. One disadvantage of dry geothermal systems is that they are only useful in certain places. In some locations, hot rock formations are more than two miles beneath the surface. Drilling holes that deep requires expensive equipment and skilled workers. In addition, drilling sometimes fractures rocks and causes small earthquakes.

Hydroelectric

People can use falling or flowing water from rivers and streams as a source of mechanical energy. Hydroelectric systems convert this mechanical energy into electrical energy by diverting flowing water through a turbine that generates electricity. For example, engineers can build a dam across a river. The dam creates a reservoir, which is an artificial lake for storing water. The water at the bottom of the reservoir is under great pressure. The pressure pushes water through the dam through a series of pipes. These pipes lead to a powerhouse where the water flows past the blades of a turbine. As the turbine spins, it generates electricity.

Hydroelectric power plants are the source of about 25 percent of the world's electricity. These systems have a long life, low maintenance costs, and do not require an input of expensive fuels such as coal or petroleum products. Hydroelectric plants also cause very little water or air pollution.

The dams required by hydroelectric systems can create better environmental conditions for some animals and plants. Such dams also create reservoirs that can provide recreation for many people. In addition, dams can help control the effects of floods by providing an even, steady flow of water out of the dam during heavy rains. On the other hand, dams can have a critical and sometimes irreversible impact on the ecological balance of a river or stream. For example, by creating a reservoir, the dam destroys the habitat of many plants and animals while causing other areas along the river to dry up. Dams also can interfere with the migration of some kinds of fish, such as salmon.

FIGURE 18.24 Hydroelectric systems convert mechanical energy from flowing water into electrical energy. The flowing water is diverted through a turbine that generates electricity.

Natural Gas

Natural gas is a fossil fuel like coal and petroleum. (For a description of fossil fuels, see Coal.) Some scientists hypothesize that natural gas formed from the partially decayed remains of ancient organisms that accumulated at the bottoms of ancient lakes, rivers, and seas. Over millions of years, these deposits became buried by sediments such as mud, sand, and rock. Gradually, the sediments accumulated in deep layers. As the layers became deeper, the pressure on the plant and animal matter increased. This pressure caused the materials to heat up. The combined pressure and heat caused chemical changes that gradually transformed the plant and animal matter into fossil fuels such as natural gas.

FIGURE 18.25 Natural gas is used to operate this appliance.

Scientists can locate, drill, and extract natural gas using many of the same methods they use to obtain petroleum. (For a description of these methods, see Petroleum). In fact, natural gas and petroleum often occur together in the same deposits. Like petroleum, producers must refine natural gas so that it is useful as an energy resource. That is because natural gas actually is a mixture of many gases including methane, ethane, propane, and butane. It also contains gases such as carbon dioxide, water, and nitrogen that producers consider contaminants.

After producers extract the natural gas, they must transport it to a processing or "stripping" plant. Those plants separate the component gases and remove contaminants. Then the gas is pressurized to reduce its volume. This refined and pressurized product is a valuable energy resource for a variety of uses.

People use natural gas to heat homes and operate appliances such as clothes dryers, stoves, air conditioners, humidifiers, and water heaters. The United States produces about 27 percent of its energy by using natural gas. Power plants that burn natural gas transform the chemical energy in the gas into electricity in much the same way that coal-burning power plants do. (For a description of these power plants, see Coal.)

Natural gas is relatively inexpensive and easy to obtain. In addition, natural gas produces more heat energy and less air pollution than any other fossil fuel. Burning natural gas also produces less carbon dioxide than does burning fossil fuels such as coal. Some scientists hypothesize that carbon dioxide contributes to global warming through a process known as the greenhouse effect. (For a description of the greenhouse effect and global warming, see Petroleum.)

To deliver natural gas to consumers, the United States contains a network of pipelines that are a total of 270,000 miles long. These pipelines cross all kinds of terrain—over mountains, rivers, and deserts and underneath cities and towns. In this way, natural gas moves from drilling sites, to processing plants, to individual users and power plants. In addition, producers can store excess supplies of natural gas in underground storage tanks and even in underground rock formations that formerly contained petroleum and natural gas. Scientists and engineers also have developed technology to convert natural gas into a liquid by reducing its temperature. Producers can then ship this liquid to consumers in other parts of the world.

One of the drawbacks of using natural gas as an energy resource is that, like coal and petroleum, it is available in limited supplies. In addition, there are dangers associated with transporting, handling, and using natural gas. For example, if the pilot light in a natural gas water heater goes out, the gas can leak into a closed space. That might create an explosion hazard. Leaking gas also might pose an asphyxiation hazard if it replaces all of the oxygen in a closed space. In general, however, manufacturers are careful to develop safe and effective natural gas products. With proper installation, adjustment, and ventilation, these appliances are safe.

Nuclear Energy

Nuclear energy is the energy stored in the nucleus of an atom. If a nucleus is unstable, it may release some of its energy. This process is called radioactive decay.

Nuclear fission: The nuclei of some heavy elements will undergo spontaneous fission (split apart) into smaller nuclei. Large amounts of energy are released during this process. The energy released is both heat and light. Some of these nuclei can be made unstable by bombarding them with neutral particles called neutrons. These are released in most fission processes so a controlled chain reaction can result. In a controlled chain reaction, the energy can be harnessed to generate electricity. (See Figure 18.26a.)

Nuclear fusion: When some light nuclei combine to make a heavy nucleus, nuclear energy is released. Nuclear fusion is the energy source for our Sun. The light nuclei have a positive charge. Large amounts of kinetic energy (which produces heat) permit the nuclei to overcome the electrical repulsion. Then they get within the range of the attractive nuclear forces. On the Sun, this hot material is held together by gravity (see Figure 18.26b).

Unfortunately, the release of nuclear energy usually produces nuclear radiation. Nuclear radiation consists of high energy that is carried by gamma rays, charged particles, and neutral particles that escape from radioactive nuclei. These nuclei are unstable, and the

Topic: Nuclear Energy
Go to www.scilinks.org
Code: physical488

a.

Nuclear Fission

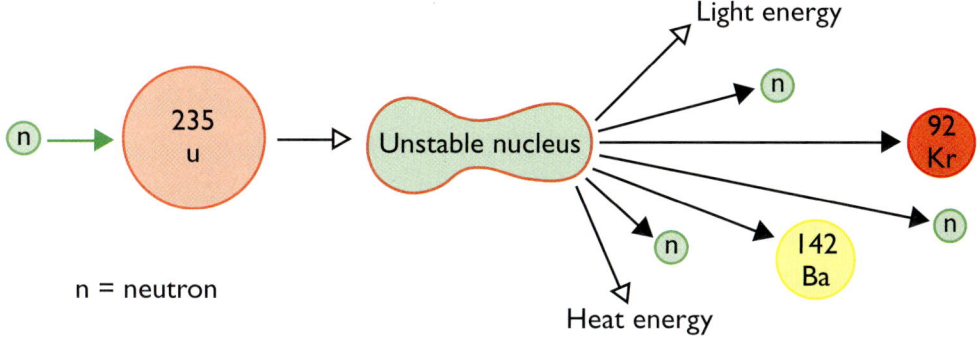

n = neutron

Source: http://library.thinkquest.org

b.

Nuclear Fusion

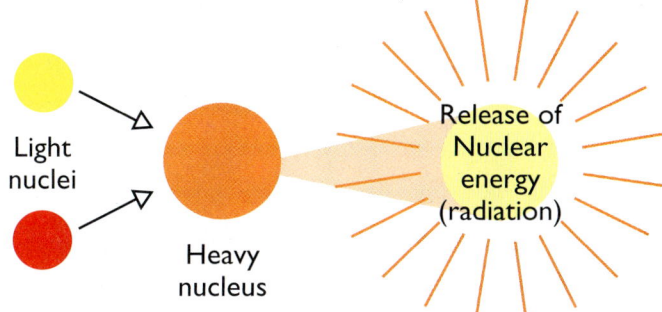

FIGURE 18.26a–b (a) When a heavy particle is bombarded with a neutron, the heavy particle splits, releasing nuclear energy. (b) When some light nuclei fused to make a heavy nucleus, nuclear energy also is released.

potential energy stored inside is occasionally released as high energy packets called gamma ray photons, or the kinetic energy of particles.

The rate of energy release from a particular radioactive sample depends only on the number of unstable nuclei it contains. As these are changed to stable nuclei, the total radiation decreases. The time required for half the nuclei to change is called the half-life. The radiation released is then half as great. Even after many "half-lives" there still will be some radiation. Nuclear half-lives range from fractions of a second to millions of years. The radiation from a short half-life sample is intense for a short time. Radiation from a long half-life sample decreases very slowly.

CHAPTER 18 Energy Benefits and Costs

Nuclear Fission Reactors

Nuclear power plants use nuclear fission systems to generate electricity. These plants contain a nuclear reactor that generates a large amount of heat. The plant uses this heat to turn water into pressurized steam which in turn drives an electrical generator. The technologies used to heat the water and drive the generator are similar to those used in coal-burning power plants. The unique part of the nuclear power plant is the nuclear reactor.

FIGURE 18.27 Nuclear power plants use nuclear fission systems to generate electricity.

Most nuclear reactors contain uranium as a source of nuclear energy. To obtain this uranium, companies must mine naturally occurring uranium ore and refine it using a series of complex and costly steps. One ton of uranium ore yields only 0.9 to 1.8 kilograms (2 to 4 pounds) of refined uranium. Companies must then form the refined uranium into fuel rods. An average nuclear reactor contains roughly 40,000 uranium fuel rods. These fuel rods are located inside the reactor in what is called the "core."

Nuclear power plants generate large amounts of electricity with relatively little fuel input. For example, a coal-burning power plant requires 2.3 million tons of coal to generate enough electricity for a city of roughly one-half million people. A nuclear power plant of the same size requires only 33 tons of uranium. In addition, unlike coal and other fossil fuels, nuclear fission does not produce large amounts of air pollution. Finally, nuclear fission systems rely on a fuel source that is relatively abundant compared to fossil fuels.

Like all radioactive materials, however, the radioactivity released from uranium can cause changes in the chemical and nuclear composition of many things. These include changes in the genetic makeup of plants, animals, and humans. Small amounts of naturally occurring radioactive energy and particles constantly bombard the organisms living on the Earth's surface. This radioactivity in small amounts has no apparent ill effect. Yet scientific studies show that large doses of radiation can cause sickness, an increased risk of cancer, or even death.

To protect people from dangerous doses of radioactivity, nuclear power plants must contain several barriers to trap and contain radioactive materials. These include safety systems and backup safety systems to prevent malfunctions, mistakes, and potential accidents. One dangerous kind of accident is known as a "meltdown." Normally, the nuclear power plant must use cooling systems to keep the core from overheating. In a meltdown, the cooling systems fail and the core becomes overheated. This overheating begins a runaway process that produces more and more heat. In the worst case, the core could become so hot that it would melt through its container and release large amounts of radioactivity into the environment.

Although there have been accidents in nuclear power plants, none has involved a complete meltdown. In 1979, an accident at the Three Mile Island nuclear plant in Pennsylvania resulted in a partial meltdown of the reactor core. In 1986, an accident at the Chernobyl nuclear power plant in the Ukraine caused an explosion, blew the roof off the reactor building, and set parts of the core on fire. Winds spread radioactive debris from this explosion over long distances including parts of Asia and Europe.

Another disadvantage of nuclear fission systems is the radioactive wastes they produce. First, the mining and processing of uranium ore produces radioactive and chemical wastes. Companies must dispose of these wastes so that they do not contaminate water or land at dumpsites. Another form of waste produced by nuclear fission systems is used uranium fuel rods. Nuclear power plants must replace their uranium fuel rods roughly every three years. Used uranium fuel rods are considered "high-level" radioactive wastes. They emit high levels of radioactivity (including energy and small pieces of matter) for many thousands of years. ("Low-level" radioactive wastes give off smaller amounts of radioactivity for shorter periods of time.)

Scientists and engineers have proposed several different kinds of waste-disposal systems for high-level radioactive waste. Possible systems include burial in deep underground sites; burial in the Antarctic or Greenland ice sheets; dumping into outer space; dumping in deep ocean sediments; processing into less harmful materials; or processing into smaller quantities that could be used as an energy resource for individual consumers. At present, the United States has several temporary storage sites for high-level radioactive waste. The government also is developing permanent storage sites at Yucca Mountain in Nevada and near Carlsbad, New Mexico.

In the United States, nuclear power accounted for about 20 percent of the energy supply produced in 2003. Many scientists predict that our use of nuclear energy will increase in the coming years. Other people, however, suggest that public dissatisfaction with nuclear energy could cause its use to decline. Advances in technology, however, are reducing and eliminating some of the disadvantages of nuclear fission systems.

Nuclear Fusion Reactors

A tremendous amount of nuclear energy is released when the nuclei of two particles join together to form a single nucleus. The process in which nuclei join or fuse together is known as nuclear fusion. (Note: Nuclear fission is a different process in which a single nucleus splits and releases nuclear energy and small amounts of matter.) Nuclear fusion occurs in the Sun and other stars and has kept them shining for billions of years.

It is obviously not possible to make a fusion reactor like that of the Sun for a power station. Confining the very hot, reacting nuclei is a major research problem that scientists are trying to solve. Modern technological systems cannot produce such conditions, nor can they control the large amounts of energy released by the process. Nonetheless, scientists continue to experiment with different systems to overcome those difficulties.

One experimental system relies on a reaction between two forms of hydrogen, deuterium and tritium. Another system relies on a reaction between two deuterium nuclei. This nucleus of hydrogen contains a neutron as well as the usual proton. Water that has this type of hydrogen nucleus is called "heavy water." Because small concentrations of heavy water are found in sea water, it is fairly abundant and easy to obtain. Scientists also are testing a system that relies on magnets and lasers to contain the nuclear fusion reaction. All of these systems, however, are in the experimental stages. They require much more testing and revising before they might produce successful results. So far, all these systems use more energy than they produce.

A system to harness the nuclear fusion process would provide a valuable, long-lasting energy resource. For example, power plants could use the heat from the fusion reaction to turn water into pressurized steam. This steam could drive a turbine that would generate electricity, similar to the way that coal-burning power plants generate electricity. Advantages of such future nuclear fusion systems include that they would produce little air or water pollution. The nuclei of both the source and the product are stable, and little nuclear radiation should result. In addition, unlike fossil fuels, nuclear fusion relies on an essentially unlimited resource. (For a description of fossil fuels, see Coal.) Disadvantages of future nuclear fusion systems include that they might require equipment that is costly to construct and operate.

Ocean Thermal Energy Conversion

The world's oceans store large amounts of heat from the Sun. Scientists and engineers have developed a kind of system called Ocean Thermal Energy Conversion (OTEC) that produces electricity using the temperature difference between the warm surface waters and the cooler, deeper waters of tropical oceans. In these systems, pumps move warm surface water through a

FIGURE 18.28 Ocean Thermal Energy Conversion produces electricity by using the temperature difference between warm surface waters and the cooler, deeper waters of tropical oceans.

device that contains a liquid such as ammonia. This liquid boils at a low temperature. Therefore, heat energy from the water causes the liquid ammonia to boil and then evaporate as ammonia gas. The pressurized ammonia gas can drive a turbine that generates electricity, similar to the way that fossil-fuel burning power plants generate electricity. OTEC systems also contain devices that pump cold water up from the ocean bottom. This water cools the ammonia gas, which then becomes a liquid. The liquid ammonia can then return to the first device where the cycle continues.

One disadvantage of OTEC systems is that they must be located offshore. Large electrical cables must transport the electricity across the distance from the offshore OTEC systems to consumers on land. Alternatively, companies also can use the electricity on site to perform such energy-intensive tasks as removing salt from seawater to produce drinking water.

OTEC systems are an attractive energy resource for countries located near the equatorial oceans. The systems require no fuel input and contain devices made from existing technologies. Construction costs for OTEC systems, however, are high—roughly two to three times as high as the construction costs for a comparable, coal-burning power plant. The expensive equipment also can be damaged by severe tropical storms and by the corrosion of metal parts in contact with seawater.

To date, small-scale OTEC plants have caused little environmental damage. There is, however, a potential for accidental spills of chemicals such as ammonia or chlorine. If these chemicals leaked into the ocean in large amounts, they could kill plants, animals, and microscopic organisms. In addition, because the systems draw cold water up from the ocean bottom, many marine organisms could be trapped and killed.

Some scientists suggest that the large-scale use of OTEC systems could change climatic conditions in some regions. For example, cold water pumped from the ocean depths could cool surface waters enough so that the entire climate of a region would change. One group of scientists has predicted that if OTEC systems were to supply all of the world's energy needs, the water temperature of tropical oceans would drop by less than one degree Celsius. But even such a seemingly insignificant change could have dramatic effects on global climate. For example, scientists hypothesize that the dramatic climate phenomenon known as "El Niño" originates from a small warming of the ocean's surface temperature in the Western Pacific of roughly one degree Celsius.

In addition, the cold, deep water contains dissolved carbon dioxide gas. When this water comes to the surface, the carbon dioxide gas escapes into the atmosphere. Some scientists hypothesize that increases in the amount of carbon dioxide and other "greenhouse gases" in the atmosphere could cause an increase in global temperatures, a phenomenon called "global warming." (For a description of greenhouse gases and global warming, see Synthetic Fuels.) Other scientists, however, argue that this hypothesis is not well supported by evidence. In any case, the amount of carbon dioxide gas released by OTEC systems is minor compared to the amount released during the burning of fossil fuels such as coal and petroleum.

Petroleum

Petroleum is a liquid fossil fuel. (For a description of fossil fuels, see Coal.) Scientists hypothesize that petroleum formed from the remains of tiny marine organisms that became buried by ocean sediments such as sand and silt. These remains accumulated over many millions of years. Gradually thick layers formed. Pressure from the upper layers caused the lower layers to heat up. This pressure and heat changed the remains of the organisms into solid, liquid, and gaseous fossil fuels. Petroleum is one kind of liquid fossil fuel.

As the liquid petroleum formed, it began to move through the layers of sediments. Sometimes the petroleum moved upward through layers of permeable sediments toward the Earth's surface. Sometimes it became trapped by a layer that it could not permeate. In this instance, the petroleum accumulated underneath the layer in a large formation called a "trap." Geologists have discovered traps in many different locations beneath the land and beneath the sea floor. (Scientists hypothesize that the traps now located beneath dry land were once located beneath seas.)

Drilling for petroleum can be expensive. Drilling companies must obtain permits and drilling rights for the areas where they want to drill. Drilling equipment and skilled workers also contribute to the high costs.

To find the best locations to drill, companies often search for clues to the existence of petroleum traps. One of the most accurate ways to locate traps is seismic technology. In this process, engineers create sound waves in Earth's surface by setting off underground explosives. Different kinds of rock layers reflect the sound waves in unique ways. An instrument called a seismograph

FIGURE 18.29 Crude oil obtained from drilling is refined to produce such products as gasoline, kerosene, and diesel fuel. Think about the advantages and disadvantages of using petroleum as an energy source.

measures the reflected waves. Engineers can then analyze data from the seismograph to make inferences about the location of possible traps. Yet, these data do not give exact answers. Many times companies will drill "dry" wells that do not reach petroleum. In fact, eight out of every 10 wells drilled turn out to be dry.

The petroleum that drilling companies extract from underground deposits is known as "crude oil." Crude oil contains many different chemicals, including some that make it unfit for use as an energy resource. Therefore, companies must process or refine the crude oil at oil refineries. Pipelines and oil tankers carry most of the crude oil from wells to refineries. At refineries, the crude oil undergoes a process that separates petroleum into components of liquid fuels such as gasoline, kerosene, heating oil, diesel fuel, and gases such as butane and propane. (We include a more complete description of these gases in the reading Natural Gas).

Refined petroleum fuels are an important and relatively inexpensive energy resource. In the United States, we consume roughly 63 percent of our refined petroleum supply for transportation. Only a small fraction of the supply is burned by power plants to produce electricity. Those plants use a method similar to that used by coal-burning power plants. (For a description of these methods, see Coal.) Unfortunately like coal, burning petroleum products contributes to air pollution such as sulfur oxides and nitrogen oxides. Those forms of pollution can be harmful to people, plants, and animals. Refined petroleum fuels, however, burn more cleanly than coal and leave no solid waste.

Some scientists suggest that the carbon dioxide produced by burning petroleum might contribute to global warming. Global warming refers to an increase in the surface temperature of the Earth. Evidence collected on the temperature of the air over landmasses and oceans and the temperature of ocean water suggests that Earth's surface temperature has increased 0.3 to 0.6 degrees Celsius during the past century.

Although scientists do not know what has caused this global warming trend, they have developed several possible explanations. According to one explanation, gases such as carbon dioxide warm the atmosphere through a process known as the greenhouse effect. In this process, gases such as carbon dioxide trap the Sun's heat in a manner somewhat like a greenhouse (recall the Heat In, Heat Out box you built in Chapter 16). Light energy from the Sun can pass through Earth's atmosphere and warm up objects on the surface. These objects then release the energy to the atmosphere as heat. Greenhouse gases are gases like carbon dioxide, water vapor, and methane, which absorb heat energy. According to the greenhouse explanation, when greenhouse gases absorb energy, the temperature of Earth's atmosphere increases.

Topic: Greenhouse Effect
Go to www.scilinks.org
Code: physical499

Other scientists contest the greenhouse explanation for global warming. They suggest that the explanation is not well supported by evidence. Instead, they favor an explanation that the global warming trend of the past century is part of the naturally occurring fluctuations in Earth's climate. Global warming and greenhouse gases are the subject of an active scientific debate. The results of the debate may influence how people view the costs and benefits of energy resources such as fossil fuels in the future.

Another disadvantage of using petroleum as an energy resource is the risk of oil spills and other accidents during shipping, refining, and delivery. Some scientists estimate that nearly six million tons of oil spill into the environment every year. Such oil spills can have devastating effects on the environment. In recent years, some oil companies have improved their equipment and techniques to reduce the risk of accidents.

FIGURE 18.30 PV cells power small items such as calculators and watches. A large PV cell like this one can produce enough electricity to help power appliances in homes or offices.

Finally, one disadvantage of all fossil fuels is their limited supply. In the case of petroleum, if we continue to use oil at our present rate, our supplies will run out in a few decades. Although there still will be petroleum in deposits that are hard to reach, extracting it would be too expensive. As a result, those countries that have large reserves might be able to control the price of oil.

Photovoltaic Conversion

Photovoltaic systems convert light energy from sunlight into electricity. These systems consist of a collection of photovoltaic cells or "PV cells." PV cells are made of thin wafers of silicon, one of the most abundant materials on Earth, commonly found in sand. Manufacturers also include small amounts of other materials such as gallium arsenide in these wafers. When sunlight strikes the surface of a PV cell, the wafer produces a small amount of electricity. A photovoltaic conversion system contains many PV cells wired together in a solar panel. By combining several solar panels in a device that tracks the Sun, a photovoltaic conversion system can produce enough electricity to run all of the electrical appliances in a home or building.

We also can use PV cells to power some small items such as calculators and watches, as well as some large items such as electrical appliances. PV cells even can power cars. (Recall the "Nuna" World Solar Challenge winner from Chapter 17.) In addition, PV systems are a valuable energy resource for locations that are difficult to supply with electricity from power lines. Many homes and businesses located in remote or isolated areas currently use PV systems. Scientists also equip satellites and spacecraft with PV systems.

PV cells generate electricity without moving, making noise, or producing pollution. Some PV cells can last up to 20 years. Unfortunately PV cells are expensive to produce because the process of refining silicon is inefficient and requires a lot of energy. In addition, photovoltaic conversion systems are not a dependable source of electricity on cloudy days or in cold climates. Even during the course of a single day, the electricity output of these systems can fluctuate. However, people can use rechargeable batteries to store any extra electricity that they do not use. That electricity can act as a backup for periods when the Sun is not shining and the solar panels are not generating electricity.

Another disadvantage of photovoltaic conversion systems is that PV cells produce a kind of electricity known as "Direct Current" or DC. Most home appliances use a kind of electricity known as "Alternating Current" or AC and cannot use DC. Therefore consumers either must purchase special appliances that run on DC or they must purchase a system known as an "inverter" that converts DC to AC electricity. Often DC appliances are more expensive than their AC counterparts. Converting entire buildings to photovoltaic conversion systems can be costly and inconvenient. As engineers and designers improve the technologies involved in photovoltaic conversion systems, however, they could eliminate many of these disadvantages.

Solar Thermal Power Systems

Solar Thermal Power Systems (STPS) are large-scale systems for generating electrical power using light energy. They consist of a field of mirrors that track the Sun. These mirrors reflect sunlight toward a device on top of a tower. The device transforms light energy from the mirrors into energy for heating water and changing it into steam. The pressurized steam drives a turbine that generates electricity in a manner similar to that used by coal-burning power plants.

STPS relies on a free and abundant energy source, sunlight. The systems also do not produce much air or water pollution. The mirrors and other equipment are expensive, however, and building the equipment can produce pollution. Another disadvantage is that STPS systems require large areas of flat land in a region that has few cloudy days such as a desert. Although large areas take away the habitat of many forms of wildlife, the total amount of land STPS systems require is much less than that needed to mine coal or uranium.

Another disadvantage of STPS is that the systems use chemicals, such as chlorofluorocarbon freon. Such chemicals can cause damage to plants and to Earth's ozone layer if they leak into the environment. (For a description of how chlorofluorocarbons affect the ozone layer, see the Sidelight: Ozone Depletion in Chapter 17.) As engineers and designers improve the technologies involved in the STPS, they might be able to eliminate many of these disadvantages.

The 1996 Summer Olympics in Atlanta, Georgia, used STPS to maintain the temperature of the pool water for all the water sports competitions. STPS was also used to provide electricity for other buildings during the Olympics.

Synthetic Fuels

Synthetic fuels are liquid and gaseous fuels that are synthesized from naturally occurring fossil fuels. (For a description of three different fossil fuels, see Coal, Petroleum, and Natural Gas.) For example, scientists and engineers have developed technologies to convert solid coal into synthetic natural gas or gasoline. Such technologies are called coal gasification and liquefaction, respectively. Other technologies can extract liquid oil from solid oil shale or tar sands.

As with naturally occurring fossil fuels, power companies can burn synthetic fuels to produce electricity. (For a description of this process, see Coal.) Synthetic gas and liquid fuels are more expensive to produce than their naturally occurring counterparts. Nonetheless, engineers might be able to refine and improve the technologies so that the costs eventually decrease. In addition, scientists estimate that we will run out of naturally occurring petroleum and natural gas much faster than we will run out of coal. Therefore, coal gasification and liquefaction might become viable alternatives for producing energy resources in the future.

The process of converting fossil fuels to synthetic fuels can pose environmental hazards to people, plants, and animals. In the process of extracting oil from oil shale, companies produce large quantities of ash. The ash is difficult to dispose of and can contain high concentrations of toxic chemicals. In other respects, however, synthetic fuels have advantages for the environment. Companies can convert coal into synthetic gas while it is still in underground deposits. This process avoids many of the problems of conventional surface mining, including the need for reclamation. (For a description of surface mining, see Coal.) The synthetic product also is cleaner than coal. When it burns, it produces less pollution.

FIGURE 18.31 Harnessing the mechanical energy of the tides can produce electrical energy.

Tides

Every day along coastlines throughout the world, the level of ocean water rises and falls in a cyclic pattern called tides. Technological systems known as tidal stations harness the mechanical energy of the tides to produce electrical energy. These systems consist of a dam that regulates the flow of ocean water into and out of coastal inlets. These inlets consist of narrow passages through which ocean water flows from the ocean to an inland bay or other body of water. The dam allows ocean water to flow into the inlet during a high tide and then traps the water. Later, during low tide, the water flows out of the dam through a system of pipes. Water flowing through the pipes drives a turbine that produces electricity.

Tidal energy is an important energy resource for some areas. Studies of stations in Germany and France suggest that tidal energy could produce electricity at an operating cost comparable to electricity-generating systems that use coal. In addition, the technologies used in tidal stations are fairly well developed and are similar to the technologies used in hydroelectric plants. (For a description of these technologies, see Hydroelectric Energy.) Unfortunately tidal stations are not feasible for all locations. Scientists and engineers have identified at least 25 places in the world where the difference between high and low tides is great enough to allow tidal stations. The advantages of tidal energy are that it does not disturb the land very much, and it does not create significant amounts of water or air pollution.

Disadvantages of tidal stations include that they might interfere with shipping and recreational activities along a coastline. Tidal stations also can damage the environment of the inlet itself; for example, by altering the migratory patterns of fish and other marine organisms. In the United States, many of these areas are classified as wetlands and fall under the protection of federal laws. Although the cost of operating the station is relatively low, the cost of construction can be high. Expensive equipment also can be damaged during severe storms or by the corrosive effect of ocean water on metal parts.

Waves

Waves contain a large amount of mechanical energy. Scientists and engineers have developed many systems to

FIGURE 18.32 Large-scale, wave-powered systems may produce as much energy as a coal-burning power plant.

convert this mechanical energy into electricity. One system uses a device containing air-filled bags to drive a turbine and generate electricity. When a wave passes over the bags, the bags compress. Pressurized air inside the bags rushes out through a series of ducts. These ducts connect to the turbine. As the pressurized air moves through the ducts, it causes the blades of the turbine to move, thus generating electricity. After the wave has passed, the bags again fill with air and return to their original shape, ready for the next wave.

Another system that converts wave energy into electricity is called a contouring raft. This system contains a series of floating rafts that are joined together by special pumps. The individual rafts move like hinged doors as waves pass beneath them. As the rafts move up and down, the pumps connecting the rafts alternately compress and expand, causing fluid inside the pumps to be forced through a series of pipes. This moving fluid provides the energy to drive a turbine and generate electricity.

Such wave-powered systems do not produce significant amounts of air or water pollution. They also do not disturb the land very much. Wave-powered systems might calm the water in an area, however, acting as wave breakers. This might be an advantage in some harbors. On the other hand, it might disrupt the habitat of some marine organisms.

Large-scale systems to harness wave energy could potentially produce as much energy as a large, coal-burning power plant. The systems would have to be strong enough to withstand turbulent weather conditions along coastlines as well as the corrosive effect of seawater on metal parts. One disadvantage is that the electricity output might fluctuate significantly according to the activity of waves at different times of the day or seasons of the year. In addition, the systems likely would have to be 7 to 11 kilometers long (roughly 4.5 to 7 miles). Some people might object to such large structures along scenic coastlines.

Wet Geothermal

Geothermal means "earth-heat." Deep inside the Earth, layers of rock are hot and molten (liquefied). In some places such as the western United States, this hot, molten rock is close to the surface. Here volcanoes occasionally bring molten rock to the surface. In other places, the molten rock heats large underground reservoirs of water. These reservoirs can form hot springs and geysers at the surface. (Geysers are fountains of steam and hot water.)

Wet geothermal systems use the heat from underground reservoirs of hot water and turn it into heat energy to heat homes and businesses, or to generate electricity. (Dry geothermal systems use heat from only the hot rocks. For more information, see Dry Geothermal.) For example, since the early 1900s, people in certain areas of Idaho have heated their homes with hot water piped from underground reservoirs or surface geysers. Alternatively, people can drill wells directly into the underground reservoirs and pipe the pressurized steam to the surface. At the surface, the steam spins a turbine and generates electricity.

The first wet geothermal system to generate electricity, located in Larderello, Italy, began operating in 1904. Today, there are more than 150 geothermal power plants around the world. In California, a plant uses steam from a geothermal reservoir to provide heat and electricity for one million homes. In Iceland, which is an island formed by volcanic activity, geothermal energy generates electricity and heats homes, factories, and swimming pools.

Wet geothermal power plants take less time to build than coal-burning or nuclear power plants. Geothermal

FIGURE 18.33 Molten rock heats large underground reservoirs of water. These reservoirs can form geysers like the one shown here.

systems also do not require a complex network of facilities and equipment. Only a small amount of land is disturbed. Such systems sometimes release a variety of gases into the atmosphere during the process of drilling holes and allowing steam to escape to the surface. The gas that is potentially the most hazardous is hydrogen sulfide. It is toxic to humans at high concentrations. It has a foul odor like rotten eggs. Residents of communities downwind of the California wet geothermal plant have complained about such gases. To deal with such problems, scientists are developing "scrubber" systems that remove some of the gases from the emissions.

An additional disadvantage of geothermal systems is that the liquids pumped to the surface can contain many contaminants, including some toxic substances such as ammonia, arsenic, and mercury. These contaminants can kill plants and wildlife and make water unfit for drinking. The California wet geothermal plant also releases hot waste water into lakes and streams. These wastes have caused temperature changes in the water that affected many aquatic organisms. Therefore scientists continue to monitor these systems and try to improve their design in order to reduce or eliminate such disadvantages.

Wind

Wind is the movement of air. Typically people experience wind when large bodies of air move. (If you studied *Investigating Earth Systems*, you might recall that scientists define wind as the horizontal movement of air in a convection cell. To review this information, see *Investigating Earth Systems*, Unit 3.) People have been able to harness the mechanical energy of wind effectively since the time of the Egyptian pharaohs. Modern windmills transform wind energy into spinning energy that can be used to turn a turbine and thus generate electricity. People can use such windmills to supply the electrical needs of individual homes, farms, and businesses. People also can cluster windmills in large groups called wind farms. These systems can generate enough electricity to supply the needs of large communities.

FIGURE 18.34 Windmills can supply electricity to homes, farms, and businesses. Clusters of windmills, called wind farms, can supply the needs of entire communities.

Wind energy is a clean, efficient energy resource for some locations. To be efficient, wind-powered systems require steady wind speeds of no less than 7.5 kilometers per hour. These winds must blow continuous throughout the year. In addition, the systems must be located in areas free of obstructions such as buildings and trees.

These constraints are disadvantages that scientists and engineers are trying to overcome. For example, some companies are developing storage systems such as rechargeable batteries. These systems would save the excess electricity from a windmill to act as a backup when the wind is not blowing, or when it is blowing too hard and the windmill must be shut down. Other companies are experimenting with different kinds of blades that could generate electricity more efficiently. Advances in such technologies could improve the efficiency and viability of wind-powered systems.

CHAPTER 19

The Power to Choose

By now you probably realize that the issues related to using energy are complex. But what does all this have to do with you? You have control over your personal actions, and every action you take involves energy. You can choose to use practices that conserve energy. Consider each item pictured in the opening photograph. How much energy does each item use?

You also can choose to share your choices and practices with others, which may influence others in your community. The choices you make about using energy have benefits and costs for you, your community, and the planet.

In this chapter, you will explore how the choices you make about your lifestyle affect your personal use of energy, whether in a positive or negative way. What you learn will increase your ability to use energy efficiently.

ENGAGE ■ Toast

EXPLORE ■ Cans of Energy

EXPLAIN ■ Complex Technological Systems

ELABORATE
■ It's Not a Simple Matter
■ Ways to Improve Our Use of Energy
■ What Is My Choice?

EVALUATE ■ The Global Picture

ENGAGE connections

Toast

Watch the video *Toast*. Immediately after viewing it, write a paragraph that begins with the following phrase, "I think the main idea of this video was _____." Then share your paragraph with some of your classmates who are seated nearby. How are the paragraphs similar? How are they different? Contribute your ideas to a class discussion of the video.

EXPLORE investigation

Cans of Energy

WORKING COOPERATIVELY

You will work in one of four groups. Each group will be responsible for producing one act of the play. Work cooperatively with members of your large group by using the cooperative techniques and skills you have learned this year. In particular, practice the Unit 4 skills.

When it comes to eating, does Al have a choice about how he uses energy? (See Figure 19.1.) Think about that question as you and your classmates participate in this investigation. You will create a play about a technological system for making and delivering cans of vegetable soup to consumers. Each part of the system requires some form of energy input. You and your classmates will prepare scripts and props for each of four different acts to explain how the system works.

Materials for the Entire Class:
- various items to use as props
- poster board
- construction paper

CHAPTER 19 The Power to Choose

- markers
- transparent tape
- glue

Materials for Each Student:
- 1 role card

Process and Procedure

1. Read the role card that your teacher assigns you.
   → This role card describes the person you will portray during the play.

2. Decide in which act of the play you will appear.
   - Act I—Production
   - Act II—Packaging
   - Act III—Distribution
   - Act IV—Consumption

3. Meet with other students who are in your same act.

FIGURE 19.1 Does Al have the power to choose how he uses energy?

CHAPTER 19 The Power to Choose

4. As a group, decide the order in which the roles appear during the act.
   → Compare the roles of all of the students in your group.
5. As a group, decide what form of energy input is needed for each role.
   → In other words, would you need energy input in the form of sunlight, gasoline, electricity, or something else to fulfill the roles of everyone in your group?
6. As a group, develop a script and props that you can use to produce your act of the play.
   → Assign group recorders to record the script and then make copies for the other group members. Be sure to include each role in the act. Your script and props should explain the roles and the energy input needed for each role.
7. Practice using the script and the props to produce your act of the play.
8. Perform your group's act along with the other acts of the play during the time your teacher arranges.

Wrap Up

After performing the play, meet with your regular team of three. Discuss these questions. Record your own answers in your notebook and be prepared to share them with the class.

1. How many different forms of energy input went into manufacturing and delivering the vegetable soup?
2. How many technological devices were involved in the manufacture and delivery of the vegetable soup?

Topic: Alternative Forms of Energy
Go to www.scilinks.org
Code: physical514

3. Which consumer made the more efficient use of energy: the one who disposed of the can in the trash or the one who used the recycling bin? Why do you think so?

4. Draw a diagram that illustrates the interactions of the energy inputs, outputs, people, and devices that were involved in your act of the play.

5. Describe whether it was difficult working cooperatively in a large group. In particular, mention the cooperative learning weaknesses and strengths you observed as you worked in your large group.

Complex Technological Systems

In Chapters 17 and 18, you learned about technological systems that had an energy input, a device, and an output. You diagramed these systems as shown in Figure 19.2. There was generally one energy input, such as wind, sun, or falling water; and one output, such as a spinning motion that you used to lift paper clips or generate electricity.

FIGURE 19.2 Technological devices transform an input into an output.

Now review the diagram you drew for the investigation Cans of Energy. In looking at the diagram, you might realize that a system that produces vegetable soup has many devices, such as farm equipment, equipment for processing the vegetables, and for making the soup. The system also has many energy inputs, such as fuel for delivery trucks and farm equipment; electricity to operate machinery; human energy to plant, grow, and harvest the vegetables; and energy from the Sun to grow the vegetables.

It might seem an impossible task to draw a diagram for this complicated system. It is not so difficult, however, if you first think of several small systems that make up the large system. For example, consider how you might diagram the vegetable soup production system. To produce vegetable soup, you need to grow and harvest vegetables, process the vegetables, and combine the vegetables with other ingredients to make soup. Each of these steps can be thought of as a system (see Figure 19.3).

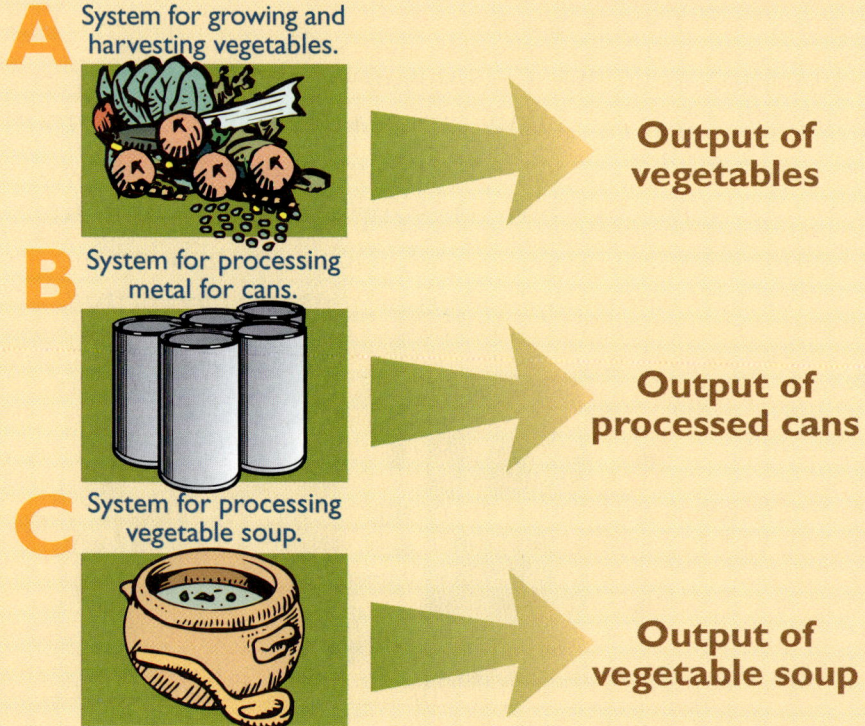

FIGURE 19.3 You can diagram a system for producing vegetable soup as three smaller systems.

Notice that the three systems shown in Figure 19.3 do not include energy inputs. To determine the energy inputs, start with the system for growing and harvesting vegetables. If you want to grow vegetables, in addition to the energy from the Sun, you need water. Sometimes that water must be pumped to the fields over long distances. This step requires energy to operate the pumps. You also need energy to operate farm equipment and energy to sustain the farm workers. Therefore, the complete diagram for the system for growing and harvesting vegetables might look like the one shown in Figure 19.4. A similar system for processing the vegetables might look like the one shown in Figure 19.5, and a system for producing the soup might look like the one shown in Figure 19.6. You can combine all three of these smaller systems to make a larger system (see Figure 19.7). The small systems are connected because the output of one system goes into another. For example, the output of the first system is raw vegetables; it is input into the second system. Likewise, the output of the second system, processed vegetables, is the input for the third system.

FIGURE 19.4 A technological system for growing and harvesting vegetables includes several kinds of energy inputs.

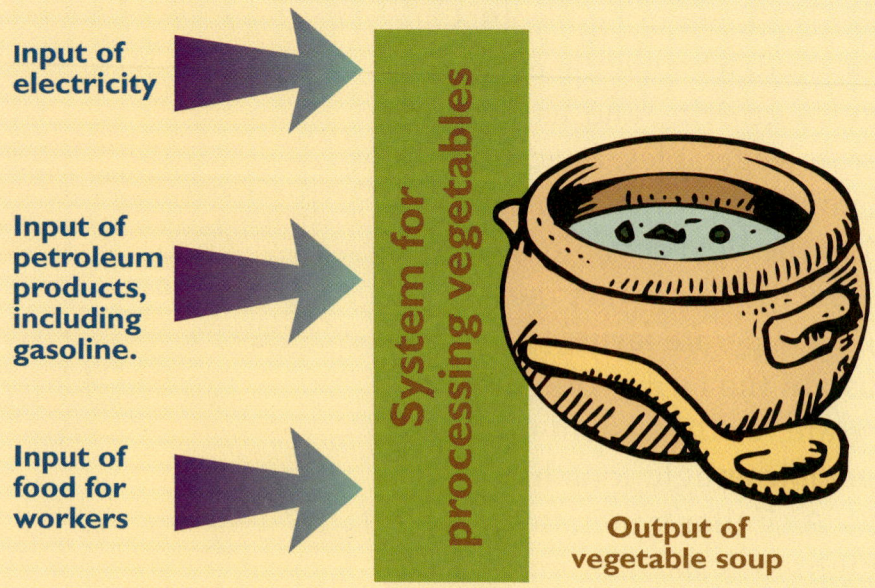

FIGURE 19.5 A technological system for processing vegetables includes several energy inputs.

Notice that Figure 19.7 does not include the energy inputs necessary to run the systems. You can add those inputs in a way that shows that all the smaller systems require energy. Figure 19.8 shows that it takes a lot of energy to make vegetable soup. It includes only one act of the play your class performed in Cans of Energy. Yet there are many steps involved in getting a can of vegetable soup to you, including mining cassiterite to produce tin

FIGURE 19.6 A technological system for producing the vegetable soup includes several energy inputs.

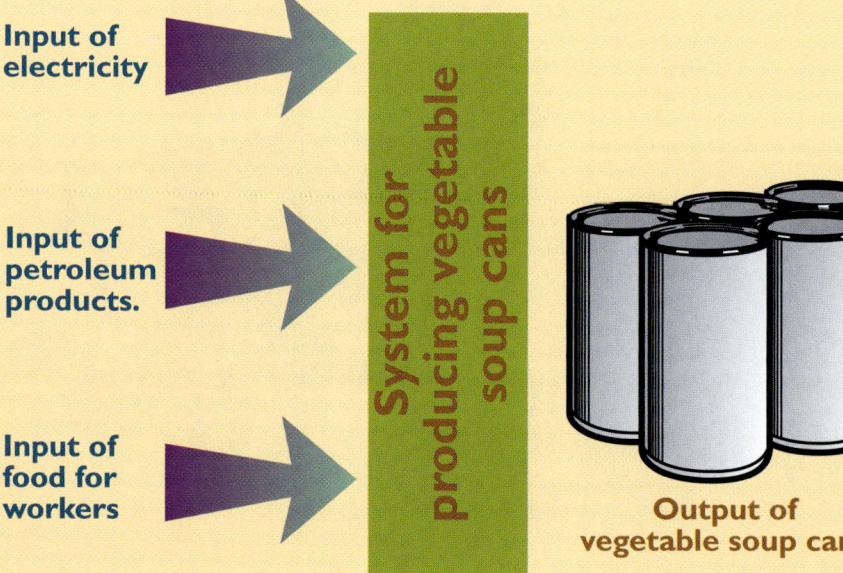

FIGURE 19.7 You can combine the three smaller systems to see what the larger system for producing vegetable soup looks like.

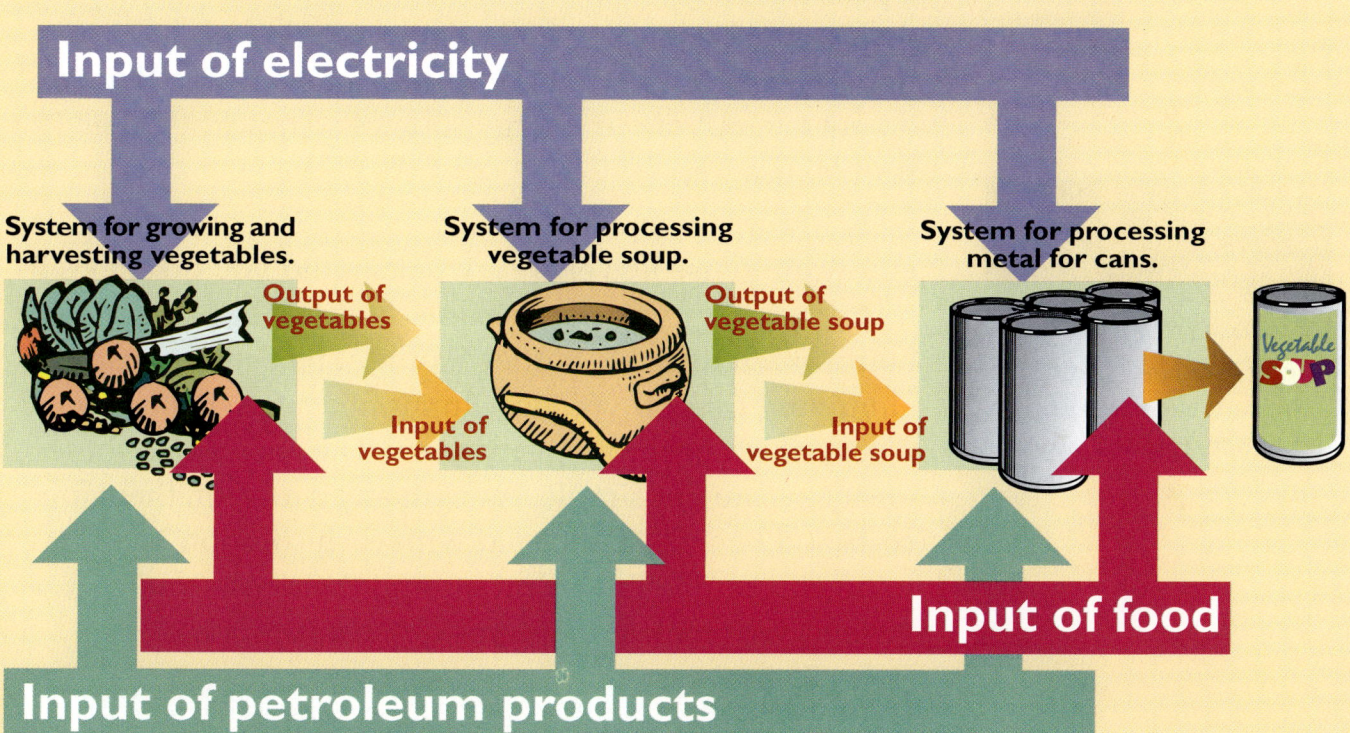

FIGURE 19.8 By adding the inputs to the diagram, you can see how all the smaller systems work together to transform energy inputs into outputs.

cans, distributing the soup to grocery stores, and purchasing the soup. Each of these systems is composed of smaller systems and each of the smaller systems requires the input of energy.

CHAPTER 19 The Power to Choose

STOP & THINK

1. Draw a diagram similar to the one for the System for Producing Vegetable Soup (see Figure 19.8), but this time depict a system for producing CDs or shoes.

The title of this unit is Limits of Energy in Systems. Do we really need to improve our use of energy? Some sources of energy, such as the Sun, appear to be limitless. Other sources of energy, however, are limited. For example, we have only limited supplies of fossil fuels such as petroleum, coal, and natural gas. We will reach a point where it costs too much, in terms of energy or dollars, to obtain these fuels. Some people feel that it is important to conserve these sources of energy so that we do not run out of our energy sources so quickly. Other people disagree.

For now, assume that saving energy is a good thing. If you want to use less energy, what should you do? Does it matter whether you purchase a soft drink in a disposable bottle or a returnable bottle? Would it be better to use aluminum cans? If you use aluminum cans, should you recycle them or throw them away? The answers to these questions are not simple. As with any problem, there are costs and benefits associated with each solution. One way to figure out possible solutions and their costs and benefits is to look at diagrams of the appropriate technological systems.

Suppose you have a system that requires energy input and produces an output. One way to reduce the amount of energy that the system uses is to reduce, improve, or change the output. For example, if you use less hot water or water that is not as hot as usual, then you can reduce the output of a water-heating system. If there is less output, then the system requires less energy

Topic: Sources of Energy
Go to www.scilinks.org
Code: physical520

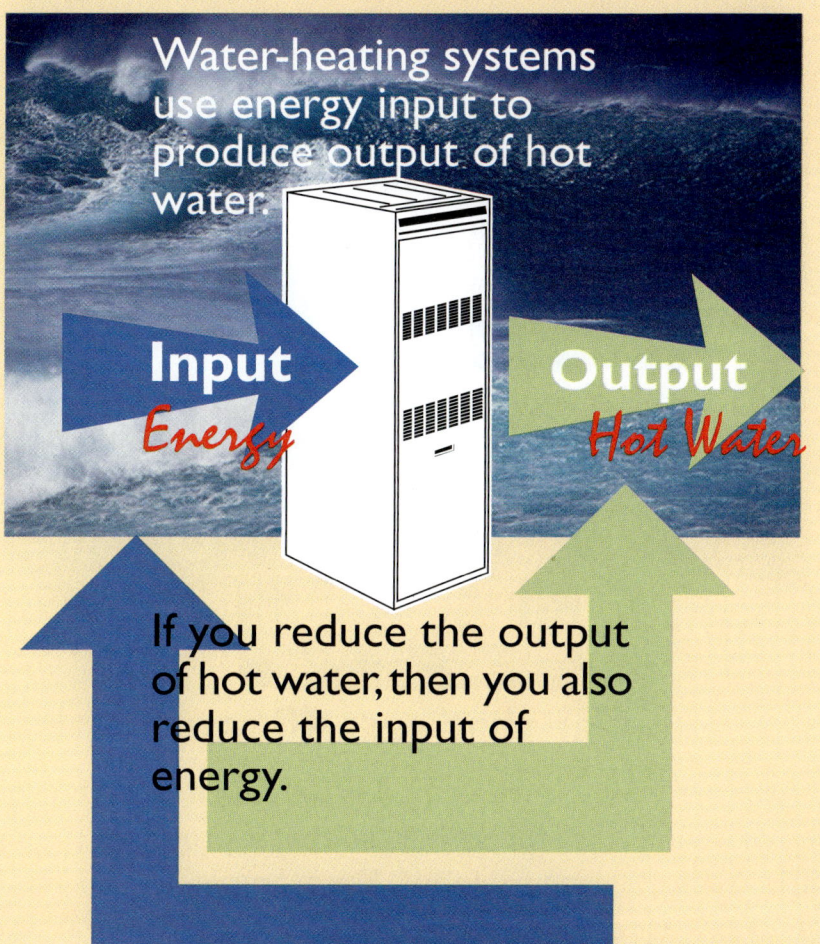

FIGURE 19.9 You can reduce the total amount of energy a system uses by reducing the output.

input (see Figure 19.9). There are many ways to reduce the output of technological systems. You might turn off lights when no one is using them, lower the thermostat setting on your home heating system, or walk to school rather than have your parents or older siblings drive you to school. All of these actions result in less energy input to technological systems.

2. What are the costs and benefits of reducing the output by (a) turning off lights, (b) setting a thermostat lower, and (c) walking rather than driving?

CHAPTER 19 The Power to Choose   521

FIGURE 19.10 You can save energy by improving the energy efficiency of a system.

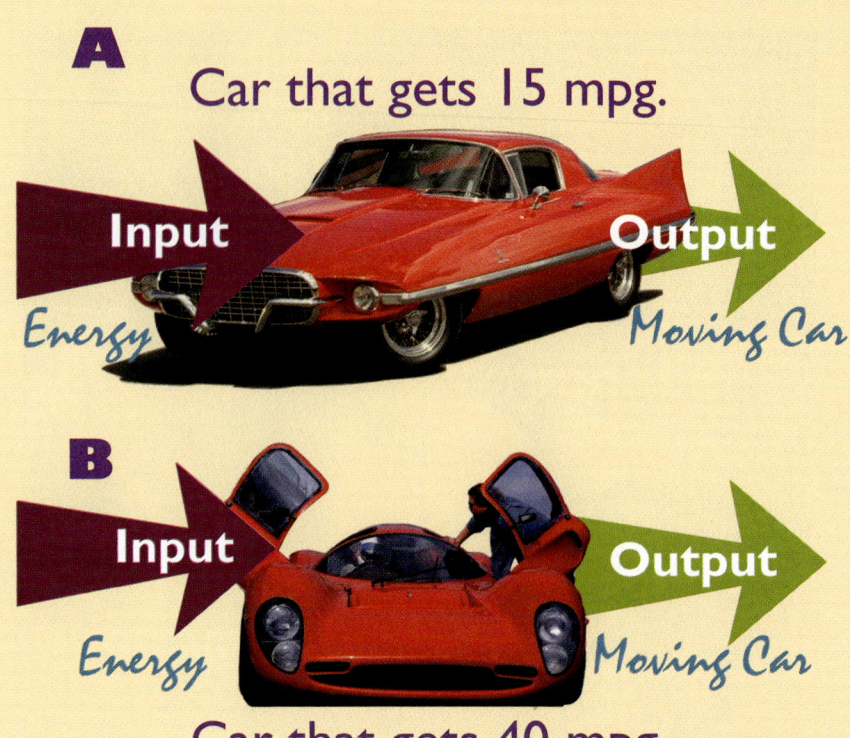

Another way to save energy is to reduce, improve, or change the input to the system so that it produces the same output with less energy input. For example, if your family uses a car that gets 40 miles per gallon rather than 15 miles per gallon, you are using a system that requires less energy input—gasoline—to get the same output—miles traveled (see Figure 19.10). Devices that waste very little energy are said to be energy efficient. Many modern washing machines, water heaters, refrigerators, and other household appliances now come with Energy Guide stickers. These stickers explain how energy efficient the appliance is.

3. What are the costs and benefits of reducing energy input by (a) using energy-efficient appliances and (b) driving fuel-efficient cars?

Another way to change a system to save energy is to eliminate some of the smaller systems or devices that make up a large system. For example, consider a

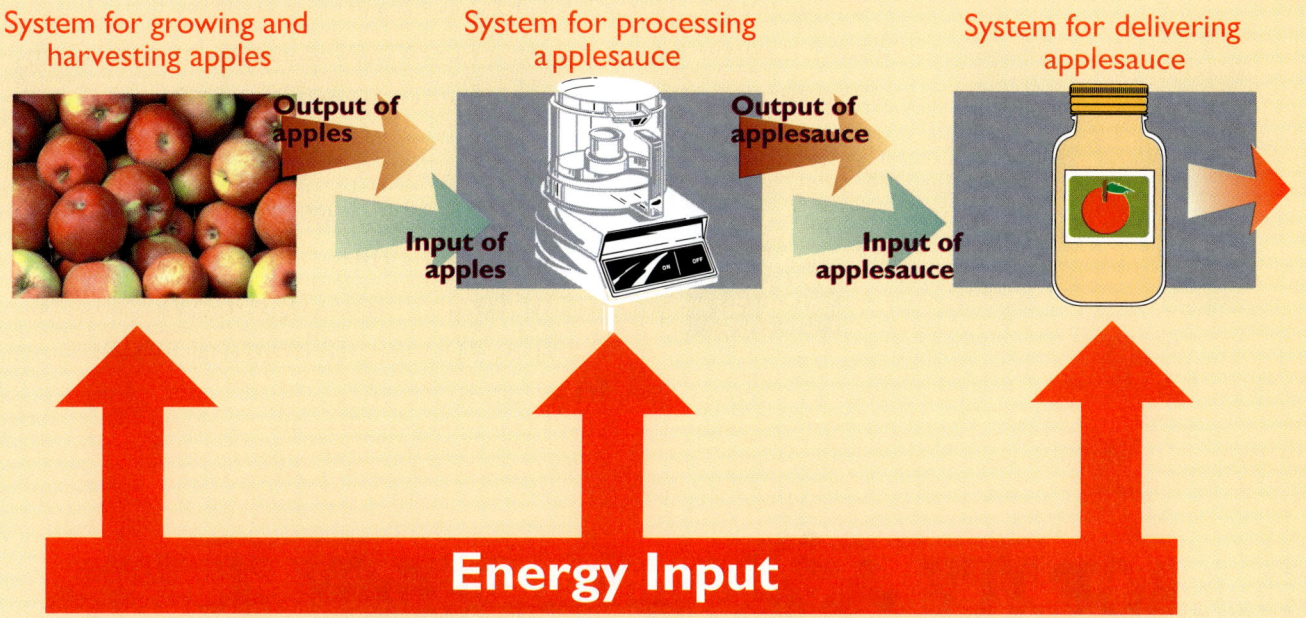

technological system for getting applesauce to consumers. It might look like the diagram shown in Figure 19.11. Each of the smaller systems requires energy input. Suppose that you eliminate the applesauce processing system by buying fresh apples rather than applesauce in a jar. In doing so, you eliminate the energy that the processing and packaging systems require. Now, however, you also require a refrigeration system to keep the apples fresh during transport, at the store, and at home.

FIGURE 19.11 How might you save energy by eliminating one or more of the smaller systems that make up this larger system?

4. What are some costs and benefits of eliminating some of the smaller systems that make up the technological system shown in Figure 19.11?

One final way to save energy is to change the input to the system. For example, you could switch from an electrical water-heating system to a solar water-heating system. This does not necessarily mean that you will use less energy, but solar energy comes from an essentially

CHAPTER 19 The Power to Choose

unlimited source. Sunlight is free, and will be available for a long, long time. Electricity can require the burning of coal or some other limited source of energy. By changing the energy input from electricity to solar energy, you are saving a limited energy source for other uses.

5. What are some costs and benefits involved in changing the energy input to a technological system such as the water-heating system in your home?

ELABORATE connections

It's Not a Simple Matter

Rosalind, Al, Marie, and Isaac are starting to apply what they have learned about energy to their personal decisions (see Figure 19.12). Yet there is one thing missing from their conversation: data. Consider the following information.

- Glass is made up largely of silicon. Silicon mining produces 385 pounds of waste for every 2000 pounds of glass produced.
- Each person in the United States uses 85 pounds of glass per year. One hundred percent of that glass is recyclable.
- It takes 30 percent less energy to make a new bottle from recycled glass than to make a new bottle from raw materials.
- Recycling aluminum cans saves about 95 percent of the energy needed to make new aluminum cans from raw materials.

FIGURE 19.12 What do you think about disposable juice containers? Should Al buy only recyclable containers?

- Each pound of raw aluminum produces 29 new cans.
- Recycling one glass bottle saves enough energy to light a 100-watt lightbulb for four hours.
- Recycling one aluminum can saves enough energy to light a 100-watt bulb for almost two and one-half hours.
- Americans recycle about 26 percent of their glass.
- Americans recycle roughly 66 percent of all aluminum cans.
- Glass is more difficult to recycle than aluminum because the glass must be separated by color.
- An aluminum can used once consumes seven times more energy than a refillable glass bottle used 10 times.
- It takes slightly more energy to recycle an aluminum can than a glass bottle.

Now participate in a class discussion to answer the question, What is the best choice for a juice container?

Topic: Waste Management
Go to www.scilinks.org
Code: physical526

Be sure you think about energy use, convenience, safety, availability, and any other factors that would influence your decision about the best container.

After discussing reusable containers with your classmates, make a chart of the costs and benefits associated with recycling aluminum cans and glass bottles. Decide whether recycling aluminum or glass is an energy-efficient decision and be ready to defend your decision.

SIDELIGHT ON TECHNOLOGY
Another Bright Idea

Most people are interested in saving money. Because energy is expensive, many people sell technological devices that are supposed to help other people save money by saving energy. And even though these new technologies are expensive, the advertisements say that if you use them, you will save a lot of money on electricity, natural gas, or gasoline. Are these products too good to be true?

Consider the example of compact fluorescent bulbs. You can find advertisements in many magazines that say that these bulbs last 10 times longer than the typical incandescent lightbulbs that you probably use in your home. The new bulbs are supposed to help

you save money because you can replace a 60-watt, incandescent bulb with a 15-watt, compact fluorescent bulb and still get the same amount of light. (Lower-wattage bulbs use less electricity.) The advertisement also tells you that even though the new bulbs

CHAPTER 19 The Power to Choose

come with a hefty price of $7 to $10 per bulb (a typical incandescent bulb costs less than $1), you can make up the difference in just a few months from savings on your electric bill.

Before you have your family rush out and buy these new energy-saving bulbs, however, you should know one thing that the advertisement does not tell you. Using the new bulbs will not necessarily save you any more money than you would save if you controlled how you use the lights in your home. In addition to turning off the lights when you don't need them, you could install lower-wattage bulbs in fixtures where you don't need bright lights. One consumer advocate, Andrew Rudin, sums up the issue this way: "In the final analysis, lightbulbs do not use electricity; people do. Efficient bulbs are not the main answer to problems involving high electric use for lighting; people are the answer." (Andrew Rudin, Energy Management Consultant, Melrose Park, Pennsylvania.)

One way people can save energy is by turning off the lights when they do not need them. Another way is to use high-efficiency lightbulbs. Imagine how much energy we would save if people did both.

You have seen how technology can fix some problems. People can fix some problems, too. Consider the costs and benefits of the compact fluorescent bulbs. Would you buy these bulbs to replace all the older bulbs in your house? Why or why not?

Ways to Improve Our Use of Energy

Long ago, people knew where everything they needed came from. If they needed food, they gathered it, grew it, or hunted for it themselves. If they needed shelter, they built it using materials in their environment—wood, stone, mud, and grass, for example. If they needed water, they dug a well or carried it from a nearby stream. Their needs were simple, but their lives involved hard work to fill their needs.

FIGURE 19.13 In what ways are these mud huts examples of how energy sources can be found in our environment? How is this structure different from those in your community? How is it the same?

Work cooperatively in your teams of three with a Manager and a Communicator. Choose a skill to review as you work. Your team can review a skill that is different from other teams. Just be sure to record the social skill and to think of a way to evaluate your use of the skill.

Today, our lifestyles are very different. Instead of spending our time growing food and making our own clothing, we can spend money to buy these things. The food, lumber, and water that we buy, however, are expensive. The trade-off is that we must work hard at our jobs to earn the money to buy things.

Why do food and other items cost so much? If you think about the play you produced in the investigation Cans of Energy, you will know the answer. The products you buy come from somewhere other than the local store. Manufacturers produce and ship the products to the store where you buy them. Every step that gets a product from the producer to the user adds to the cost of the product. Every additional step requires more energy input into the system.

In this investigation, you and your teammates will use what you know about delivery systems to help you use energy wisely. You will diagram a system that delivers a product such as a CD or a CD player to a consumer such as you. Then you will identify three ways you could improve the use of energy in the system. To improve the system, you could eliminate steps, change the energy input, modify the device, or even change the output. The solutions are up to you. For each solution, however, you must explain at least one benefit and one cost. Finally, you will decide whether you think anyone actually would use your solution.

Materials for Each Team of Three:
- 1 set of markers
- 1 sheet of newsprint

Process and Procedure
1. As a team, decide on 1 product.

→ Make sure that you choose a product you know something about so that you can draw a reasonably accurate delivery system.

→ Notebook entry: Record your choice.

2. Identify the parts of the system that deliver this product to the consumer.

   → Notebook entry: Record the parts you will need to produce, package, distribute, and consume the product.

3. Diagram the parts of the delivery system on a large sheet of paper.

   → Record on the diagram the form of energy input into each part of the delivery system.

4. Decide on 3 different methods to improve the use of energy in the delivery system for this product.

   → Each Team Member should contribute 1 different way.

   → Notebook entry: Record all of the different ways.

5. Beside each proposal, put an "I" if the method reduces, improves, or changes the energy input into the system; an "O" if the method improves or changes the output from the system; and a "D" if the method eliminates steps or otherwise changes parts of the devices within the system.

6. Identify at least 1 benefit and 1 cost of using each method.

   → Notebook entry: Record the benefits and costs for each method.

7. Discuss the benefits and costs of each method and decide whether you think anyone actually would use this method to improve the use of energy in the delivery system.

FIGURE 19.14 What are the steps in the delivery system for a product such as a CD? Many stores now sell used or recycled CDs. How does this change the delivery system of this product?

FIGURE 19.15 What components do you think are involved in the delivery system for this product and other products pictured in this activity?

- You might want to rate each of the benefits and costs on a scale from 1 to 10, with 10 being the greatest benefit or the greatest cost to help you decide.
- Notebook entry: Record your decisions.

8. Select which method your team thinks is best for improving the use of energy in the system.
   - Make sure you choose a method that someone actually might use.
   - Notebook entry: Record the solutions you chose.

9. Explain your team's diagram and methods to the rest of the class.
   - Be ready to explain which solution your team thinks is best for improving the use of energy in the system. Also describe the social skill you reviewed and how you rate yourself on the use of this skill.

Wrap Up

Listen to each team's presentation of its delivery system and the method they chose for improving the use of energy in the system. After each presentation, discuss the advantages and disadvantages of the solution each team proposes. As a class, decide whether you actually would use that solution.

What Is My Choice?

By now, you should know about some of the problems related to using energy. You also should know about some of the benefits and costs of different solutions to energy-related problems. Furthermore, you have information about different energy sources that can help you evaluate different solutions. Are you ready to make some personal choices?

Read the five situations that follow. For each situation, decide which action you should take in order to be the most energy efficient. In addition, list the benefits and costs of each action that you choose. When you finish, look at your answers. Decide whether your answers show that you chose to use energy efficiently.

Write a paragraph that explains your decisions.

After you have answered the previous questions, discuss as a class what you could do to help others in your school learn to use energy efficiently. Could you start a school-wide recycling project? Could you ask the cafeteria manager to replace disposable plates and utensils with reusable ones? Think about what actions you could promote to save energy. Then organize yourselves and promote those actions. Before you begin, you should know that it is not always easy to get people to change their behavior. When you plan a project for your school, you need to plan ways to motivate other students and teachers to take part in the project. What do you think would motivate the people in your school to participate in a recycling project or to begin washing and reusing cafeteria plates and utensils? Think about this issue as you plan your project.

Topic: Recycling
Go to www.scilinks.org
Code: physical531

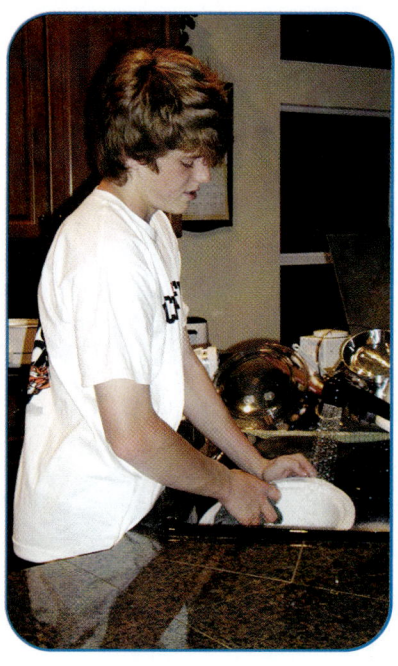

Situation: How should you wash dishes?
Action
   a. I wash the dishes by hand under a steady flow of hot water and use cool water instead of hot water to rinse the dishes.
   b. I run the dishwasher whenever I need to and set the water-heater thermostat at the minimum recommended temperature.

    c. I use paper plates and cups made out of recycled paper to minimize the number of dishes I have to wash.
    d. I wash the dishes by hand using a sink full of soapy water; then I rinse the dishes in a sink full of clean water.
    e. I run the dishwasher when it is full and make sure the water-heater thermostat is set on the high setting.
    f. I wash the dishes by hand, turning on the hot water each time I need to rinse and turning the hot water off after each dish.

Situation: How should you bathe?
Action

    a. I take three-minute, hot showers every day.
    b. I take big warm, bubble baths.
    c. I take 15-minute, hot showers every other day.
    d. I wash my hair in the sink under a steady flow of warm water and sponge bathe every day.
    e. I take 15-minute, warm showers every day.

Situation: How should you get to and from school?
Action

    a. A family member drives me to school.
    b. I ride the school bus.
    c. I car pool with several of my neighbors or friends.
    d. I ride my bike.
    e. An adult gives me a ride on a motorcycle.

Situation: If you had extra money, which product would you buy?
Action

    a. A pair of in-line skates.
    b. A new CD player.

c. A ticket to a concert.
d. New clothes.

Situation: You live in a region that has a cold climate. Sometimes you feel too cold in your home. What should you do?

Action

a. I turn the thermostat up, although never over 72°F (22°C). I use an extra sweater if I am still cold.
b. I never turn the thermostat up past 68°F (20°C). I also turn on a space heater to quickly warm up the room I am in.
c. I put another log on the fire.
d. I turn the thermostat up to 78°F (25°C) during the day and down to 65°F (18°C) when I am in bed at night.
e. I take a long, hot shower.

SIDELIGHT ON CAREERS

Lighting Designer

Imagine what your life would be like if Thomas Edison had not invented the lightbulb. What would you do when the Sun goes down at night? Candles or a kerosene lamp might be enough light if you were sitting in one place, working on a small project. What if you wanted to play tennis at night, go to the mall, or watch a play at an outdoor theater? A lighting designer's job is to think about and create ways to light different kinds of spaces efficiently.

Megan Strawn, a lighting designer for a company in Seattle, Washington, has done lighting designs for theaters, TV stations, office buildings, schools, parks, and retail establishments. For each of these spaces, she decided what kinds of lights and light fixtures should be installed. She also did layouts of the spaces to determine where the lights should be placed.

Lighting designers must attend engineering school to learn their trade. Megan has a bachelor's degree in English Literature and a bachelor's degree in Architectural Engineering with an emphasis in lighting. Her English degree helps her to communicate her lighting design ideas clearly. Her engineering degree helps her to understand how light interacts with a building's architecture. In addition to her own knowledge about lighting and architecture, Megan must work cooperatively with architects, other engineers, and building contractors. As Megan explains it, "Buildings are incredibly complex and require the expertise of several different fields of design and engineering."

In this unit, you have been learning about the costs and benefits of using energy. Lighting designers also are aware of the costs and benefits of using energy to create light. According to Megan, 20 to 25 percent of the energy consumed in a building is used for lighting. About 15 to 20 percent of the heat removed from a building comes from lighting. Lighting designers can minimize the amount of wasted light by using energy-efficient light sources and carefully planning the placement of light fixtures. Sometimes time clocks are added to help ensure that

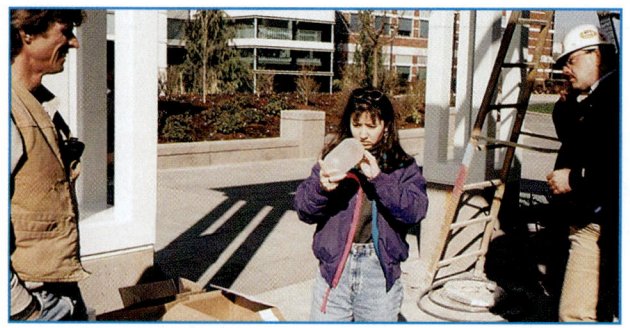

lights are not left on when they are not needed.

Lighting designers also are conscious of how lighting can affect people's moods. "Lighting has a more important role than just helping you see when the Sun goes down. Lighting has everything to do with how a space is perceived," says Megan. Sometimes lights can be too dim or too bright. Light can be used to transform an unpleasant space into one that is more comfortable. Proper lighting can lower stress and help to increase productivity. As Megan says, "Light can be used to emphasize elements of a building or make them go away. When used in the theater, light can evoke moods and responses from the audience."

As advances are made in lighting technology, lighting designers will take advantage of those advances. For example, more efficient forms of lighting, such as smaller light sources that put out a large amount of light, are already in use. In the future, fiber optic lighting may be more widely used. In the past, architects or interior designers handled lighting design. Now, lighting design has become its own consulting field.

The Global Picture

You have just considered several actions that you could take to improve your use of energy. At this point, you might be asking the question, Can one person make a difference? It might seem that the actions of one person do not matter in the global energy system. Yet what would happen if you combined the actions of six billion people?

Figure 19.16 is a photograph of the Earth at night. Each point of light represents the output of light energy at a particular location on the Earth's surface. Study this photograph and answer the following questions. (You can use the descriptions at the bottom of the photograph to identify particular locations.)

1. What patterns do you observe in the photograph? (Identify at least two.)
2. Compare the photograph with a map of Earth. Identify the location of the six continents in the photograph of Earth at night.
3. What do you see in the photograph that you did not expect to see?
4. Can you identify your community in the photograph?

Topic: Producing Light
Go to www.scilinks.org
Code: physical535

When you finish answering these questions, record your impressions about the photograph in your notebook. Include your feelings about the amount of energy being used in the global energy system. Describe whether your power to choose can change this photograph in any way. Then write a newspaper article, poem, essay, or short story that describes how your decisions about using energy might affect the global energy system. Be prepared to share your writing with the class.

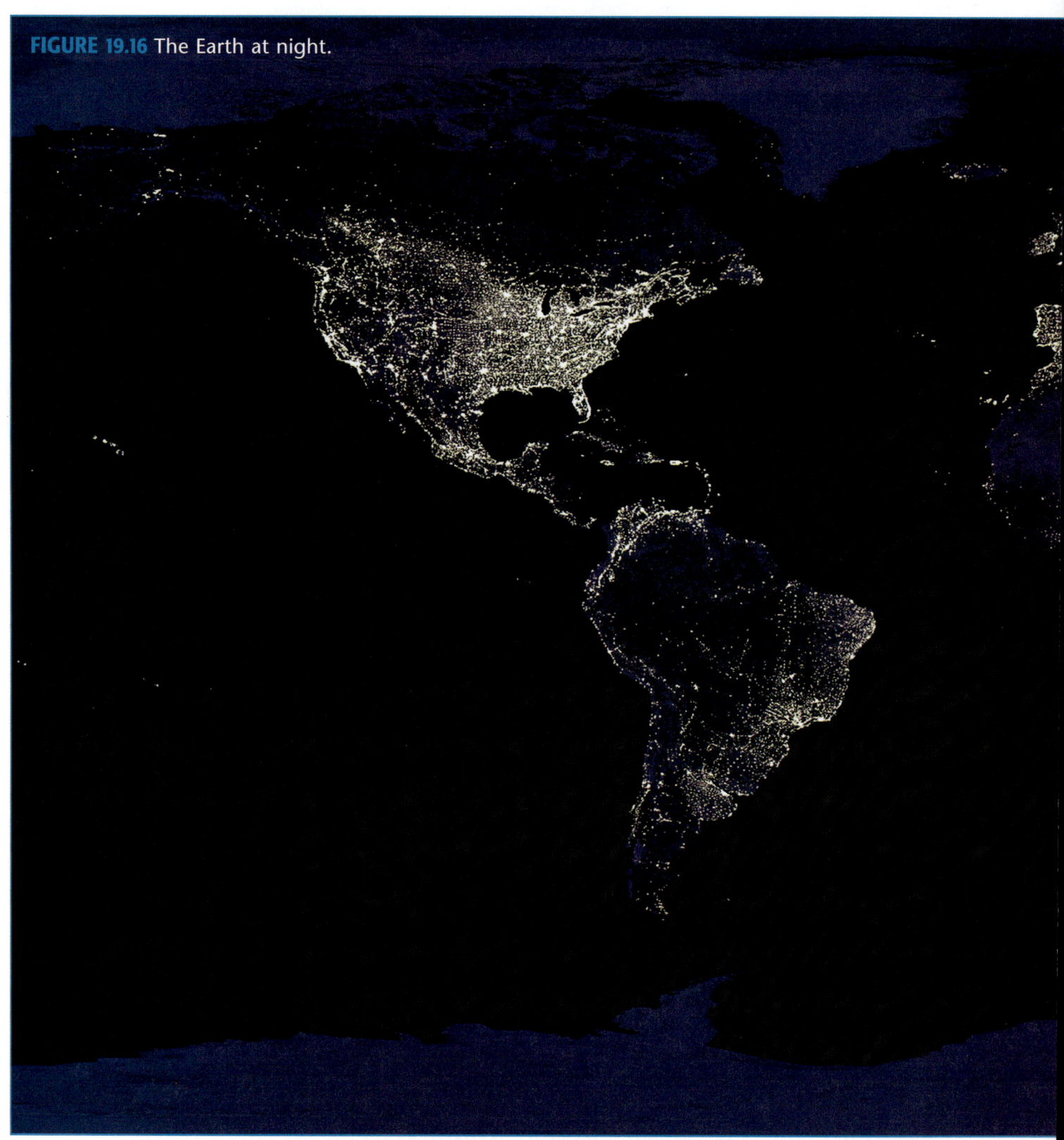

FIGURE 19.16 The Earth at night.

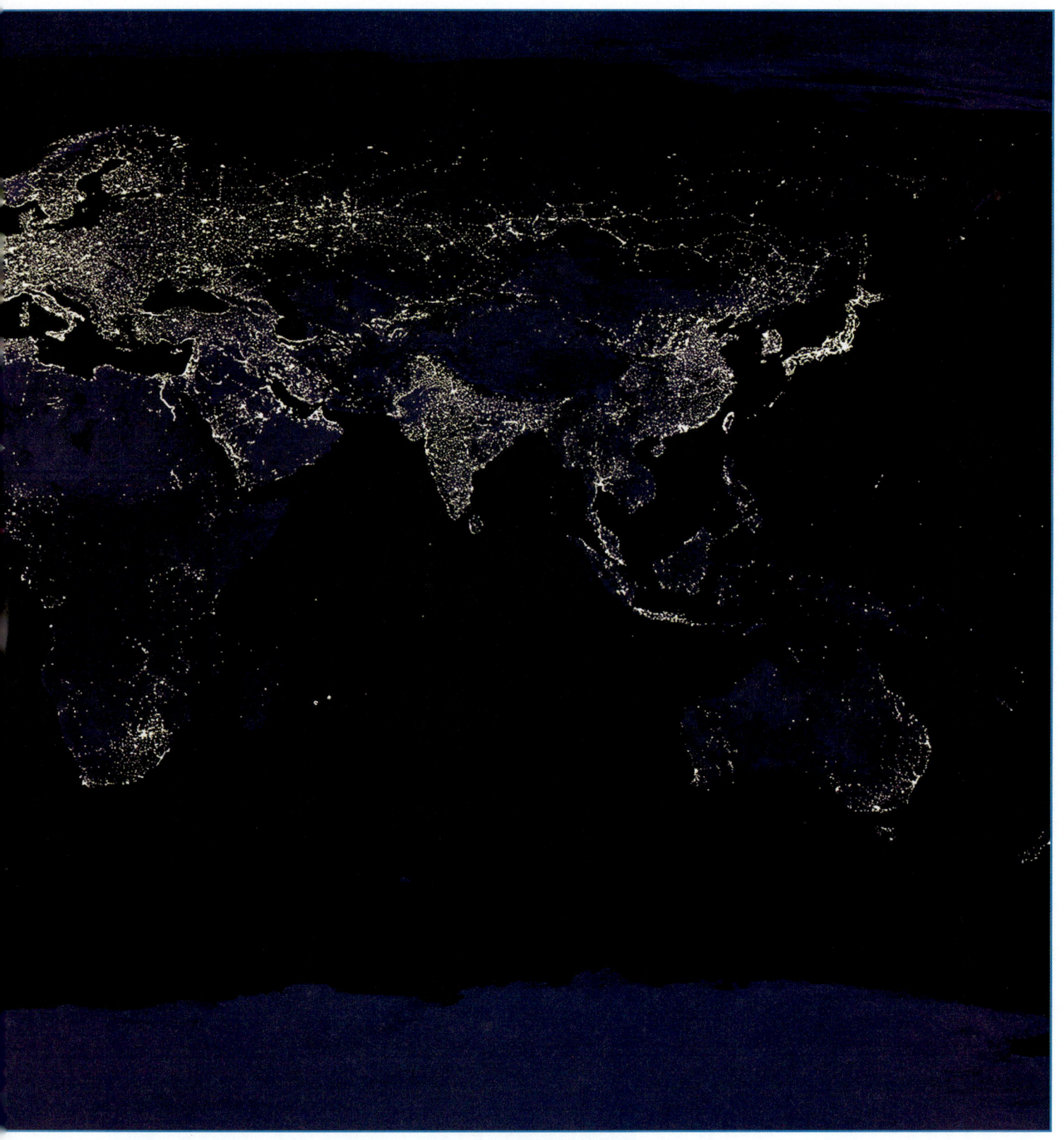

CHAPTER 19 The Power to Choose

537

A tapestry of city lights and rural fires announces our presence on this planet. In contrast to daytime images, where only natural features are easily visible, the activities of humankind at night are readily traced in this mosaic of images from U.S. Air Force weather satellites. Much of the light leakage to space corresponds to street and building lights in urbanized regions, especially in Europe, North America, and eastern Asia. Also nicely etched are transportation features such as the Trans-Siberian railroad, the main railroad through central China, the spoke pattern centered on the hub of Moscow, and Interstate Highway 5 along the western coast of the United States. The delimiting effects of geographical features such as the Nile River, the Sahara Desert, the Himalayas, and the Australian Outback are also apparent. In the tropics, the major sources of light are controlled fires—the result of grassland burning, slash-and-burn agriculture, and clearing of forests. The frequency of these fires depends on season, but in the present image they are prominent throughout the highlands of Southeast Asia, the sub-Saharan savannas, and East Africa. Other lights arise from huge burn-offs of natural gas associated with oil wells. Gas flares show clearly in Indonesia, the Tashkent region of the Soviet Union, Siberia, the Middle East, North and West Africa, and northwestern South America. In the Sea of Japan, the large blotch of light emanates from a fishing fleet that hangs multitudes of lights on its boats in order to lure squid and saury to the surface. The only natural source of light is the aurora over Greenland. Aurorae (Northern Lights) occur when high-energy particles from the sun enter polar magnetic regions, creating currents that cause the upper atmosphere to glow like a gigantic fluorescent tube.

The image is replete with lessons in geography, economics, anthropology, and environmental science. And for astronomers, who find themselves limited to increasingly remote and expensive sites, it also illustrates their constant battle against the "light pollution" that damages observations of faint stars and galaxies. But is there not a wider loss? The image testifies that hundreds of millions of people today have no dark sky and are thus denied the nighttime universe. No longer do they know the exquisite thrill of a meteor shooting across the sky, nor the humility brought on by the resplendence of two thousand stars wreathed by the Milky Way. At a time when the very survival of our species depends on finding a common vision, we have wrapped Earth in a glowing fog.

CAUTIONARY NOTES: There are several aspects of this image that the discriminating viewer should keep in mind. First, it is based on a mosaic of about forty individual photographs, each of which has its own distortions. These photos have been approximately reduced to a common scale, but the final image corresponds to a Mercator projection only to about 5% accuracy. The individual photos were also taken with a variety of exposures and under varying moonlight. Several additional processing steps then led to this poster, with the result that quantitative light intensities in different regions can only roughly be judged. The photos in the mosaic were taken at various times and seasons over the period 1974–84; this affects in particular the occurrence of tropical fires, which are highly seasonal. The aurora included in the image is an artist's rendering based on satellite photography. Its positioning is only meant to be suggestive—aurorae in fact usually occur in a ring centered on the north or south magnetic pole. Finally, although most dark regions are truly lacking in light, sections of some photos were clouded out, and suitable photos were not available for a few regions such as many remote islands and portions of South China and southwestern Africa.

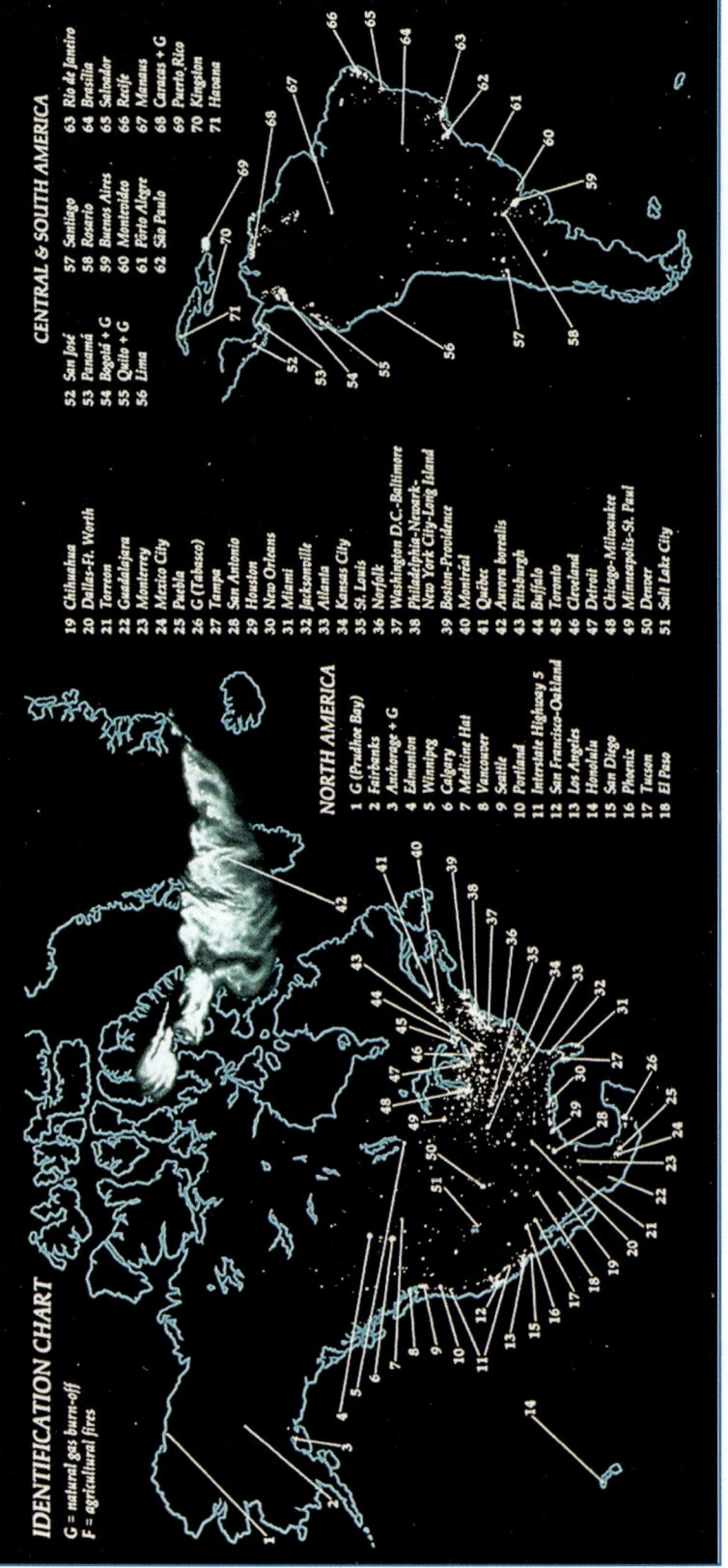

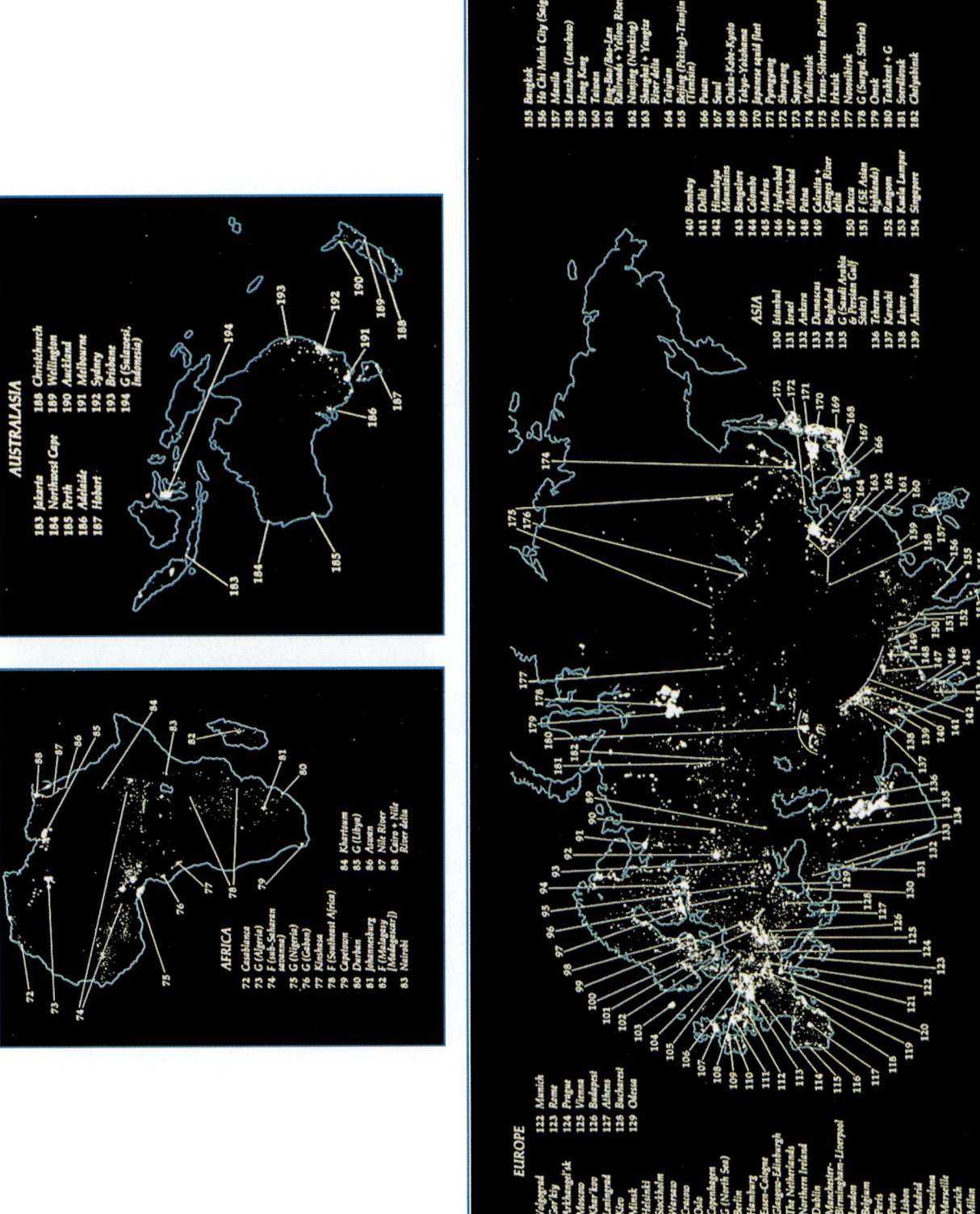

Chapter 19 The Power to Choose

HOW TO: 1 Make a T-Chart

At the beginning of each unit, Ros, Al, Isaac, and Marie introduce a new skill. For the first unit, the skill is *Show caring and respect for others and their ideas.* When you and your teammates practice that skill, what does your team look like? What does it sound like? In your teams, you will create a T-chart to describe what each skill looks and sounds like. If you have made T-charts in other classes, this will be practice for you. If you already have experience, help your teammates learn how to make T-charts.

The following steps will take you through the process of making a T-chart for the skill *Show caring and respect for others and their ideas.* Although the examples in this How To are specific for that skill, you can use these steps to make T-charts for many other skills. Therefore, you can refer to this How To whenever you need a reminder.

1. Hold a brainstorming session about what the good use of the skill *Show caring and respect for others and their ideas* would look like or sound like.
   - List as many of "looks like" and "sounds like" descriptions as you can.
   - Does everyone agree? If not, discuss those you do not agree on and compile a list that everyone can accept.

2. To make a chart like the one pictured in Figure H1.1, draw a line across the top of the paper and another line down the middle of the page, taking up about 10 lines. Title your chart "T-Chart for *Show caring and respect for others and their ideas.*"
   - Each student should make this chart in his or her notebook.

FIGURE H1.1 Label your T-chart with the skill you are charting.

How To Make a T-Chart 541

3. Label the left column "Looks Like" and the right column "Sounds Like."
4. Refer to your brainstorming list and fill in each column with your team's ideas of what *Show caring and respect for others and their ideas* looks and sounds like. Our example is pictured in Figure H1.2.
    ➡ Be sure to use your team's ideas, not ours.
    ➡ Is everyone participating? Have you included everyone's ideas?
5. Refer to this chart throughout Unit 1 to refocus yourself and your teammates on the unit skill. You can also use it to remind yourself what the unit skill looks and sounds like.
    ➡ As you begin other units and activities, your teacher may ask you to complete a T-chart for each new skill. When you do, try to follow the steps in this How To. If you have questions on making a T-chart, remember that your teammates are there to help you understand, and the Communicator can consult with Communicators from other teams. The more T-charts you make, the better you will become, and T-charts will get easier. Good job!

Show caring and respect for others and their ideas

Looks Like	Sounds Like
1. Use your teammates' name.	1. Saying "Hi, Jim."
2. Move into your teams quickly and quietly.	2. Careful—not running into people.
3. Stay with your team.	3. Sitting or standing quietly next to your teammates.
4. Listen to other's thoughts and ideas.	4. Not interrupting when another person is talking.

FIGURE H1.2 Your T-chart should have a column for what each skill looks like and sounds like. What other ideas did your team come up with for this skill?

How To Make a T-Chart

HOW TO: Construct a Data Table

While doing an investigation, you often must keep track of many pieces of information: numbers, measurements, or other observations that might tell you how long something takes, how much of something you have, or where something is. For example, in the investigation Star Tracers, you record the time your partner needs to successfully trace a star pattern.

Usually, we refer to the information collected in a scientific investigation as **data**. The dictionary defines the term "data" as "information, especially information organized for analysis or used as the basis for a decision." That definition fits the information you collect in all of your investigations, because you always analyze your information, or data, in some way. (Please note that the term "data" is actually the plural of the word "datum," which means one piece of information. Therefore, we say "data are," not "data is." That usage may sound funny to you at first because people often use the word incorrectly in conversation.)

Before you begin collecting your data, you need to plan how to keep them organized. A **data table** is a helpful tool to use when you need to organize data. If your data are organized from the start, then you will have an easier time making sense of them during and after your investigation.

Just what is a data table, and how will it help you organize information? A data table is a chart that has rows and columns. The rows and columns form boxes, called "cells," and each piece of data fits into one cell. The columns separate the data into different categories. Each row contains the information you need for each event or record. A data table can help you organize all kinds of information so that you can see patterns and find answers to questions.

The following steps will take you through the process of designing a data table for the investigation Star Tracers. Even though the examples in this How To are specific to that investigation, the steps apply to almost any data table for any investigation. Therefore, you can refer to this How To any time you need to construct a data table.

1. Read all of the steps in the investigation.

2. Decide on the problem you are trying to solve or the question(s) you are trying to answer.
   → The problem in the investigation Star Tracers is to successfully trace a star.

3. Decide what data you must collect so that you will know if you are solving the problem or answering the question. (In other words, what do you need to track during the investigation?)
   → In Star Tracers, you need to track how successfully and how quickly you and your teammate trace a star. Therefore, you have five things to record in your data table: the name of the person attempting the trial, the trial number, the time it took to complete the trial, any difficulty you or your partner had, and how successful each of you was at tracing the star. The five things you need to record are called variables. Variables are those things that can change during an experiment.

4. Draw your data table with the appropriate number of columns and rows for the data you are to collect.
   → In this case, you will need five columns and 10 rows in your data table. How do you know? First, the instructions tell you there are five variables that you need to keep track of in your data table: name, trial number, time, difficulty, and success. Those variables will become the labels of the five columns.

How To Construct a Data Table

→ Second, the instructions say that each team member should attempt to trace the star five times. Because there are two team members and you will each attempt to trace the star five times, there should be 10 rows (2 team members times five attempts each equals 10 total trials.)

5. Label the columns of your data table, including the units you will use for measurement.
   → Figure H2.1 shows how you might label your data table for the investigation Star Tracers. Note that the five variables—name, trial #, time, difficulty, and success—are the main labels of the five columns. Also note that the units appear in parentheses. Once you put the units in parentheses as part of the column label, you do not have to label each number in the column. For example, the phrase "in seconds" tells you that the numbers in the third column represent a certain number of seconds, not minutes or hours.

6. Complete as much of the data table as you can before you begin the investigation.
   → If you can fill in any parts of the data table ahead of time, then you will save time and effort during the investigation.
   → In the data table for Star Tracers, you can fill in the rows in the name column and the trial number

Star Tracers Data Table				
Person attempting the trial	Trial #	Time (in seconds)	Describe difficulty if any	Success? YES or NO
Ros	1			
Ros	2			
Ros	3			
Ros	4			
Ros	5			
Al	1			
Al	2			
Al	3			
Al	4			
Al	5			

FIGURE H2.1 To save time and effort, you can fill in as much of the data table as possible before you begin the investigation.

How To Construct a Data Table

column in advance. Because you know who will be participating, and how many trials each will take, you can fill in all the cells in columns 1 and 2 before you begin the investigation.

7. Give your data table a title.
   - The title of the data table should relate to the investigation or experiment for which you are collecting the data. You might give the data table in Figure H2.1 the title "Star Tracers Data Table."

8. Begin the investigation and record the beginning measurement or information in row 1.
   - Figure H2.2 shows how you might record the first time measurement, as well as any difficulties and whether your teammate was successful.

9. Collect the rest of your data and record each data point in the appropriate cell in the data table.
   - Keep in mind that your data table will extend for all five trials for both team members and that your time readings might be different from those in the Star Tracers Data Table shown in Figure H2.3.

As you complete other investigations in this program, you will have to construct your own data tables. In each case, try to follow the steps described here. Use the example from the investigation Star Tracers to help you visualize what a new data table might look like. If you have questions as you construct your data tables, remember that you may always ask your teammates for help. You also might ask your team's Communicator to consult with the Communicators from other teams. With practice, however, the steps should become familiar and easy for you to follow.

Star Tracers Data Table				
Person attempting the trial	Trial #	Time (in seconds)	Describe difficulty if any	Success? YES or NO
Ros	1	2 minutes	She gave up	No
Ros	2			
Ros	3			
Ros	4			
Ros	5			
Al	1			
Al	2			
Al	3			
Al	4			
Al	5			

FIGURE H2.2 Begin entering your data in row 1.

How To Construct a Data Table

Star Tracers Data Table				
Person attempting the trial	Trial #	Time (in seconds)	Describe difficulty if any	Success? YES or NO
Ros	1	2 minutes	She gave up	No
Ros	2	2 minutes, 10 seconds	She went out of the track 3 times	No
Ros	3	3 minutes, 30 seconds	She took forever	Yes
Ros	4	3 minutes	She almost went out of the track	Yes
Ros	5	3 minutes, 3 seconds	Same thing	Yes
Al	1	10 seconds	He quit	No
Al	2	8 seconds	Hand wouldn't move	No
Al	3	15 seconds	He quit	No
Al	4	5 seconds	He broke the lead	No
Al	5	11 seconds	He crumbled the paper	No

FIGURE H2.3 As you collect additional data, continue to record data in the appropriate cells.

How To Construct a Data Table

HOW TO: Construct a Bar Graph

A good graph is worth a lot of words! Graphs are useful tools because they provide "pictures" of information that often communicate more clearly than words. We organize graphs in a particular way so we can easily read the information the graphs contain. Graphs also help us find patterns in the data we have collected.

There are certain steps you should follow in constructing a graph, just as there were certain steps you followed in constructing your data tables. The following steps will help you construct your graphs for the class data in the investigation Threading the Needle.

1. Review the data you have recorded.

> To construct this sample graph, review the class data for the investigation Threading the Needle. Those data should be recorded in the class data table that your teacher has on the chalkboard or on an overhead transparency.

2. Draw the horizontal axis and the vertical axis for the graph.
   - Graphs have two lines, one that runs horizontally across the page and one that runs vertically from top to bottom. Those lines have special names. The line that runs across is the horizontal axis. The line that runs up and down is the vertical axis. The point where these two lines, or axes, meet is the place where the graph begins. (The term "axes" is the plural of the term "axis.")
   - It helps if you draw your axes on graph paper, because graph paper provides evenly spaced lines and squares. Use a ruler to draw straight lines. Draw the vertical axis close to the left-hand edge and the horizontal axis close to the bottom edge of the graph paper. Make the lines longer than you think you need. This will make your graph easy to read.
3. Label each axis with the headings in the data table.
   - The labels on each axis tell what variables you investigated. (A variable is anything that can change in an experiment.) You label each axis for a different variable so that you can see the relationship between the two variables. The variables for this graph are the number of students and how many successes they achieved during the investigation Threading the Needle.
   - Label the vertical axis "Number of Students" and the horizontal axis "Number of Successes," as the characters have done.

4. Set up the number scales on each axis.
- Both axes of a graph often have a sequence of numbers called a number scale. You read the numbers on the horizontal axis from left to right and those on the vertical axis from bottom to top. In your graph for Threading the Needle, the numbers on the horizontal axis are the number of times students were successful in threading the eyebolt, while the numbers on the vertical axis are how many students had a certain number of successes. Write a zero just outside the left-hand corner of your graph, where the vertical and horizontal axes meet.
- To decide what numbers to use in your number scales, look at your class data table. Determine the number of successes that most of the students in the class had. Count the number of students who had that number of successes, and write that number next to a line at the top of the vertical axis. Also from the class data table, determine the greatest number of successes that any one student had, and write that number in a space at the far right on the horizontal axis. Number as many of the lines on the vertical axis and the spaces on the horizontal axis as you think you need to plot your data accurately. (Compare the number scales on the graphs in the following figure. Notice that Isaac did not number every line and space.) Leave equal spaces between whole numbers. The objective is to make a big, clear picture of the class data from the investigation

How To Construct a Bar Graph

Threading the Needle. Try not to squeeze the graph down toward the bottom or over to the far left. Fill the entire page with your graph. The number scale on one axis does not have to be the same as the number scale on the other axis. The difference between the numbers next to each other on an axis must be the same, though. Look at the following outlines of graphs that Al, Rosalind, Marie, and Isaac drew. Do you think all of their number scales are okay? Why or why not?

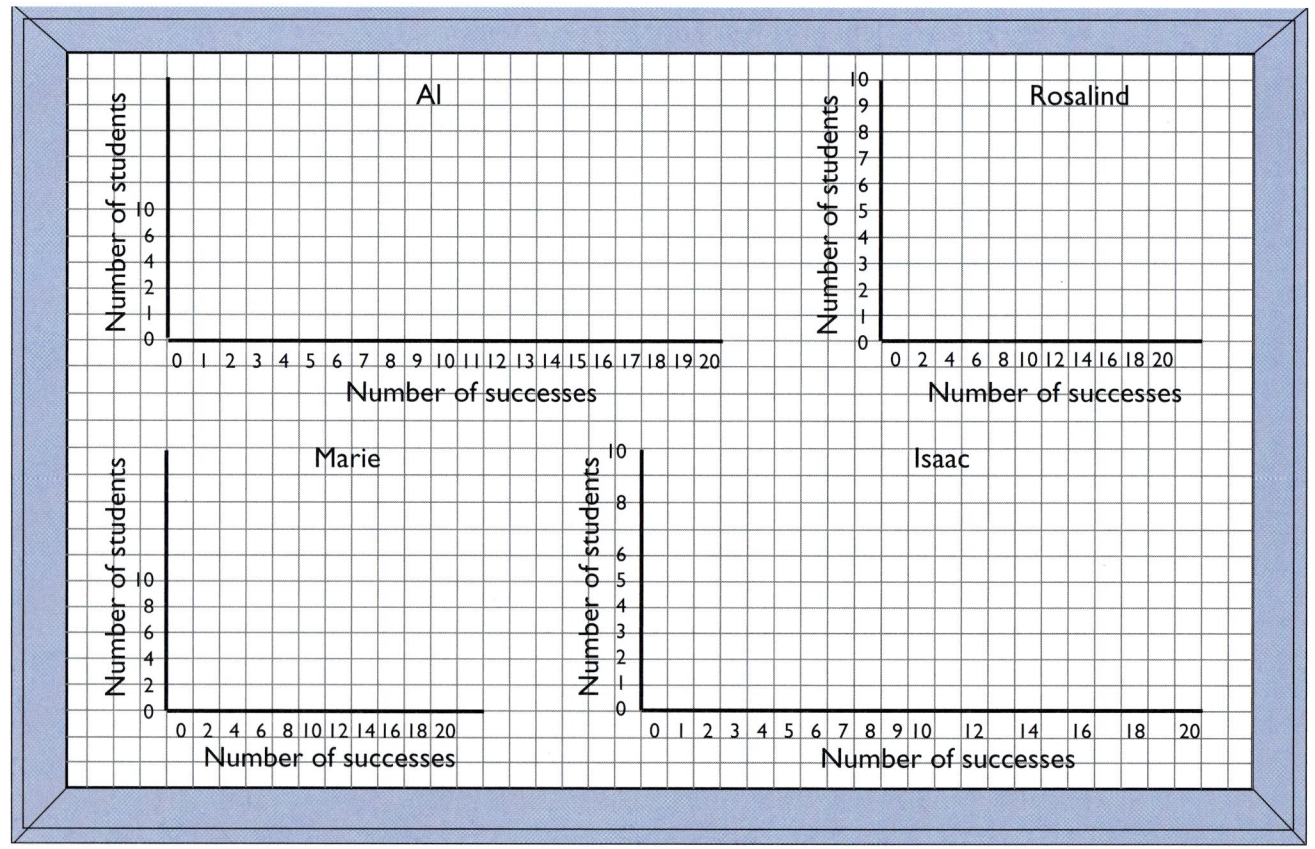

How To Construct a Bar Graph

5. Plot the data on your graph by doing the following:
- Locate the first number on the horizontal axis.
  → The first number on the horizontal axis is one. That stands for one success in threading the eyebolt. Look at your class data table. How many students were successful one time?
- Trace your finger up the column above the label to the place on the vertical axis that shows where that piece of data fits.
  → Let's say that three students were successful in threading the eyebolt one time.
- Draw a horizontal line at the correct height to make the top of the bar.
  → In this case, draw a line across the first space (#1) at the third line (#3) on the vertical axis.
- Color in the bar from that line down to the horizontal axis.
- Next, count the number of students who were successful at threading the eyebolt two times and draw a bar that represents those students. Then count the number of students who were successful at threading the eyebolt three times and draw a bar that represents those students. Continue this way until you have drawn bars for all the data in your class data table. Note that some numbers on the horizontal axis might not have a bar at all. There would be no bar if no one had that particular number of successes.
- Repeat the parts of Step 5 for all the pieces of data in the data table.

How To Construct a Bar Graph

6. Finally, give your graph a title.
   - The title should tell the reader something about the purpose of the graph. You might title this graph something simple like "Class Graph of Data," or you could be more elaborate and call it "Our Class Success Story in Threading an Eyebolt."
   - When you are finished, you will have a beautiful bar graph of what happened in your class in the investigation Threading the Needle. To see more clearly the pattern that the bars in the graph make, connect the tops of the bars with a continuous, smooth line. Your line should not dip and bulge dramatically with the bars. The objective is to round out the bars to make a smooth curve. What is the general shape of the curve that connects the tops of the bars?
   - Whenever you need to construct a bar graph in future investigations, refer to the steps in this How To. Even though the example is specific to the investigation Threading the Needle, the general steps will help you draw any bar graph.

HOW TO: Have a Brainstorming Session

Having a brainstorming session can be a lot of fun. During a brainstorming session, you can voice any idea that comes to your brain, no matter how crazy it might seem. Any idea that you think might provide a solution is one you can propose. Sometimes the ideas that seem really far-fetched at first can lead to other ideas that really work! The main purpose of a brainstorming session is to create a "storm in your brain" so that you become really creative.

Marie: "I wonder where the word 'brainstorming' came from."

Ros: "The teacher said it had something to do with 'divergent thinking'."

Isaac: "According to the dictionary, to diverge means 'to go or extend in different directions from a common point, to branch out; to differ, as in opinion or manner; to depart from a set course or norm'."

(*Isaac is reading from a dictionary.*)

Al: "Good! That means I can diverge and express my own ideas, even if they are different from yours."

As Ros indicated, the process of brainstorming involves divergent thinking, a term that comes from the root word "diverge." Isaac defined the term for you, and Al interpreted it correctly. During a brainstorming session, you have permission to be as different in your thinking from the others in the group as you like. Your goal is to come up with as many different ideas as possible. After you finish your brainstorming session, you can return to reality and use the ideas you thought of to solve a problem or answer a question.

Before you begin your brainstorming session, decide how you will record your ideas. Remember that no editing is allowed during the brainstorming session. The editing comes later. When you are ready to begin your brainstorming session, follow these guidelines:

1. State any idea about the topic that comes to your mind.
2. Record everyone's ideas. Don't judge whether the ideas are good or bad. Write them all down.
3. Keep thinking of ideas for as long as you can or until your teacher tells you the time is up.
4. If you can't think of a new idea, try to add something to an idea that is already on the list. Do not change the idea that is on the list; just add a new twist or a new way of looking at that idea.
5. If you are working in a group, take turns. Be sure that each person has a turn to suggest ideas.

→ Remember that some people might need more time than others to think before suggesting something.

→ After you finish your brainstorming session, look at your list and decide which ideas might be better than others for solving the problem or answering the question. Now is the time to edit your ideas. You should have a great list of ideas to choose from!

HOW TO: 5 Determine Averages

Averages can be very useful numbers. An average gives you information about what you can expect to happen most of the time in a given situation. For example, weather forecasters use averages in their daily weather reports. You might hear a weather forecaster say, "Today's high temperature was 72 degrees Fahrenheit. The average high temperature for this time of year is 81 degrees Fahrenheit." Such a statement tells you that the temperature today was not as high as you might expect for this time of year. The weather forecaster might go on to tell you what weather patterns caused the unusual temperature.

Part A—Procedures for Averaging Numbers

How do weather forecasters come up with an average high temperature for a particular day? First, someone had to record the high temperature on that particular date for a number of years. (You must have at least two numbers to find an average.) Let's use the date of July 25 as an example. Figure H5.1 shows the data that weather forecasters in a town in Ohio collected over a 10-year period.

Year	Temperature (in degrees Fahrenheit [°F])
1992	82
1991	78
1990	85
1989	83
1988	78
1987	83
1986	81
1985	79
1984	81
1983	80

FIGURE H5.1 This table shows the high temperature recorded on July 25 from 1983 through 1992 for a town in Ohio.

How To Determine Averages

From the data table, you can see that the high temperature was exactly 81°F only two times during the 10-year period. The rest of the time, the high temperature varied between 78°F and 85°F. How did the weather forecasters determine the average high temperature for July 25 if the temperature was never the same from one year to the next? Follow these steps and you will see how they found the average temperature.

1. Find the sum by adding all the numbers together.
   - The sum of high temperatures on July 25 from 1983 to 1992 is: 82 + 78 + 85 + 83 + 78 + 83 + 81 + 79 + 81 + 80 = 810°F.

2. Divide the sum by the number of temperatures that you have.
   - In this case, divide the sum by 10 because you have data for 10 years, each year from 1983 through 1992: 810°F ÷ 10 = 81°F.

If you look carefully at the numbers and at this procedure, you will understand more about what an average tells you. In this case, the average high temperature will not tell you exactly what the high temperature will be on any given day, but it gives you a good idea of what the temperature is likely to be. If you use the example of July 25 in the Ohio town, the high temperature every year on July 25 should be somewhere near 81°F. It might be a few degrees higher than 81°F and it might be a few degrees lower. Occasionally it might be much different if, for example, the region experienced a heat wave or an unusually cool spell. Yet most of the time the daily high temperature will be near 81°F.

How many years of data do the weather forecasters use to determine the average high temperature in your town or city? Usually weather forecasters base their averages on more than 10 years of data. Often they have records going back more than 100 years. The more years of data they have, the more accurate their average will be. Can you figure out why that is true?

Part B—Averages for Your Personal Reaction Time

In the investigation Your Personal Reaction Time, you measure the distance that the meter stick drops before you catch it. To find the average, follow the same steps as in Part A of this How To.

1. Find the sum by adding all the numbers together.
   → Before you start adding numbers to determine an average, you must decide for what purpose you want the average. In this case, you want to find one average, the falling distance, or the distance the meter stick dropped before you or your partner caught it.

2. Divide the sum by the number of pieces of data you have.
   → To determine the average, you will divide each sum by 3 because you added three numbers together to get each sum. For example, let's say that the first time, your falling distance was 0.60 meters, the second time it was 0.50 meters, and the third time it was 0.40 meters. Your sum would be 1.50 meters (0.60 + 0.50 + 0.40). You then divide 1.50 meters by 3 to determine the average falling distance because you attempted the trial three times (1.50 meters ÷ 3 = 0.50 meters).

3. Repeat Steps 1 and 2 until you have determined all averages necessary for the activity.

How To Determine Averages

HOW TO: Conduct a Research Project 6

Your task in the investigation Enabling the Physically Challenged is to learn enough about one physical challenge so that you can redesign an environment to make it more accessible to people who have that physical challenge. To find out about a particular challenge, you will do some research. There are two keys to conducting good research: (1) choose a topic that interests you and (2) get organized! If you go about your research in an organized manner, then it will be much easier to put your information together to complete your task. This How To has some tips that will help you conduct your research.

How is conducting research different from completing other assignments? According to the *American Heritage Dictionary*, to research a topic, such as a physical challenge, means to study that topic thoroughly. Therefore, conducting research is different from simply reading an article in a magazine or looking up something on the Internet. Also, when you conduct research, you do more than just read. You gather as much information from as many sources as you can. Then you put the information together in an organized way to accomplish your task.

Because all Team Members will find out something about the physical challenge your team selects, you should divide up the responsibilities. The Tracker should make sure that each Team Member is responsible for a specific part of the research and that teammates are not duplicating efforts. Be sure to check in with one another often, so that you know you are on the right track and that others know what information you are finding. That way, too, you can find out if you need to change your plan before you waste too much time.

Tips for Conducting Research

Part A—Choosing Your Topic

Tip #1: List several topics (in this case, physical challenges) that you might research.

Your team probably has a head start on this part because you already conducted a brainstorming session. The purpose of your brainstorming session was to decide on a challenge that you want to accommodate in the environment you chose. If your team has not conducted a brainstorming session yet, continue with Part A. If your team has already chosen a physical challenge, go to Part B.

To get started, list at least five possible physical challenges that you could accommodate in the environment you will redesign. Then rank order those challenges and circle your first and second choices. (You should always have at least two choices, because you might not be able to find enough information about your first choice.)

Stop now and conduct your brainstorming session. Make a list of five or more physical challenges that you could study. You might write your list on BLM HT6.1, Planning for My Research on a Physical Challenge.

Part B—Getting Organized

Tip #2: Think about your topic before you read anything.

You might ask yourselves these questions about the physical challenge you have chosen:

- What do we already know about this condition?
- What do we need to know about the physical challenge?
- What sources of information might have some information about the physical challenge?
- Is there anyone at school or in our neighborhoods who has this physical challenge?

If you organize your thoughts first, then it will be easier to locate and organize information.

You might organize your research by writing each of those questions, in addition to others you choose, at the top of one sheet of paper. Then as you find information that answers one of those questions, write it on that sheet of paper. That way you can stay organized as you go.

As you read and talk to people, stay open to new ideas. You might think of new questions about the physical challenge that you would like to answer. You can start a new page for each question that you find interesting.

Stop now and write some things you already know about the physical challenge your team chose and some of the questions you would like to answer. You might write what you already know and your questions on BLM HT6.1, Planning for My Research on a Physical Challenge.

Part C—Finding Information

Tip #3: Use more than one source for information.

Sources of information you might want to consider include

- a person who has the physical challenge you chose;
- a hospital or medical supply company;
- a school for the visually or hearing impaired, a hospital, or a nursing facility;
- a school, community, university, or medical library;
- the World Wide Web; and
- yourself, if you can gain experience with the physical challenge.

This section will discuss each of these sources and how you might access them. You might want to list your sources on BLM HT6.2, Gathering Information for My Research.

A person who has the physical challenge you want to research

If you know of someone who has the physical challenge you want to research, you might set up an interview. Even if one of your team members knows the person, you should still organize a formal interview. You might interview the person directly and/or interview a member of the physically-challenged person's family. Before you arrange for an interview, review the following points:

- Call or personally contact the person and ask if he or she is willing to talk with you.
- In your initial call, be sure to state your name and explain that you are conducting a research project for your science class. Explain that you want to find out how you could make an environment more accessible to a physically challenged person.
- Ask the person how much time he or she can spend with you and arrange for a specific time and place to meet.
- Once you arrange for an interview, be sure you write down the date, time, and place.
- Write a list of questions that you plan to ask the person. Be sure the questions are specific to the physical challenge and to the environment you are redesigning. You might share your list of questions before the interview so the person can be thinking about answers.
- Be on time for your interview. Dress neatly and speak politely.
- Ask direct questions such as, "Could you use a desk such as this one?" and show the person a photograph or a drawing of a desk from the environment you are redesigning or provide a description of the environment. Then ask, "If so, how would you use it? If not, how would you change it?"

- Be sure to show respect for the person you are interviewing. Do not treat him or her any differently from anyone else you might talk with. Remember, this person has important information to share with you.
- Thank the person for taking the time to talk with you. Offer to share your team's report if the person is interested.

A hospital or medical supply company that has relevant equipment

You might contact a medical or hospital supply company that has special equipment for a person with the physical challenge you are studying. When you contact such a company, you should

- State your name and the reason for your call;
- Ask if the company carries equipment that a person with the specific physical challenge might use;
- Ask about the store hours and when someone might be available to help you find out about specific equipment, such as a prosthesis (artificial limb) or a hearing aid;
- Set up an appointment with someone from the company;
- Arrange for transportation ahead of time to be sure you arrive on time for your appointment;
- Take tape measures or meter sticks to measure the equipment. For example, you might need to know the height and width of a wheelchair, walker, or crutches; and
- Be sure to thank the person who helps you.

A school for the visually or hearing impaired, a hospital, or a nursing facility

If the physical challenge you chose involves visual or hearing impairment, then you might visit a school and talk with staff and students. Staff at local hospitals or nursing facilities also might provide you with information or access to someone who has the physical challenge you are researching. Be sure you follow the same procedures outlined previously for conducting interviews and making appointments. Always let the contact person know who you are and why you are calling. Always be on time for any appointments you make, and thank everyone who helps you.

A school, community, university, or medical library

Libraries contain a lot of information that might help you with your research project. In addition to printed material, libraries contain card catalogs, on-line catalogs and databases, the *Reader's Guide to Periodical Literature*, encyclopedias, medical dictionaries and medical reference materials, audiovisual materials, and pamphlet files. If you have not used such resources before, ask for help from your media specialist, librarian, teacher, or a friend who has used them.

Before you use reference materials in a library, you need to know enough about your topic to identify a few key words. Key words are important words that are related to the topic you want to study. Sometimes, you can find key words by reading an article in an encyclopedia or in a magazine about your topic. You also can come up with key words by interviewing someone. For example, if you chose the challenge of cerebral palsy, you might first look up the words "cerebral palsy" in an encyclopedia or dictionary. There you might find words such as "birth defect," "lack of coordination," and "muscle weakness." The words inside the quotation marks are all possible key words because they tell you something about the cause and effects of cerebral palsy.

You need to have some key words in mind before you can effectively use the card catalog, an on-line database, or the *Reader's Guide to Periodical Literature*. You might list the key words you will use on BLM HT6.2, Gathering Information for My Research.

The World Wide Web

If you have access to the World Wide Web, you may find a wide range of information about the physical challenge that you want to research. You may find personal information from people who have the challenge, medical information about the condition, and even historical data.

In order to conduct searches on the Web, it is important to have key words in mind before you begin. (See the information about libraries.)

When you find interesting information on the Web, remember that you can print it so that you can refer to it later. If you don't print it, remember to write down the Web site addresses that you use. They will be important information for your bibliography.

Remember to disconnect from the Internet when you finish.

Yourself, if you can gain experience with the physical challenge

You should assume the physical challenge only within the environment you plan to redesign so that your direct experience will help your team complete its task. Depending on what physical challenge you chose, you might use a wheelchair, walker, or crutches; wear a blindfold; tie your hands behind your back; navigate on only one foot; or write with the opposite hand. Decide on a reasonable length of time to assume the physical challenge, and be sure you have taken safety precautions so that you do not injure yourself or someone else in the process. Keep track of problems you encounter, how you solve them, and your feelings about the physical challenge.

Stop now and list the sources you will use to start your research. Then list a few key words that will help you find out more. You can add to your list of key words as you do your research. Refer to BLM HT6.2 for help in organizing your information.

Part D—Organizing your information.

Tip #4: Reread your task and the questions you want to answer. Decide which information you have collected relates directly to accomplishing your team's task.

When you have collected enough information, your team needs to put together all the information and then decide which information helps your team accomplish its task. First, have each Team Member organize his or her information from most important to least important in terms of accomplishing the team task (redesigning the environment to accommodate the physical challenge you have been researching). Second, ask each Team Member to read or describe the three most important pieces of information he or she found. Decide how you will keep a record of each Team Member's information. You might use BLM HT6.3, Organizing My Information, as a guide.

Continue sharing information until you have enough to begin work on redesigning the environment to accommodate the physical challenge you chose. Team Members can share additional information as you complete the investigation.

HOW TO: Read a Thermometer

Thermometers are important tools for scientists and students alike because they measure temperature. Thermometers can measure the temperature of many liquids, gases, and solids. You probably are most familiar with thermometers that measure air or body temperature.

The simplest kind of thermometer is a hollow glass rod with a bulb at one end that is filled with a liquid. Sometimes this liquid has a metallic silver color; other times, the liquid is red. Small changes in temperature cause the liquid in the thermometer to expand or contract. As the thermometer gets warmer, the liquid expands and moves up the glass column. As the thermometer gets cooler, the liquid contracts and shrinks back toward and into the bulb.

The glass rod of a thermometer has marks on it that translate the height of the liquid column into temperature. Those marks are called the scale of the thermometer. Depending on their intended uses, the scales on a thermometer can show degrees of temperature in metric units (°C) or in English units (°F). Some thermometers show both degrees Celsius and degrees Fahrenheit in scales that appear on opposite sides of the glass rod. In the investigation Heat In, Heat Out, you will use thermometers that have a metric scale.

Scales on metric thermometers do not always look the same. Sometimes, a metric scale might have one long mark for every 5 degrees Celsius with four short marks in between, as shown in Figure H7.1. Each of the four short marks stands for an interval of 1 degree Celsius. Sometimes, as in Figure H7.2, a scale might show long marks every 10 degrees Celsius. On this type of scale, there are four short marks, then one medium mark, and then four more short marks between the long

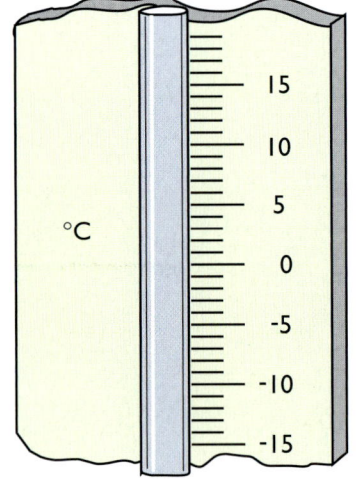

FIGURE H7.1 This is one example of a metric scale you might find on a thermometer.

570 How To Read a Thermometer

marks. Each of the medium marks stands for an interval of 5 degrees Celsius and each of the short marks stands for an interval of 1 degree Celsius. Some thermometers might have scales that are different from these two examples. Before you use any thermometer, look at its scale and decide what each mark means.

To read a thermometer, you must first look at the scale and then rotate the thermometer slowly from side to side until you see the silver or red column of liquid. (That might take practice. If you have trouble seeing the liquid column, try moving the thermometer into better light or holding it against a background of a different color.) The point on the scale where the liquid column ends indicates the temperature. You must read the scale to get the temperature reading. Figure H7.3 shows a thermometer that reads "45 degrees Celsius."

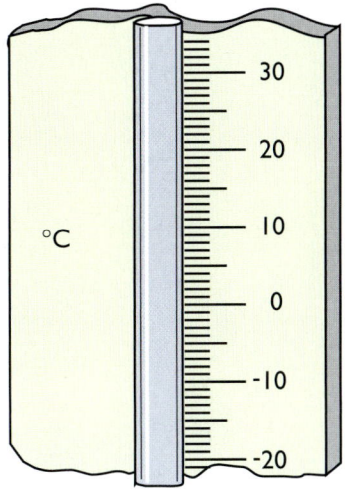

FIGURE H7.2 Some metric thermometers have scales that mark every 10 degrees Celsius.

 Always use caution when working with a glass thermometer. If you drop it or bump it against a hard surface, the thermometer might break. If a thermometer does break, tell your teacher immediately.

- If you break a thermometer that contains a red liquid, sweep up as many pieces of broken glass as you can; do not pick up the pieces of broken glass by hand. Dispose of the broken thermometer by placing it into a double bag and sealing the bag. You can dispose of the bag along with the regular trash or place it in the special container for broken glass.
- If the thermometers you are using contain a silver liquid called mercury, tell your teacher immediately, but do not pick up or sweep up any of the broken glass or the silver liquid yourself. Let your teacher take care of the spill. Mercury is a hazardous material.

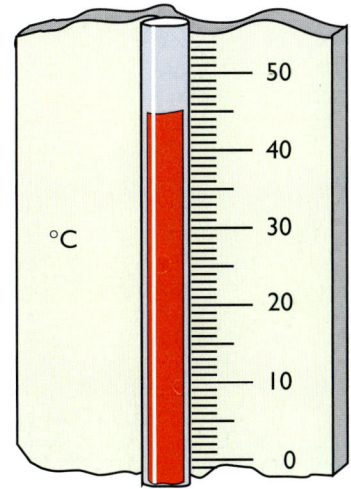

FIGURE H7.3 The scale on this thermometer indicates that the temperature is 45 degrees Celsius (45°C).

HOW TO: Conduct a Fair Test
8

In the investigation Collecting the Sun, your assignment is to design a fair test to compare the water-heating systems each team in your class designed. What does it mean to conduct a fair test? By definition, a fair test is one that shows no favoritism or bias in the results. In a fair test, you must treat all elements to be tested (in this case, the different water-heating systems) the same so that there is no bias toward one element (such as aluminum foil) or another.

Consider a different example that describes a fair test. Suppose you are a botanist, a scientist who studies plants. You want to breed a new type of bean plant that you think could alleviate world hunger. You want this new bean plant to grow in places that get a lot of sun, but not much water. You have a collection of five different bean seeds, and you are ready to conduct your investigation to find out which bean seed will produce a plant that will grow best in sunny, dry conditions.

First, you plant each different bean seed in the same sized pot in the same amount of soil at a depth of two inches. However, each pot has a different type of soil: Pot #1 has topsoil with fertilizer added; Pot #2 has clay soil; Pot #3 has commercial potting soil; Pot #4 has dirt from your backyard; and Pot #5 has sandy soil. Then you place Pots #1 and #2 in a sunny greenhouse, Pots #3 and #4 in the shade outdoors, and Pot #5 in your living room window where the sun shines on the pot about two hours every morning. You water each pot every three days with 250 mL of water. After six weeks, the bean plant in Pot #1 is the tallest by about 6 inches and has twice as many leaves. You conclude that the bean seed in Pot #1 is the best, and you decide to produce a lot of those seeds to sell to people who live in sunny, dry climates. Is this a valid conclusion? In other words, was this a fair test of the five different bean seeds? Why or why not?

In deciding your answer, go back to the definition of a fair test. In a fair test, you treat all the things to be tested the same so that there is no bias toward one thing or another. Were all the bean seeds treated the same, or did some of the seeds receive more favorable treatment than the others? Name the similarities and differences in how the bean seeds were treated during the test.

You probably noticed that the major differences were the types of soil and the amount of sunlight that each plant received. The conditions that were the same were the size of the pot, the amount of soil, the depth the seeds were planted, the amount of water, and frequency of watering. Those similarities and differences are known as **variables**. The variables are factors that can possibly change within an investigation or from one investigation to another. (How To #2, How to Construct a Data Table, introduced the term "variable." You might review that discussion if you do not remember what variables are.)

To conduct a fair test, you need to **control variables**. That means that you need to keep all variables the same except the one you want to test. That includes all the variables involved in the entire test, from start to finish. For the bean seed test, you listed the variables you needed to control: the size of the pot, the amount of soil, the type of soil, the depth you plant the seeds, the amount of sunlight the seeds and plants receive, the amount of water they receive, and how often you water the seeds and plants. If you control all of these variables and grow all of the bean plants under the same sunny, dry conditions, then you should be able to choose which seed will produce the bean plant that will grow best in a sunny, dry climate.

In the investigation Collecting the Sun, you design a fair test to compare the water-heating systems each team in your class designed. As you design your test, keep the test of the bean seeds in mind. You might follow these steps in designing and conducting a fair test.

1. Identify the question you are trying to answer.
    → In the investigation Collecting the Sun, you want to know which water-heating system heats water best.

2. Identify the variable that will help you answer the question from Step 1.
    → How will you know which water-heating system will heat water the best?
    → You will know by the temperature of the water in each system. Therefore, the variable that will help you answer the question about the water-heating systems is the temperature.

3. Identify what measurements or observations you will make of the variable you identified in Step 2.
    → Decide ahead of time how you will describe and record the temperature. How many temperature samples will you take during the test?

4. Identify all of the variables that might affect the outcome of your test.
    → Some variables that might affect the outcome of your test include the amount of time the water-heating system has been collecting sunlight; another is the type of materials used in the system. Those are not the only variables, of course. Come up with your own, complete list.

5. Plan how you will control all of the variables.
    → Think about how you will make sure that you treat each water-heating system the same. Remember, you do not want to show any bias or favoritism toward any of the systems, even though you might have an idea how your test will turn out.

How To Conduct a Fair Test

6. Conduct your test according to your plan.
7. Record your results in your data table.
   → Remember to leave room in your data table to record any comments about surprising things that happen, or notes about which variables were difficult to control and why. For example, on the day you conduct your test, maybe the Sun is covered by clouds. Is the variable of cloud cover one that you can control? You might want to include notes about whether you think the clouds affected the results of your test. Such notes will help you explain the results of your test.
8. Report the results of your test.
   → In your report, explain exactly what happened during the test. Note the temperature of the system. Be as detailed as you can about what happened.
9. Draw your conclusions.
   → State answer to the question you identified in Step 1. Which water-heating system collects the Sun's heat the best? Be sure you explain why you think that system is best. Use the test results that you recorded in Step 8.

This How To might help you complete the remaining investigations in this program. Refer to it each time you need to control variables and conduct a fair test.

HOW TO: 9 Use a Bunsen Burner

A Bunsen burner is a special kind of heat source that burns fuel gas. It was designed in 1855 by a German scientist named Robert Bunsen. The Bunsen burner was a more convenient and efficient heat source for a laboratory than either stoves or candles, which were the only other heat sources available at that time. The Bunsen burner has changed very little since Mr. Bunsen invented it.

Part A—Parts of a Bunsen Burner

The Bunsen burner has very few parts (see Figure H9.1). It has a fuel gas inlet, which is connected to the gas jet with rubber tubing. The burner has a movable ring near the base, which has openings called air ports. (The air ports look like small holes around the base of the Bunsen burner.) The air ports allow air to enter the barrel of the burner and mix with the fuel gas. Once ignited, the air and gas mix in the barrel to produce a flame at the top of the barrel.

You can change the size of the air port openings by rotating the ring. By changing the size of the air ports, you can control how much air enters the barrel of the burner. The amount of air entering the Bunsen burner is important because fuel gas cannot burn without air. If there is not enough air mixing with the gas, the gas will not burn completely, and you will not get a very hot flame.

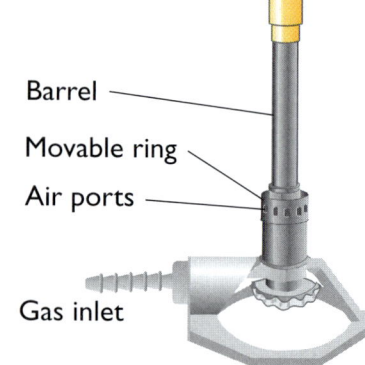

FIGURE H9.1 A Bunsen burner has few moving parts and is easy to use. A Bunsen burner burns with an open flame, though, so you must always be careful when you use one.

Part B—Safety Cautions When Using a Bunsen Burner

Always be careful when lighting and using a Bunsen burner. Wear eye protection at all times. Be sure you do not put your face near the top of the barrel at any time. Tie back long hair, roll up your sleeves, and remove scarves, ties, or jewelry that might hang down into the flame. Remember that the flame of a Bunsen burner burns very hot; do not play around during an investigation that requires the use of a Bunsen burner or any heat source.

Strike the match before you turn on the gas jet. As soon as you hear the gas, light the burner. The gas should not be on longer than a few seconds before you light it. If you have trouble lighting the burner, turn off the gas and try again. Do not keep the gas flowing while you light a new match. If you fail to light the burner after a couple of attempts, ask your teacher for help.

When you have finished with the burner, be sure that you turn off the gas jet completely so that no gas escapes into the room.

Part C—How to Light and Adjust a Bunsen Burner

Follow these steps in lighting and adjusting a Bunsen burner:

1. Review the safety cautions in Part B.
2. Read through all the steps for lighting the burner and be sure you understand what to do in the proper order.
3. Put on your safety goggles.
4. Tie back long hair, roll up your sleeves, and remove any scarves, ties, or jewelry that might hang down into the flame.
5. Turn the ring near the base of the burner so that the air port openings are almost closed.
   ⟹ The air port openings should be open just slightly so that a little air flows through them.
6. Strike a long, wooden match and hold the flame near the top of the barrel of the Bunsen burner.
   ⟹ You should strike the match first so that when you turn on the gas, you can light the burner right away. You can hold the flame of the match near the top of the burner until the gas is actually on.

7. Turn on the gas jet to a low setting and move the flame of the match to the side of the barrel, just below the top.
   → The burner should light immediately and probably will burn with an orange flame. The orange flame is a sign that the gas in the burner is not burning completely.

    If the burner does not light after a few seconds, turn off the gas and try again. Do not keep the gas flowing while you light a new match. If you fail to light the burner after a couple of attempts, ask your teacher for help.

8. Adjust the flow of air and the flow of gas until you see a blue flame.
   → When the burner is burning properly, you will see a blue flame that has two distinct parts (see Figure H9.2). To get the blue flame, slowly open the air ports to add more air to the fuel. You might need to adjust the flow of gas from the gas jet as well. Keep adjusting the air port openings and the gas flow until you get a steady, blue flame.
   → The gas should burn quietly, also. If you hear a roaring sound, then there is probably too much gas coming into the barrel. Reduce the flow of gas and possibly close the air port openings a little, too.

9. When you have finished with the Bunsen burner, turn off the gas jet completely and close the air ports.
   → The flame should go out quickly.

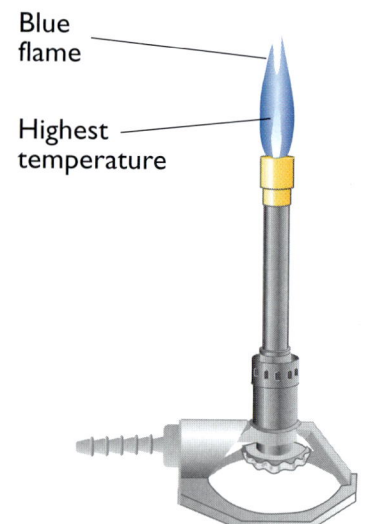

FIGURE H9.2 When you have adjusted your Bunsen burner properly, you should see a flame that looks like this.

How To Use a Bunsen Burner

acceleration: Acceleration is an increase in speed during a given time interval.

adhesive forces: The attractive forces between unlike particles. In this book, adhesive forces refers to the forces between particles of water and the particles that make up a paper boat.

afterimage: An afterimage occurs when a spot exists in your vision after you have seen a bright flash of light. Afterimage refers to the desensitization of the retina to certain colors so that you see only the complementary colors for a brief period of time.

alchemy: The branch of ancient science concerned with changing one material into another using an elusive material known as philosophers' stone.

alcohol consumption: Alcohol consumption is the act of consuming or drinking alcohol.

atoms: Atoms are the small particles that make up materials. The philosopher Democritus named these small particles atoms.

average: If you made a list of every classmate's height, added the heights together, and divided by the total number of people on your list, this would give you the average or the mean height for students in your class.

axis (plural, axes): An axis is a horizontal or vertical line in a graph on which we write numbers or labels. Most graphs have both a horizontal and a vertical axis, and the place where these two lines meet is where the graph begins.

bell-shaped curve: Bell curves show the range of diversity in various limits for a group of organisms.

benefit: A benefit is an advantage or something that promotes well-being according to what a person identifies as a need or a desired outcome.

blind spot: In the eyeball, the portion of the retina that contains no rods or cones is called the blind spot. You do not see images that fall on the blind spot.

blood alcohol content (BAC): BAC is a measure of how much alcohol is in a person's blood. This is measured as a percentage.

carcinoma: (car sih NO muh): A carcinoma is a cancer that occurs in tissues such as skin or linings of organs.

chemical energy: Energy that holds particles of an object together.

cohesive forces: The attractive forces between like particles are called the cohesive forces. In this book, cohesive forces refer to the attractive forces between water particles.

concave: A lens shape in which the edges are thicker than the center is called concave.

cones: Cones are specialized cells in the retina of the eye that help you see colors.

constraints: Human factors and other things that affect the criteria a designer sets for a product are called constraints. Constraints are limits a designer might encounter when trying to fulfill the criteria.

continuous motion rate: This refers to the speed at which you must present a series of pictures in order to perceive continuous motion.

controlling variables: This refers to the steps you take to keep all variables in an experiment constant except for the one that you want to test.

convex: A lens shape in which the edges are thinner than the center is called convex.

cooperative learning: Working cooperatively with other students to complete a task as a team.

cornea: The clear part of the coat of the eyeball that covers the iris and pupil at the front of the eye.

cost: A cost is a disadvantage or something that causes a loss of one kind or another according to what a person identifies as a need or a desired outcome.

criteria: In this book, criteria refers to technology and design. Criteria is the plural form, criterion is the singular. Criteria are the goals that designers set for what a product will do, look like, act like, or be like. A product then is judged by whether or not it fulfills all the criteria.

D/s ratio: This is the value derived from dividing the distance a person is from a screen of lines by the space between the lines. This value represents how far a person can stand from a screen to see a clear picture.

data: The data are the measurements or information that you collect during an investigation.

data table: A data table is a chart that helps a person keep track of observations or when making observations.

deoxyribonucleic (dee OK see RY boh noo KLEE ik) acid (DNA): These are particles that make up chromosomes. Components of DNA make up genes.

dermatologist (der mah TAL uh jist): A dermatologist is a medical doctor who specializes in problems of the skin.

diversity: In this book, the term "diversity" describes a variety in individual abilities. The word means "differences."

driving under the influence (DUI): A person driving under the influence is driving while intoxicated with alcohol or other drugs.

dry geothermal: A system that transforms the heat energy of hot rocks into electricity.

echolocation: The ability to use sounds, instead of vision, to locate objects.

electrical energy: Energy produced when electrons move from one atom to another.

electrons: Electrons are tiny particles in atoms.

energy: Energy is the ability to change the motion of an object or its physical state. It is often referred to as the ability to do work.

energy of motion: Energy of motion is energy that is released when something is in motion.

energy of position: Energy of position is the energy something has because of its position. This is also called potential energy.

ergonomics (er goh NAH miks): This is the branch of science that studies how people interact with the products, facilities, equipment, environments, and procedures they use at work and in their homes. It is the science of human factors.

fair test: When you control all variables except the one you wish to study, you are said to be conducting a fair test.

flicker-fusion frequency: This frequency is how fast a series of flashes must be presented to achieve the appearance of continuous motion.

focal length: The distance between the surface of the lens and the point on the eye where the distant image is the sharpest.

force: A push or a pull that causes something to move or change its speed or direction is called a force.

fovea (FOH vee ah): This is the area of the retina that contains the most rods and cones. You can see an image more clearly if it falls on the fovea.

function: In this book, function refers to what a product can do. Function is a criterion.

galvanometer (gal van OHM ih ter): A galvanometer is a device that can measure small amounts of electricity. A simple galvanometer consists of a compass inside a coil of wires. The needle of the compass is a small magnetic bar. When electricity flows through the coil of wire, the needle will move.

generator: A generator is any system that generates electricity. Power plants contain large generators known as turbines. You also can construct a much simpler generator using a coil of wire and a magnetic bar. In either case, electricity is produced or generated as magnets move past coils of wire.

graph: A visual representation of the results of your data.

hardness: The property of a material that is measured by how firm it is.

horizontal axis: The line in a graph that runs across the page from left to right.

human factors: Human factors are the differences in individual human beings that must be considered in setting standards, such as sizes, to ensure a proper "fit."

if–then statements: A logical statement that first states what a model is and then the results you would expect from an experiment.

input: Something that goes into a system is called input. In the investigation Falls and Waterfalls, the input to the beaker of water was the water coming into the system through the tubes and raising the water level in the beaker.

interlacing: In television, when the electron gun activates every other line of pixels from the top of the screen to the bottom then returns to the top to activate the remaining lines.

iris: The colored part of the eye that is a muscle that opens or closes to control the amount of light coming into the eye.

kinetic energy: The energy of movement.

lens: A lens is a piece of transparent material with curved surfaces that refracts light that passes through it.

light: Light is something that makes things visible. It is a type of electromagnetic radiation to which the eyes react.

limit: The term "limit" means boundary or something that you cannot go beyond.

material: This refers to the makeup of a product or thing.

melanin (MEL uh nin): Melanin is a dark brown pigment in the skin that absorbs sunlight, both visible light and ultraviolet radiation.

melanoma (mel uh NO muh): A melanoma is a type of skin cancer that can spread to other areas of the body. Melanomas involve the cells that give the skin its color.

motor: A motor is a system that transforms electrical energy into the mechanical energy of a spinning shaft. People can attach this spinning shaft to many other devices (for example, the blades of a fan) to produce many different kinds of output. A motor contains coils of wire inside stationary magnets; the coils are attached to the shaft. When electricity flows through the coils, the shaft spins.

NTSC standards: The National Television Systems Committee (NTSC) is the organization that sets the standards we currently use to produce TV pictures in the United States and Japan.

nucleus (NOO klee us): This is the part of an atom that contains the protons and neutrons.

operational definition: An operational definition is a standard definition of how you measure something.

output: Something that comes out of a system is called output.

particle model (or particle theory): This model presents the idea that all materials are composed of particles. The model is an estimate of what one of these particles might look like.

perception distance: The perception distance is the distance you travel during your perception time. This is the first phase in the stopping process.

perception time: The perception time is the time that elapses between hearing and perceiving.

peripheral vision (per IF er al): This is the ability to see around you without turning your head and while looking straight ahead.

persistence of vision: Persistence of vision is when the human visual system retains an image it sees for a very short time after that image is no longer on the retina.

petroleum: A liquid form of a fossil fuel.

phenomena: Phenomena means happenings or occurrences. It is the plural of phenomenon.

philosopher: A philosopher is a person who thinks about things and develops explanations for them.

photon: Ultraviolet packets of energy.

pitch: In this book, pitch refers to the shape and twist of a certain propeller.

pixels: The term "pixels" refers to small sections that appear as dots on a grid that covers the TV or computer screen. Pixels come in three colors: red, green, and blue.

potential energy: Energy that an object has because of its position.

property: A property of a material can be its look, feel, function, or other unique characteristic.

propulsion: The force that makes an inanimate object move forward.

pupil: A clear opening in the center of the iris that lets light into the eye.

range: A range defines the inner and outer limits of a characteristic such as peripheral vision. Ranges exist in individuals as well as within populations.

reaction distance: This is the distance you travel during your reaction time, before you begin to stop. This is the second phase of the stopping process.

reaction time: The time it takes for one to react after perceiving an event.

research: Research means to study a particular topic thoroughly. When you research, you gather information about the topic from as many sources as you can.

retina: The retina is the innermost layer of the eyeball, and contains the rods and cones.

rods: Rods are specialized cells in the retina of the eye that help you see in the dark.

scientific model: A representation or model of an object or scientific phenomena that scientists cannot observe directly.

skidding distance: During the skidding time, you travel a distance known as the skidding distance. This is the third and final phase of the stopping process.

skidding time: The skidding time is the time that passes between when you first try to stop and when you finally come to a complete stop.

solar radiation: A process by which energy from the Sun can be transported.

spectrum: The spectrum is the range of frequencies of visible light from red to violet.

standard: Set of guidelines that industries use to maintain product consistency.

stratosphere: The stratosphere is a layer of the atmosphere that is about 25 km (15.5 mi) above the surface of the earth.

sunburn: Exposure to the Sun's UV radiation that can cause damage to the cells of our skin.

surface tension: This refers to the special way that particles of water are cohesively attracted to each other.

system: A group of objects that a person has isolated in order to study characteristics of properties.

technology: Technology refers to designing products that fit people and help them solve problems.

testability: This refers to the component of models that classify them as science or non-science. If a model is testable and does not include elements of magic, superstition, faith, or other components that are impossible to test, the model is classified as science.

theory: An explanation for a phenomenon that is based on a set of observations.

total stopping distance: This refers to the distance you travel during the total stopping time. You can derive the total stopping distance by adding the perception distance, the reaction distance, and the skidding distance.

total stopping time: This is the entire time it takes you to come to a complete stop. You derive the total stopping time by adding the perception time, the reaction time, and the skidding time.

translucence: This is the property of a material that is measured by how much light passes through it.

ultraviolet radiation: Sunlight is composed of many different kinds of light. Ultraviolet radiation, abbreviated UV radiation, is a type of invisible light that contains a great deal of energy and can damage the skin.

variable: A variable is anything that can change in an experiment or investigation. A variable can affect the results of your experiment.

vertical axis: The line in a graph that goes from the bottom of the page to the top of the page is called the vertical axis.

viscosity (vis KOS i tee): This refers to the property of a material that is measured by how easily it flows or pours.

Artists, Photographers, and Photography Suppliers, Third Edition

Illustrations: Cukjati Design; LaurelTech; Pizzazz Productions; Mary Snyder
BSCS Photographers: Carlye Calvin, Alex Fenlon, Melissa Richie, Mark Schoenenberger
Scanning and Prepress: Advanced Graphics & Printing

Introduction Unit: Opener (**Fiberoptic wires, Beakers, Laptop computer** PhotoDisc; **Runners** Comstock; **Students studying, Girls on slide, Students in line** Corbis)
CHAPTER 1: Opener NASA, Johnson Space Center; **Colored yarn, 1.2, 1.3 (Market), Mural art, Colored powder dyes, 1.4, 1.11** Corel; 1.1 (**Chocolates, CDs**), 1.3 (**Sushi**) PhotoDisc
CHAPTER 2: Opener EyeWire; **Students building robots** LEGOLAND

Unit 1: Opener (**Rock climber, Karate punch, Gymnast** PhotoDisc; **Mountain biker, Snowboarder, Basketball player, Swimmer** EyeWire)
CHAPTER 3: Opener Comstock; **Life cycle of stars, Wolf-Rayet star, Glowing eye of NGC** NASA; **Looking through binoculars, Bat** Corel; **Binoculars** Pioneer Research; **Girls talking on phone** Corbis
CHAPTER 4: Opener (**Snowflakes**) Corel; **Bhutanese schoolgirls** Corel
CHAPTER 5: Opener PhotoDisc; **Televisions,** Carlye Calvin; **5.1 Early Television Foundation; 5.9** EyeWire; **5.10, Video camera** PhotoDisc
CHAPTER 6: Opener EyeWire; **Winter road, Summer road, Traffic lights, 6.2, City street at night, Stop sign, Speed limit signs, Studebaker** Corel; **6.3** Comstock; **Cell phone, Hands-free phone, Single car accident** PhotoDisc; **6.16** Carlye Calvin
CHAPTER 7: Opener Mr. Erik Weihenmayer, Comstock; **7.1** (**Seal, Fire engine**) Corel, **Trumpet player, sea, jackhammer,** Carlye Calvin

Unit 2: Opener PhotoDisc
CHAPTER 8: Opener BSCS; **General Santa Anna** © 2003 www.clipart.com; **Dr. Chien-Shiung Wu** The Corbis-Bettmann Archive; **8.8** Corel
CHAPTER 9: Opener BSCS; **Prism** PhotoDisc; **9.4, 9.5, 9.6** (**Mountain meadow, Ocean waves, Clouds**) Corel; **9.6** (**Fire**) Comstock; **9.7** © 2003, www.clipart.com
CHAPTER 10: Opener Comstock; **10.3, 10.6, 10.8, 10.9** (**Syncom IV-3, Rocket, Niagara Falls**) Corel; **10.4, Alchemists painting** The Corbis-Bettmann Archive; **10.9** (**Atlantis docked at Mir**) NASA, Johnson Space Center
CHAPTER 11: Opener Visuals Unlimited; **11.1** PhotoDisc; **11.3** Visuals Unlimited/ J. Daley; **11.10** © 2003 www.clipart.com; **11.14** Courtesy of the New York State Museum

Unit 3: Opener (**Baby** DigitalStock; **Woman, Eye chart** PhotoDisc; **Satellite dish, Circuit board, Beakers, Circuit board tunnel** Corel)
CHAPTER 12: Opener BSCS; **12.3** Tom Stack/B. August; **Family Breakfast,** Carlye Calvin
CHAPTER 13: Opener Stabi-Craft Marine; **Boats in Sidelight, 13.3, 13.15** Corel; **Toy store** © 2003, www.clipart.com; **Child on tricycle** Brand X Pictures; **Ergonomic keyboard** PhotoDisc; **Playground** Courtesy of Goric Marketing Group USA, Inc., www.goric.com; **LEGOLAND** LEGOLAND

CHAPTER 14: Opener PhotoDisc; **14.2 (Satellite)** USGS; **14.2 (Bank, Pueblo, Arch, Cathedral)**, **14.4, 14.5, 14.6, 14.7 (Hikers)**, **14.9a, c, d** Corel; **14.3** © Wally Hage, Doll House Lady Miniatures. Photo by Charles Webley Edwards; **14.7 (Students)**, **14.8** Comstock; **14.9b** The Corbis-Bettmann Archive

CHAPTER 15: Opener Carlye Calvin; **15.1** Corel; **Bus with wheelchair lift, Van cross section** The Braun Corporation; **15.4** The Corbis-Bettmann Archive; **Computer** PhotoDisc; **Underwater vehicle** OAR/National Undersea Research Program (NURP), U.S. Navy

Unit 4: Opener (Castle Geyser) Corel; **(Windmills, Sun, Oil drill)** Comstock

CHAPTER 16: Opener Comstock; **16.1 (Dancers, Flower, Plane)**, **16.2, 16.3, 16.4, 16.5, 16.6, 16.11 (Lizard)**, **16.13** Corel; **16.1 (Computer)** DigitalStock; **16.7** The Corbis-Bettmann Archive; **16.11 (Pouring antifreeze)** Tom Stack; **16.16a** Carlye Calvin

CHAPTER 17: Opener (Model of Mars Pathfinder NASA Glenn Research Center; **Solar windmill** Sandia National Laboratories; **Helios** NASA; **Roof-top solar panels** PowerLight Corporation); **17.4, 17.5, 17.6** Visuals Unlimited/K. E. Greer; **17.7, Ice break up with birds, 17.12** Corel

CHAPTER 18: Opener NEG Micon; **18.6, Paper clips, 18.10, 18.19 (Electrical generators, Dam), 18.22, 18.23, 18.28, 18.31, 18.32** Corel; **18.11, Environmental activists** Comstock; **Rancher, 18.21** PhotoDisc; **Family** Brand X Pictures; **18.19 (Windmills)** NEG Micon; **18.24, 18.27, 18.29, 18.30, 18.34,** Carlye Calvin; **18.33** Jeff Vanuga

CHAPTER 19: Opener Comstock; **Students discussing play** Comstock; **19.13, 19.15, Colored Pencils, screws** Corel; **19.16,** NASA Goddard Space Flight Center; **pages 538–539** © 1985 W. T. Sullivan, III

Contributing Writers and Consultants, Third Edition

Clyde R. Burnett, *Fritz Peak Observatory, Rollinsville, Colorado* (Science Content)

Debra Hannigan, *BSCS, Colorado Springs, Colorado*

Sherie McClam, *Boulder, Colorado*

The Third Edition Revision was funded by a grant from Kendall/Hunt Publishing Company, Inc., Dubuque, Iowa.

Program Reviewers, First Edition

Michael R. Abraham, *University of Oklahoma, Norman, Oklahoma* (Science Content, Instructional Model)

Thomas Anderson, *University of Illinois, Champaign-Urbana, Illinois* (Reading)

Albert A. Bartlett, *Professor Emeritus, University of Colorado, Boulder, Colorado* (Science Content)

Clyde R. Burnett, *Fritz Peak Observatory, Rollinsville, Colorado* (Science Content)

Elizabeth Beaver Burnett, *Fritz Peak Observatory, Rollinsville, Colorado* (Science Content)

Kallene Casias, *Turman Elementary School, Colorado Springs, Colorado* (Cooperative Learning)

Audrey Champagne, *SUNY, Albany, New York* (Instructional Model)

Aileen Dickey, *Wildflower Elementary School, Colorado Springs, Colorado* (Cooperative Learning)

Peter Drotman, *Centers for Disease Control, Chamblee, Georgia* (Science Content)

Richard A. Duschl, *University of Pittsburgh, Pittsburgh, Pennsylvania* (Nature of Science, Science Content)

Diane Ebert-May, *Northern Arizona University, Flagstaff, Arizona* (Science Content)

Timothy Falls, *Meadows Elementary School, Novi, Michigan* (Safety)

Robert J. Francis, *GM Hughes Electronics, Los Angeles, California* (Science Content)

Terry Gerbstadt, *KRDO, Channel 13, Colorado Springs, Colorado* (Science Content)

Jerald Harder, *Aeronomy Laboratory, National Oceanic and Atmospheric Administration, Boulder, Colorado* (Science Content)

Henry Heikkinen, *University of Northern Colorado, Greeley, Colorado* (Science Content)

Werner Heim, *The Colorado College, Colorado Springs, Colorado* (Science Content)

Jane Heinze-Fry, *Cornell University, Ithaca, New York* (Science Content)

Sheryl Hobbs, *Carmel Middle School, Colorado Springs, Colorado* (Cooperative Learning)

Martin Hudson, *Hughes Aircraft, Denver, Colorado* (Science Content)

Jack Lochhead, *Ventures in Education, New York, New York* (Instructional Model)

James McClurg, *University of Wyoming, Laramie, Wyoming* (Science Content)

Joseph D. McInerney, *BSCS, Colorado Springs, Colorado* (Science Content)

Verjanis Peoples, *Grambling University, Grambling, Louisiana* (Equity)

E. Joseph Piel, *Professor Emeritus, SUNY, Stony Brook, New York* (Science Content)

Belinda Rossiter, *Baylor College of Medicine, Houston, Texas* (Science Content)

Kathleen Roth, *Michigan State University, East Lansing, Michigan* (Instructional Model)

Frank Tallentire, *Aerospace Engineer, Retired, Littleton, Colorado* (Science Content)

Lynn Williams, *University of Oklahoma, Norman, Oklahoma* (Nature of Science)

Advisory Board Members, First Edition

Elliot Asp, *Littleton Public Schools, Littleton, Colorado*

Randall Backe, *Kansas State University, Manhattan, Kansas*

Pat Barry, *Wilbur Wright Middle School, Milwaukee, Wisconsin*

Bonnie Brunkhorst, *California State University, San Bernardino, California*

Herbert Brunkhorst, *California State University, San Bernardino, California*

Janet Carlson Powell, *Spectrum Science Education, Boulder, Colorado*

H. Mack Clark, *Air Academy District #20, Colorado Springs, Colorado*

Mary Doyen, *Rocky Mountain Center for Health Promotion and Education, Northglenn, Colorado*

Linda Ganatta, *Timberview Middle School, Colorado Springs, Colorado*

April Gardner, *University of Northern Colorado, Greeley, Colorado*

Cynthia Geer, *University of Cincinnati, Cincinnati, Ohio*

Merton Glass, *University of South Florida, Tampa, Florida*

Johnnie P. Hamilton, *Franklin Intermediate School, Chantilly, Virginia*

Debbie Hill, *Eagleview Middle School, Colorado Springs, Colorado*

David Housel, *Oakland Public Schools, Waterford, Michigan*

Roger Hubley, *Pleasant Run Middle School, Cincinnati, Ohio*

Paul DeHart Hurd, *Professor Emeritus, Stanford University, Palo Alto, California*
Candace Julyan, *Technical Education Research Centers, Cambridge, Massachusetts*
David Kennedy, *State Department of Education, Olympia, Washington*
Joyce Kerce, *W. D. Sugg Middle School, Bradenton, Florida*
Keith Kester, *The Colorado College, Colorado Springs, Colorado*
Julie Kropf, *Hollenbeck Middle School, Los Angeles, California*
Thomas Liao, *SUNY, Stony Brook, New York*
Thomas Lord, *Indiana University of Pennsylvania, Indiana, Pennsylvania*
Susan Loucks-Horsley, *The NETWORK, Andover, Massachusetts*
Glenn Markle, *University of Cincinnati, Cincinnati, Ohio*
James McClurg, *University of Wyoming, Laramie, Wyoming*
Sam Milazzo, *University of Colorado, Colorado Springs, Colorado*
Francesca Mollura, *Academy of Liberal Arts and Sciences, Kansas City, Missouri*
Cathy Oates, *Challenger Middle School, Colorado Springs, Colorado*
Michael Padilla, *University of Georgia, Athens, Georgia*
Rita Patel-Eng, *SUNY, Stony Brook, New York*
E. Joseph Piel, *Professor Emeritus, SUNY, Stony Brook, New York*
Tracy Posnanski, *University of Wisconsin-Milwaukee, Milwaukee, Wisconsin*
Douglas Reid, *Southridge Middle School, Fontana, California*
Josina Romero-O'Connell, *Eagleview Middle School, Colorado Springs, Colorado*
Rochelle Rubin, *Instructional Materials Center, Waterford, Michigan*
Charlotte Schartz, *Kingman Elementary School, Kingman, Kansas*
Dick Sevits, *Panorama Middle School, Colorado Springs, Colorado*
M. Gail Shroyer, *Kansas State University, Manhattan, Kansas*
Elayne Shulman, *Classroom Consortia Media, Metuchen, New York*
Barbara Spector, *University of South Florida, Tampa, Florida*
John Staver, *Kansas State University, Manhattan, Kansas*
John Swaim, *University of Northern Colorado, Greeley, Colorado*
Robert Tinker, *Technical Education Research Centers, Cambridge, Massachusetts*
David Trowbridge, *University of Washington, Seattle, Washington*

Project Advisors and Consultants, First Edition

Clyde R. Burnett, *Fritz Peak Observatory, Rollinsville, Colorado* (Content Reviewer)
William D. Gillan, *IBM, Boca Raton, Florida* (Corporate Advisor for Design Study)
Martin Guttmann, *IBM, Boca Raton, Florida* (Corporate Advisor for Design Study)
Ann Haley-Oliphant, *Mainville, Ohio* (Contributing Author)
Jerald Harder, *NOAA Aeronomy Laboratory, Boulder, Colorado* (Content Reviewer)
Norris Harms, *Arvada, Colorado* (Evaluation)
A. W. Harton, *IBM, Atlanta, Georgia* (Corporate Advisor for Design Study)
James McClurg, *University of Wyoming, Laramie, Wyoming* (Curriculum Development)
Ann Primm, *Knoxville, Tennessee* (Contributing Author)
James R. Robinson, *Boulder, Colorado* (History)
M. Gail Shroyer, *Kansas State University, Manhattan, Kansas* (Implementation)
Dave Somers, *Colorado Springs, Colorado* (Editor)
Terry G. Switzer, *Fort Collins, Colorado* (Contributing Author)
Luise Woelflein, *Washington, DC* (Contributing Author)

Project Advisors and Consultants, Second Edition

Karen Bertollini, *BSCS, Colorado Springs, Colorado*
Clyde R. Burnett, *Fritz Peak Observatory, Rollinsville, Colorado*
Michael J. Dougherty, *BSCS, Colorado Springs, Colorado*
Sariya Jarasviroj, *BSCS, Colorado Springs, Colorado*
Mark Johnson, *Gustavus Adolphus College, Saint Peter, Minnesota*
Sam Milazzo, *University of Colorado, Colorado Springs, Colorado*

Field-Test Sites, First Edition
Primary Site Centers and Affiliated Schools

California
Almeria Middle School, Fontana, California, 1990–91
Southridge Middle School, Fontana, California, 1990–92
Coordinated by Herbert Brunkhorst (Site Coordinator) and Carol Cyr (Graduate Assistant, 1990–91) and Cynthia Peterson (Graduate Assistant, 1991–92), based at California State University, San Bernardino, California.

Colorado
Carmel Middle School, Colorado Springs, Colorado, 1990–92
Challenger Middle School, Colorado Springs, Colorado, 1990–92
Colegio Los Nogales, Bogota, Colombia, South America, 1991–92

The Colorado Springs School, Colorado Springs, Colorado, 1990–92
Desert School, Rock Springs, Wyoming, 1991–92
Eagleview Middle School, Colorado Springs, Colorado, 1990–91
East Junior High School, Rock Springs, Wyoming, 1991–92
Gorman Middle School, Colorado Springs, Colorado, 1990–92
Panorama Middle School, Colorado Springs, Colorado, 1990–92
Smiley Middle School, Denver, Colorado, 1991–92
Timberview Middle School, Colorado Springs, Colorado, 1990–91
White Mountain Junior High School, Rock Springs, Wyoming, 1991–92
Coordinated by BSCS staff, based in Colorado Springs, Colorado.

Florida
Clearwater Comprehensive School, Clearwater, Florida, 1990–91
Harllee Middle School, Bradenton, Florida, 1991–92
Lincoln Middle School, Palmetto, Florida, 1991–92
Sixteenth Street Middle School, St. Petersburg, Florida, 1990–92
Southside Fundamental School, St. Petersburg, Florida, 1990–92
W. D. Sugg Middle School, Bradenton, Florida, 1990–92
Coordinated by Barbara Spector (Site Coordinator) and Merton Glass (Graduate Assistant), based at University of South Florida, Tampa, Florida.

Kansas/Nebraska
Chapman Middle School, Chapman, Kansas, 1990–92
Dawes Junior High School, Lincoln, Nebraska, 1991–92
East Junior High School, Lincoln, Nebraska, 1991–92
Fort Riley Middle School, Fort Riley, Kansas, 1990–92
Kingman Middle School, Kingman, Kansas, 1990–92
Murdock Elementary School, Kingman, Kansas, 1990–92
Norwich High School, Kingman, Kansas, 1990–92
Norwich Junior High School, Kingman, Kansas 1990–92
Pound Junior High School, Lincoln, Nebraska, 1991–92
Coordinated by John Staver (Site Coordinator) and Randall Backe (Graduate Assistant, 1989–91) and Ronald Krestan (Graduate Assistant, 1991–92), based at Kansas State University, Manhattan, Kansas.

New York
Roy W. Brown Middle School, Bergenfield, New Jersey, 1991–92
Longwood Junior and Senior High School, Middle Island, New York, 1990–91
Longwood Middle School, Middle Island, New York, 1990–91
Mount Sinai Middle School, Mount Sinai, New York, 1991–92
Shoreham-Wading River Middle School, Shoreham, New York, 1990–91
Southampton Intermediate School, Southampton, New York, 1991–92
Tremont School, Mount Desert, Maine, 1990–91
Coordinated by Thomas Liao (Site Coordinator) and Rita Patel-Eng (Graduate Assistant, 1989–91) and Cynthia Anderson (Graduate Assistant, 1991–92), based at State University of New York, Stony Brook, New York.

Ohio
Dater Junior High, Cincinnati, Ohio, 1990–91
McCord Middle School, Worthington, Ohio, 1991–92
Perry Middle School, Worthington, Ohio, 1991–92
Pleasant Run Middle School, Cincinnati, Ohio, 1990–92
Coordinated by Glenn Markle (Site Coordinator) and Cynthia Geer (Graduate Assistant), based at University of Cincinnati, Cincinnati, Ohio.

Secondary Site Centers and Affiliated Schools

Arizona
Lee Kornegay Junior High School, Miami, Arizona, 1991–92
Tso Ho Tso Middle School, Fort Defiance, Arizona, 1991–92
Williams Middle School, Williams, Arizona, 1991–92
Coordinated by Diane Ebert-May (Site Coordinator) and Alison Graber (Graduate Assistant), based at Northern Arizona University, Flagstaff, Arizona.

California
Hollenbeck Middle School, Los Angeles, California, 1990–91
Coordinated by Andrea Gombar, based at Los Angeles Unified School District, Los Angeles, California.

Colorado
Bookcliff Middle School, Grand Junction, Colorado, 1991–92
East Middle School, Grand Junction, Colorado, 1991–92
Fruita Middle School, Grand Junction, Colorado, 1991–92
Mount Garfield Middle School, Grand Junction, Colorado, 1991–92
Orchard Mesa Middle School, Grand Junction, Colorado, 1991–92
West Middle School, Grand Junction, Colorado, 1991–92

Coordinated by Kathleen Kain (Site Coordinator) and Rebecca Johnson (Field-Test Teacher), based at Mesa County Schools, Grand Junction, Colorado.

Michigan
Isaac E. Crary Middle School, Waterford, Michigan, 1990–92
Detroit Country Day School, Birmingham, Michigan, 1990–92
Stevens T. Mason Middle School, Waterford, Michigan, 1990–92
John D. Pierce Middle School, Waterford, Michigan, 1990–92
Coordinated by Rochelle Rubin, based at the Instructional Materials Center, Waterford, Michigan, and David Housel, based at Waterford Public Schools, Waterford, Michigan.

Missouri
Academy of Arts & Sciences, Kansas City, Missouri, 1990–91
Coordinated by Francesca Mollura, based at the Academy of Arts & Sciences, Kansas City, Missouri.

North Carolina
Farmville Middle School, Farmville, North Carolina, 1991–92
Coordinated by Brenda Evans, based at the Department of Public Instruction, Raleigh, North Carolina.

Pennsylvania
Davis School at IUP, Indiana, Pennsylvania, 1991–92
Freeport Junior High School, Freeport, Pennsylvania, 1990–92
Milton Hershey School, Hershey, Pennsylvania, 1991–92
North Hills Junior High School, Pittsburgh, Pennsylvania, 1991–92
Coordinated by Thomas Lord (Site Coordinator) and Terry Peard (Assistant), based at Indiana University of Pennsylvania, Indiana, Pennsylvania.

Wisconsin
Lundahl Junior High, Crystal Lake, Illinois, 1991–92
North Junior High, Crystal Lake, Illinois, 1991–92
Richfield Senior High, Richfield, Minnesota, 1991–92
Wilbur Wright Middle School, Milwaukee, Wisconsin, 1990–92
Coordinated by Jean Moon (Site Coordinator, 1989–90) and Craig Berg (Site Coordinator, 1991–92) and Tracy Posnanski (Graduate Assistant), based at University of Wisconsin-Milwaukee, Milwaukee, Wisconsin.

Coordination, Text Design, Electronic Production and Prepress
LaurelTech Integrated Publishing Services, Manchester, New Hampshire

Public Support
National Science Foundation

Private Support
Kendall/Hunt Publishing Company, Inc., Dubuque, Iowa
Science Kit & Boreal Laboratories, Inc., Tonawanda, New York
IBM Educational Systems, Atlanta, Georgia

Index

A

AC. *See* Alternating Current
Acceleration, 153–161
Accident. *See* Car accident
Acid rain, 483
ACS. *See* American Cancer Society
Activity skills, 22
Adams, Thomas, 191
Adhesive forces, 321
Afterimage, 90–94
    color and, 92, 94–96
    defined, 90
    diversity of, 94–95
Air, 216–217, 245
Airplane design, 373–378
Air pollution, 464
    biofuels and, 479
    coals and, 480
    petroleum contributes to, 498
    sulfur and nitrogen oxides, 498
Alchemy, defined, 245
Alcohol, 152–153, 168. *See also* Drunk driving
Alcohol consumption, 168, 171
Altamont Pass, 466
Alternate Forms of Energy Web site code, 514
Alternating Current (AC), 501
Alternative fuel, 173
Aluminum cans, recycling, 524–525
American Automobile Association, 391
American Cancer Society (ACS), 427, 429
American flag, 92
Analog TV. *See* Color TV
Anaxagorus, 224
Animal senses, limits and diversity in, 72–73
Animation, continuous motion through, 111–113
A&P, 298
Apple sauce, 523
Aristotle, 217, 245

Athletic shoes design, 369
Atmosphere, 431
Atmospheric scientist, 431
Atom, 232, 409
    defined, 225
    nucleus of, 488–489
Atoms and Elements Web site code, 225
Automobile, 172–173. *See also* Car accident
Automobile safety laws, 151
Average, determining, 561–563
Axis, 86, 93, 554–557

B

Backpack design, 370
Balloon, 232–235, 333–337
Bar graph
    horizontal and vertical axes, 554–555, 557
    how to construct, 553–558
    number scales, 555–556
    title of, 558
Bats, 72–73
Becquerel, Antoine, 16
Behavioral constraints, 355
Bell-shaped curve, 81, 88–90
    defined, 84
    and setting standards, 165–168
    use of, by clothing manufacturers, 86–87
    value of, 85–88
Bend, 209
Benefits, 461–465
    defined, 462–463
    of energy, 447–509
    of wind farms, 466–467
Biofuels, 474, 477–479
Biogas, 479
Biosphere, 262
Black Jack gum, 191
Black lines, 118
Black lung disease, 482
Blind spot, 96–97

Blood Alcohol Content (BAC), 169–171
Boat(s)
    building, 338–340
    common criteria of, 324–325
    gas, 333–337
    mast, 329
    parts of, 327–337
    propeller, 330–333
    propulsion, 326
    sails, 329–330
    small-scale, 318–322
    technology and, 323
    types of, 324
Bounceability, 188–190, 192
Bounty, 294, 297
Bradbury, Ray, 392
Brainstorming session, 559–560, 565
Brawny, 298
Breathalyzer, 171
"Bubble TV," 103
Bunsen, Robert, 578
Bunsen burner, 578–580
    how to light and adjust, 579–580
    parts of, 578
    safety cautions when using, 578–579
Butane, 486, 498

C

Cancer. *See* Skin cancer
Car accident
    cellular phones and, 151
    drunk driving, 168–172
    energy conserved during, 162
Carbon dioxide, 483, 486–487, 499
Carcinomas, 426
CBS system, 106–107
Cell model, 229
Cells, in data table, 544
Cellular phone safety, 151–153
Celsius, 572–573
CFC. *See* Chlorofluorocarbon
Chemical energy, 408

Chemistry, basis for modern, 245
Chernobyl nuclear power plant, 492
Chicle, 191–192
Chien-Shiung Wu, 202–203
Chinese philosopher, 214–215
Chlorine, 431
Chlorofluorocarbon (CFC), 431–432, 502
Christiansen, Ole Kirk, 354–357
Christiansen (GKC), Godfried Kirk, 357
Clock, 219
Clothes, manufacturing, 86–87
Coal, 474, 480–483, 520
    greenhouse effect, 483
    open pit coal mine, 482
    power plants, 483
    restoration project, 481
Cognitive psychologist, 18
Cognitive scientist, 20
Cohesive force, 279, 321–322
Color, 94–98
    process of, 94–95
    producing, 8–11
Color TV, 103–106
Columns, in data table, 544
Commercial, 287–315
Communicator, 23–24
Complementary color, 94–95
Compost, 445
Concave lens, 65, 70
Cones, 94–95. *See also* Rods and cones
Constraints, 308–309, 458
    behavioral, 355
    costs, time, budget, and human factors, 382–383
    criteria and, 371–372
    evaluate toy based on, 352
    mental, 355
    physical, 354
Continuous motion
    at movies, 113–115
    through animation, 111–113
Continuous motion rate, 113
Control variable, 60, 575
Convex lens, 64, 70
Cooperative learning, 18–25
    defined, 21
    teamwork, 18, 21
Corn, 478
Cornea, 69, 96

Costs, 461–465
    energy, 447–509
    of wind farms, 466–467
Crick, Francis, 17
Criteria, 308, 458
    constraints and, 371–372
    as limitation to diversity of design, 371
Crude oil, 497–498
Crutzen, Paul, 431
Curie, Marie, 15–16, 225
Curie, Pierre, 15–16

D

Dalton, John, 225
Dam, 485, 504
Data, defined, 544
Data collection, 5–8
Data table
    defined, 544
    how to construct, 544–552
    label, 548
    sample outline of, 547
    title of, 550
Da Vinci, Leonardo, 207
DC. *See* Direct Current
Deforestation, 479
Delivery system, 528–530
Delta, 297
Democritus, 225, 409
Deoxyribonucleic acid (DNA), 17
Dermatologist, 425–427
Design, 317–357
    airplane, 373–378
    athletic shoes, 369
    backpack, 370
    criteria as limitation to diversity of, 371
    doors, 368
    furniture, 370
    furniture for doll house, 364
    human factors as, constraints, 354–355
    masters of, 381–393
    process of, 343–346
    with shapes, 361–368
    similarity and diversity in, 368–372
    toothbrush, 353–356
    toy, 347–352
    vehicle, 371
    window, 367
Design Process Cards, 341

Design Web site code, 318
Deuterium, 493
Devonian fossil, 270
Diesel fuel, 498
Digestor, the, 445
Digital cable service, 105–106
Digital camera, 105
Digital signal, 105
Digital TV, 105–106
Direct Current (DC), 501
Dishes, washing, 531–532
Distance-to-space ratio, 119–123, 130
Diversity, 45–73
    in height, 85–88
    limits and, in animal senses, 72–73
    limits in peripheral vision, 70
    in peripheral vision among humans, 79–80
    of popcorn, 82–83
    range of limits and, 75–97
    similarity and, in design, 368–372
    use, to set standards, 133–173
DNA. *See* Deoxyribonucleic acid
Door design, 368
Driver's license, 167
Driving Under the Influence (DUI), 170–172
Drunk driving, 168–172
Drunk Driving Web site code, 170
Dry geothermal, 474, 507
    advantages and disadvantages of, 484
    defined, 483–484
"D-to-s" ratio. *See* Distance-to-space ratio
DTV, 105
Duck, 73
DUI. *See* Driving Under the Influence
Duplo bricks, 357
DuPont, 193–194

E

Earth, 215–217, 535–539
Earth heat. *See* Dry geothermal
Echolocation, 73
Einstein, Albert, 259
Elaborate, 19–20
Electrical energy, 409–410
Electrical water-heating system, 523–524

Electricity, generating, 448–453
    benefits and costs, 461–465
    systems of, 473–474
Electricity Current Web site code, 454
Electricity Web site code, 448
Electric wheel motor, 173
Electron beam, 104–105
Electron gun, 104–105, 131
Electron microscope, 225
Electrons
    atoms and, 225
    defined, 104–105
    water and, 409–410
Electrostatic precipitators, 483
Element, 245
Elements, 215–217
"El Niño," 496
Empedocles, 216–217
Energy, 404
    benefits and costs, 447–509
    forms of, 405–410
    from fossil fuels, 480–481
    from garbage, 444–445
    geothermal, 401
    improve use of, 527–532
    mechanical, 505–506
    saving, 520–527
    from Sun, 414–415, 419–445
    in systems, 399–417
    tracing flow of, 416–417
Energy Guide sticker, 522
Energy of motion. *See* Kinetic energy
Energy system(s), 477–509
    biofuels, 477–479
    coal, 480–483
    dry geothermal, 483–484
    hydroelectric, 484–485
    natural gas, 486–488
    nuclear energy, 488–489
    nuclear fission, 488
    nuclear fusion, 488–489
    Ocean Thermal Energy Conversion (OTEC), 494–496
    petroleum, 486, 496–500
    photovoltaic conversion, 500–501
    Solar Thermal Power System (STPS), 502
    synthetic fuel, 503
    tidal stations, 504–505
    waves, 505–506
    wet geothermal, 483, 507–508
    wind, 508–509
Energy Transformation Web site code, 469
Energy Web site code, 404
Engage, 19–20
Engineer, 248
Environment, evaluating, 383–385
Ergonomics, 353–354
Ergonomics specialist, 353
Ethane, 486
Ethanol, 478
Evaluate, 19–20
Explain, 19–20
Explore, 19–20
Eye
    composed of, 69
    concave lens, 70
    convex lens, 70
    cornea, 69, 96
    development of, 68–72
    eyelid, 69, 96
    farsighted, 67, 70–71
    fovea, 69–71
    iris, 69, 96
    lens, 69–71, 96
    model of, 65–72
    nearsighted, 67, 70
    optic nerve, 69, 96–97
    peripheral vision, 70
    pupil, 69, 96
    retina, 69–71, 96
    rods and cones, 69–71
    sclera, 69
Eyelid, 69, 96
Eye Web site code, 62

F

Fahrenheit, 572
Fahrenheit 451, 392
Fair test
    defined, 574
    how to conduct, 574–577
    steps in determining and conducting, 576–577
Falling distance, 141–142
Farsighted eye, 67, 70–71
Federal Communications Commission (FCC), 107
Feng-Shui, 214
Fermentation, 478
Fiarkoski, John, 444–445
Final Factor, The, 134–135
Fire
    build, 407–409
    as element, 215–217, 245
Fire system, 408
Flicker-fusion frequency, 116–118, 123, 130
Flipbook, 111–113
Flour mills, 406
Flowability, 197
Flow chart, 341–342
Fluorescent light bulb, compact, 526–527
Fly, 73
Focal length, 65
Force
    adhesive, 321–322
    cohesive, 279, 321–322
    defined, 160
    in solids versus liquids, 237
    in water, 276
Forms of Energy Web site code, 405
Fossil fuel, 480–481, 520
Fovea, 69–71
Franklin, Rosalind, 17–18
Frog, 73
Fuel
    alternative, 173
    biofuel, 474, 477–479
    diesel, 498
    fossil, 480–481, 520
    synthetic, 474, 503
Fuel rod, 490
Function, 24, 172, 193, 229, 359, 361, 370–372, 454
Furniture design, 364–365, 370

G

Galaxy, 229
Galileo, 153, 161, 187
Galvanometer, 448–454
Gamma ray photons, 489
Gamma rays, 225
Garbage, energy from, 444–445
Gasoline, 498
    as cause of air pollution, 463
    as fuel for boat, 333–337
    leaking, 488
Generator, 454–456
Generators Web site code, 467
Georgia Pacific (*Mr. Big*), 293
Geothermal energy, 401

INDEX 595

Geysers, 483, 507. *See also* Wet geothermal
Glass, recycling, 524–525
Global warming, 496
    carbon dioxide and, 483, 487
    petroleum and, 498–499
Gloop, 251–257
Goldfish, 73
Goldring, Winifred, 270
Graph. *See* Bar graph
Graph paper, 554
Gravity, 274–276
Greek philosophers, 215–216
Greenhouse effect, 483, 487, 499
Greenhouse gas. 496, 499. *See* Ozone
Gregoire, Marc, 194
Guinness Book of World Records, 194
Gum, 192–193

H

Handicapped. *See* Physically challenged
Hardness, 194, 196–197
HDTV, 106
Heat, 411–414. *See also* Solar-heating system; Water-heating system
Heat energy, 499
Heating oil, 498
Heavy water, 493
Height, diversity in, 85–88
Heraclitus, 216
Herpes simplex virus, 424
Hi-Dri, 293
Horizontal axis, 554–555, 557
How Are Motion and Speed Related Web site code, 137
How Do You Identify Matter Web site code, 190
Hubble snapshot, 47
Human factors, 308
    defined, 125
    as design constraints, 354–355
    science of, 353–354
    setting standards and, 125–126
    types of, 354–355
Hybrid-electric engine, 173
Hydroelectric, 474, 484–485
Hydrogen, 431, 493
Hyper Car, 172–173

I

Ice, 237
Icicle, 417
If–then statement, 259–261
Incandescent light bulb, 526–527
Industrial chemist, 431
Infrared light, 209
Infrared radiation, 424
Input, 407–408, 515
Interlacing, 130–131
Ions, 225
Iris, 69, 96
Isaac Newton Web site code, 259

J

Jeans, 86–87
Job Squad, 294, 296–298
Journey to the Moon, 392

K

Kerosene, 498
Kinetic energy, 162
    defined, 406
    example of, 45
    of particles, 489

L

Label, in data table, 548
Lawrence, Ernest, 203
Leak-free model, 265–272
Leaking gas, 488
Lego, 356–357
Legoland, 357
Lens, 66, 69–71, 96. *See also* Concave lens; Convex lens
Library, 569
Light
    energy-efficient sources of, 534
    in eye, 69–71
    inside prism, 208–209
    traveling, 65
    of waves and particles, 228
Light bulb, 526–527
Lightening, 140
Lighting designer, 533–534
Limit(s), 45–73
    and diversity in animal senses, 72–73
    diversity of, in peripheral vision, 70
    range of, and diversity, 75–97
    use, to set standards, 99–131
Linotype, 388
Lopez de Santa Anna, Antonio, 191

M

Magnet, 455
Making Paper Towels Web site code, 293
Manager, 23–24
Mast, 329
Masters of design, 381–393
Material, 187, 188, 190, 192–197, 199, 201, 204, 205
Mechanical energy, 505–506
Melanin, 428–429
Melanoma, 427
Mental constraints, 355
Mering, Jacques, 17
Metal, 215
Methane, 478, 486, 499. *See also* Natural, gas, types of
Metric scale, 572
Model
    leak-free, 265–272
    shaping, 248–250
    See also Scientific model
Molecule model, 232
Molecules, 225, 431
Molina, Mario, 431
Motion, particles in, 237. *See also* Continuous motion; Kinetic energy
Motor, 455
Mud hut, 528
Mystery box, 220–223

N

National Television Systems Committee (NTSC), 129–131
Natural gas, 474, 486–488, 520
    advantages and disadvantages of, 487
    stripping, 486
    types of, 486
    use of, 478
NBC system, 107
Nearsighted eye, 67, 71
Neutrinos, 225

Neutron, 225, 409
New England Journal of Medicine, The, 151–152
Newton, Sir Isaac, 16–17, 225
    First Law of Motion, 160
    Second Law of Motion, 161
    theory of gravity, 259–260
    Web site code for, 259
Niels, Bohr, 225
1984, 393
Nitrogen
    as part of natural gas, 486
    as part of reactive molecules, 431
Nitrogen oxides, 498
Noise, 179–181
Noise Pollution Web site code, 179
Nonscience, science versus, 262–264
Nontranslucence, 196
NTSC. *See* National Television Systems Committee
Nuclear energy, 488–489
Nuclear fission, 474, 488–489, 493
Nuclear fission reactor, 490–493
Nuclear fusion, 474, 488–489, 493
Nuclear fusion reactor, 493–494
Nuclear power plants, 490–492
    accidents in, 491–492
    Chernobyl, 492
    Three Mile Island, 491
Nuclear radiation, 488–489
Nucleus, 225
Number scales, 555–556
"Nuna" 419, 501

O

Observations, 230–231
Ocean Thermal Energy Conversion (OTEC), 474, 494–496
Operational definition, 59, 60, 189–190
Optic nerve, 69, 96–97
Optimal TV viewing distance, 123–125
Organism, ancient, 480–481, 486
Orwell, George, 392–393
OTEC. *See* Ocean Thermal Energy Conversion
Output, 407–408, 515, 518
Oxygen atoms (O), 431
Oxygen molecules (O_2), 431
Ozone
    depletion of, 431–432
    "hole," 432
    greenhouse gas (O_3), 426
    molecules of, in path of sunlight, 429
    Solar Thermal Power Systems (STPS) and, 502

P

Paleontology Web site code, 270
Paper, for paper towels, 293
Paper clip, 456–461
Paper towels, 288–291
    brands of, 293–294
    consumers of, 291–301
    linting and running, 297
    paper for, 293
    ratings of, 299–300
    recommendations, 297–298
    test for absorbency, 295
    test for strength, 295–296
Particle model
    defined, 229–230
    how, works, 232–236
Particles
    kinetic energy of, 489
    in motion, 236–239
Particle theory, 228
Pathmark, 298
Perception distance, 136, 138, 144–148
Perception time, 136
    average human, 146, 164–165
    driver, 152
Peripheral vision, 62, 79–81
    defined, 60
    diversity of limits in, 70
    measurement of, 80–81
Persistence of vision, 107–122, 130
Petroleum
    air pollution caused by, 498
    defined, 496
    disadvantages of, 499–501
    drilling for, 486, 497–498
    limited amount of, 520
    refined, fuels, 498
Phenomenon, 128, 248
Philosopher's stone, 245
Philosophy and Alchemy, 245
Phosphor, 104
Photon, 426, 489
Photovoltaic (PV) cells, 500–501
Photovoltaic conversion, 474, 500–501

Physical constraints, 354
Physically challenged, 385–387, 390–391
Physical Properties of Matter Web site code, 200
Phytoplankton, 432
Pitch, 407
Pixel(s), 104–105, 118
Planet, 229
Plunkett, Roy, 193–194
Polystyrene-butadiene co-polymer, 205
Polyvinyl acetate, 191–192
Popcorn, 76–79, 82–83
Potential energy, defined, 406
Pourability, 197
Predictions, 258–261
Prism, 208–209
Problem solving, technological, 340–347
Processing, 24–25
Proctor & Gamble, 294
Producing Light Web site code, 535
Product, describing function of, 371
Propane, 486
Propane molecule, 232
Propeller, 330–333
Properties of Sound Web site code, 181
Property, 187–205, 193
Propulsion, 326
Proton, 225
Protons, 409
Pupil, 66, 69, 96. *See also* Lens
PV cells. *See* Photovoltaic cells
Python, 73

Q

Quartz crystal, 219
"Queen of Nuclear Physics." *See* Chien-Shiung Wu

R

Radiation, 424
Radioactive decay, 488
Radioactive waste, 492
Radio wave, 105
Range
    defined, 62
    of limits and diversity, 75–97
Rattlesnake, 73

Reach toothbrush, 354
Reaction distance, 136–138, 144–148
Reaction time, 137, 168
   driver, 152
   with high Blood Alcohol Content (BAC), 171
   personal, 139–143, 563
Reactive molecule, 431
Reader's Guide to Periodical Literature, 569–570
Recycle, 520, 524–526
Recycling Web site code, 531
Redelmeier, Donald A., 151
Refineries, 498
Research project, 564
   choose topic for, 565
   find information, 566–571
   get organized, 565–566
   interviews, 567–569
   organize information, 571
Reservoir, 483, 485, 507
Resolution, 105
Restoration project, 481
Retina, 107
   blind spots and, 96
   color and, 94–96
   defined, 69–71
Rock, molten, 507
Rocky Mountain Institute, 172
Rods and cones
   blind spots and, 97
   color and, 94–95
   functions of, 69–71
Roles, 23–25
Rowland, F. Sherwood, 431
Rows, in data table, 544
Rudin, Andrew, 527
Rutherford, E.J., 225

S

Safety, 29–39
   contracts, 33
   cooperative learning and, 33–39
   guidelines for, 459
   using Bunsen burner, 578–579
Safety in the Science Classroom Web site code, 32
Safeway (*Marigold*), 298
Sail, boat, 329–330
Sailboat, 323
Sapodilla tree, 191
Satellite dish, 106

Scale, metric, 572
Scavenger Hunt, 26–27
Science
   versus nonscience, 262–264
   and technology, 388–389
Science safety. *See* Safety
Scientific Biographies Web site code, 225
Scientific explanations, 207–217
Scientific model, 219–245, 228–229
   criteria for, 280
   defined, 229, 230
   using, to test and predict, 247–281
   See also Model
Sclera, 69, 96
ScotTowels, 298
ScotTowels Junior, 294
Scott Paper Co., 294
Scrubber system, 508
Scrubbles, 483
Seat belts, 151
Seeds, 224
Segre, Emilio, 203
Seismograph, 497–498
Shadow mask, 105
Shapes, designing with, 361–368
Shock, avoid receiving, 451
Signal, TV, 106–107
Silicon mining, 524
Similarity, in design, 368–372
Skidding distance, 136–138, 164
Skidding time, 137
Skills. *See* Activity skills; Social skills; Unit skills
Skin cancer, 424
   defined, 426–430
   protection from, 427–428
   types of, 426–427
   See also Sunburn
Social skills, 21–23
Solar cars, 419
Solar cells, 173
Solar Cells Web site code, 421
Solar collector, 445
Solar/compost water-heating system, 445
Solar-heating system, 441, 444
Solar radiation, 424
Solar sources, 419–445
Solar system, 229–230
Solar Thermal Power System (STPS), 474, 502

Solar water-heating system, 523–524
Solid, 237
Solving problems, systems for, 437–438
Soochow Girls School, 202
Sources of Energy Web site code, 520
Spectrum, 208
Spectrum Web site code, 208
Speed limits, 162–164
SPF. *See* Sun Protection Factor
Spinner, constructing, 109–111
Standards
   bell-shaped curve and setting, 165–168
   defined, 128
   for safe toy design, 349
   use diversity to set, 133–173
   use limits to set, 99–131
Star, 47, 229
Star tracers, 46–50
Steam, 482
Stickiness, property of, 205
Stopping, phases of, 135–139
Stopping distance, total
   defined, 137–138
   equation for, 136
   setting speed limits, 162–164
Stopping time, 137–138
STPS. *See* Solar Thermal Power System
Stratosphere, 431–432
Stratospheric ozone. *See* Ozone
Stripping, 486
Subatomic particles, 203
Sulfur dioxide, 483
Sulfur oxides, 498
Sun, energy from, 414–415, 419–445
Sunburn
   defined, 423–425
   scale for determining risk, 430
   skin cancer and, 426–427
Sun detector, 421
Sundial, 219
Sun Protection Factor (SPF), 424
Sunscreen, 427, 429
Supersonic transport, 431
Surface tension, 279, 321–322
Synthetic fuel, 474, 503

System(s)
  balance of heat input and heat output, 411
  defined, 400
  energy in, 399, 404–405, 417
  forms of energy in, 405–410
  input and output of, 407–408
  limits of energy in, 395–539
  for solving problems, 437–438

T

T-chart, 42–43, 540–543
Team Member, 23–24
Technological systems, 515–524
Technology, 382–383, 437
  boats and, 323
  energy saving, 526–527
  Hyper Car, 172–173
  problem solving, 340–347
  science and, 388–389
  scientific models involving, 236
  transportation and, 323
  TV and, 105–107
Teflon, 194
Television Technology Web site code, 105
Temperature, 412. *See also* Global warming
Temperature(s)
  determining average, 561–562
  measuring, 187
  of water, 502
Test. *See* Fair test
Testability, 262
Tetrafluoroethylene, 194
Thales, 215
Theory, 228, 231
Thermocouple, 472
Thermometer
  how to read, 572–573
  to measure air temperature, 187
Thermostat, 533
Thompson, J.J., 225
Threading Needle, 50–53
Three Mile Island nuclear power plant, 491
Tibshirani, Robert J., 151
Tidal stations, 504–505
Tides, 474, 504–505
Tire manufacturing, 204
Title
  of bar graph, 558
  of data table, 550

Toothbrush design, 353–354
Toy design, 347–352
Tracker, 23–24
Translucence, 194–196
Transportation, 323, 419
Tritium, 493
TV, 100
  black lines, 118
  commercials, 287–315
  for the future, 127–131
  National Television Systems Committee (NTSC), 129–131
  optimal viewing distance, 123–125
  pixels, 118
  video cards and, 129
TV pictures, 101–106
TV screen, 103–106
Twenty Thousand Leagues under the Sea, 392

U

Ultralight cars, 173
Ultraviolet light, 209
Ultraviolet photons, 426
Ultraviolet (UV) radiation, 424
  absorbed by melanin, 429
  ozone and, 432
  protection from, 427–428
  skin cancer and, 426–427
United States Geological Survey, 481
Unit skills, 22–23
Universe, 229–230
Uranium, 490
Using Models Web site code, 229
UV radiation. *See* Ultraviolet (UV) radiation

V

Variable, 58–59, 575. *See also* Control variable
Vegetable soup, 516–519
Vehicle, 390–391
Vehicle design, 371
Verne, Jules, 392
Vertical axis, 554–555, 557
Video camera, 105
Video card, 129
Villumien, Rasmus, 14–15
Viscosity, 194, 197–198
Visible light, 424

Visible Light Web site code, 64
Vision, persistence of, 107–122. *See also* Peripheral vision
Vision test, 167
Viva, 293–294, 296–298
Volcano, 414–415

W

Waste. *See* Radioactive waste
Waste Management Web site code, 526
Watch, wrist, 219
Water, 215–217
  boiling, 237
  cohesive force, 279, 321–322
  contained in natural gas, 486
  droplets of, 250
  forces in, 276
  particles, 272–279
  properties of, 245
  surface tension, 279, 321–322
  temperature of, 502
  See also Heavy water; Reservoir
Water-heating system, 439–444, 520–521
Water particles, 274–276, 321–322
Water reservoir. *See* Reservoir
Water vapor, 499
Watson, James, 17
Wave-powered systems, 505
Waves, 474, 505–506
Wegener, Alfred, 14–15
Wet cell, 471
Wet geothermal, 474, 483, 507–508
Wheelchair, 390–391
White light, 94–95
Wind, 474
Wind farms, 466–467
Windmill, 508–509
Window design, 367
Wolf-Rayet star, The, 47
World Solar Challenge, 419, 501
World Wide Web, 570
Wrigley, Jr., William, 192

Z

"Zero emission vehicle" 173